FinFET Devices for VLSI Circuits and Systems

FinFET Devices for VLSI Circuits and Systems

Samar K. Saha

CRC Press
Taylor & Francis Group
Boca Raton London New York

CRC Press is an imprint of the
Taylor & Francis Group, an **informa** business

First edition published 2021
by CRC Press
6000 Broken Sound Parkway NW, Suite 300, Boca Raton, FL 33487-2742

and by CRC Press
2 Park Square, Milton Park, Abingdon, Oxon, OX14 4RN

ISBN: 978-1-138-58609-3 (hbk)
ISBN: 978-0-367-51556-0 (pbk)
ISBN: 978-0-429-50483-9 (ebk)

Typeset in Times
by Deanta Global Publishing Services, Chennai, India

In loving memory of my parents
Mahamaya and Phani Bhusan Saha

Contents

Preface

Silicon integrated circuits (ICs) have enormously impacted modern society, serving as the foundation of the Internet, social media, and interconnected network of networks or *Internet of Things* (IoT). The emerging Internet technologies offer people-to-people, people-to-machine, and machine-to-machine communications enabling appliances and services to provide notifications, security, energy-saving, automation, telecommunications, healthcare, computers, entertainment, and so on. The network of networks is integrated into a single ecosystem to create smart environments with a shared user interface. This ongoing progress of smart environments and integrated ecosystems is made possible due to the continuous miniaturization of metal-oxide-semiconductor (MOS) field effect-transistor (FET) or MOSFET devices, thus, providing low-cost, high-density, fast, and low-power ICs. However, the performance of MOSFETs in the design and manufacturing of "smart" electronic products necessary to create smart networks or "smart things" to enable smart environments and integrated ecosystems has approached its limits due to the fundamental physical limitations such as the short-channel effects (SCEs). Shrinking MOSFET device length in the decananometer regime degrades device performance including degradation in the subthreshold swing and decrease in device turn-on voltage. As a result, the scaled MOSFETs cannot be turned off easily by lowering the gate voltage leading to excessive leakage current. And, because of SCEs, the device characteristics become increasingly sensitive to process variation that imposes a serious challenge to continued scaling of planar-MOSFETs to the nanometer nodes. Furthermore, at gate length below 22 nm, the sub-surface leakage paths are weakly controlled by the gate irrespective of gate oxide thickness and their potential barriers can be easily lowered by drain bias through the enhanced electric field coupling to the drain. Thus, to surmount the challenges of continuous scaling of MOSFETs, the fin field-effect transistor or FinFET has emerged as the real alternative to continue scaling and manufacture ICs to create smart things and enable smart environments and integrated ecosystems. This book presents the basic features and operating principles of FinFET devices required for the understanding of design and manufacturing of very large scale integrated (VLSI) circuits and systems.

A large volume of research articles along with a handful of books are available on device technology and modeling of FinFETs. Most of the research articles are meant for experts in the field. On the other hand, the available books on FinFETs are either dedicated to device modeling for IC design or a collection of research articles on research and development. Thus, the available literature does not provide the fundamental principles of FinFET device operation and adequate background for the understanding of the newly adopted mainstream device technology for beginners as well as practicing engineers and experts transitioning to FinFET device technology. After working for over 30 years in the field of semiconductor process and device architecture and device modeling in industry, and over 20 years of teaching device and process physics and device modeling courses in academia, I felt the need for a comprehensive book that presents the fundamentals of FinFET device electronics for the understanding of the design and manufacturing of FinFET ICs at the nanometer nodes. This book provides readers the basic architecture and theory of FinFETs to continue shrinking devices to the fundamental scaling limits for VLSI manufacturing technology. Starting from the basic semiconductor electronics, this book presents the principles of operation and modeling of FinFETs. Thus, this book is useful for beginners as well as experts in the field of microelectronics device and design engineering to understand the theory and operation of FinFET devices.

This book is intended for the researchers and practitioners working in the area of electron devices as well as senior undergraduate and graduate students in electrical and electronics engineering programs. However, the presentation of the materials is such that even an undergraduate student not well-acquainted with semiconductor physics can understand the basic concepts of FinFETs.

Chapter 1 presents an introduction to FinFET devices as the alternative to the mainstream MOSFETs and planar-CMOS technology for VLSI circuits and systems at the nanometer node. The chapter overviews the scaling constraints of the mainstream MOSFETs beyond the 22-nm node due to SCEs; discusses the scalable alternative planar-MOSFETs and non-planar-FinFET devices for VLSI circuits and systems beyond the 22-nm node; and presents the advantages of the multiple-gate ultrathin-body FinFET devices in surmounting the SCEs for VLSI circuit manufacturing in the sub-22-nm regime. Furthermore, a comprehensive history of the emergence and development of FinFETs for non-planar-CMOS technology is presented. Chapter 2 presents a brief overview of the basic semiconductor electronics and *pn*-junction operation as background materials for the understanding of FinFET devices.

Chapter 3 presents the basic structure and operation of multiple-gate MOS capacitor systems as the foundation for the development of FinFET device theory. Analytical expressions for multiple-gate MOS capacitor systems are derived to discuss the accumulation, depletion, and inversion mode operations of multiple-gate MOS capacitor systems. A unified surface potential function is developed to analyze the characteristics of multiple-gate MOS capacitors applicable to FinFET devices. Also, a unified inversion charge expression is derived to account for the substrate doping effect in multiple-gate MOS capacitors that can be used for FinFET current calculation.

Chapter 4 provides an overview of FinFET device architecture, process technology, and typical fabrication processes for the manufacturing of FinFETs in non-planar-CMOS technology. Fabrication process flow for FinFETs on bulk-silicon substrate and on SOI substrate is overviewed and the complexities and benefits of each technology are highlighted.

Chapter 5 presents the basic theory of FinFETs, formulation of surface potentials, and presents the electrostatic behavior of long-channel devices. A set of simplifying assumptions are used to derive a continuous drain current expression for long-channel devices applicable to all regions of device operation. In addition, the regional drain current expressions for linear, saturation, and subthreshold operations are derived from the continuous drain current expression for intuitive analysis of device performance. Chapter 6 presents the small geometry effects in FinFETs for accurate characterization of real device effects. The mathematical formulation for SCEs, including V_{th} roll-off, DIBL, quantum mechanical effects, low-field mobility, velocity saturation, and channel length modulation and output resistance is presented.

Chapter 7 discusses the physical mechanisms and mathematical formulation of the different components of leakage currents in FinFET devices during operation in VLSI circuits and systems. These leakage currents include the subthreshold leakage currents due to the close proximity of the drain to the source, the gate-induced drain and source leakage currents due to band-to-band tunneling, source-drain pn-junction leakage currents, and gate tunneling currents.

Chapter 8 overviews the parasitic resistance and capacitance elements of FinFETs. The parasitic resistances include the contact resistance, spreading resistance, and source-drain extension resistance components of the raised source-drain series resistance as well as the gate resistance. The parasitic capacitances discussed are overlap, fringe, and the source-drain pn-junction capacitances. Chapter 9 presents an overview of the major challenges of the FinFET process, device, and circuit design.

Chapter 10 presents an overview of the present state-of-the-art compact models for common multiple-gate FinFET devices. Device model includes a core model for large geometry devices and models for short-channel devices to accurately analyze the physical and geometrical effects on real devices. The model includes current-voltage and capacitance-voltage formulations for FinFET devices. Furthermore, a process variability model is formulated to estimate the effect of dopant fluctuations in FinFET devices.

An extensive set of references is provided at the end of each chapter to help the readers identify the evolution and development of FinFETs and FinFET manufacturing technology for ICs at nanometer nodes

Samar K. Saha
January 2020

Author

Samar K. Saha received a PhD in Physics from Gauhati University, India, and MS in Engineering Management from Stanford University, CA. Currently, he is an Adjunct Professor in the Electrical Engineering Department at Santa Clara University, CA, and Chief Research Scientist at Prospicient Devices, CA. Since 1984, he has worked in various technical and management positions for National Semiconductor, LSI Logic, Texas Instruments, Philips Semiconductors, Silicon Storage Technology, Synopsys, DSM Solutions, Silterra USA, and SuVolta. He has also worked as a faculty member in the Electrical Engineering Departments at Southern Illinois University at Carbondale, IL; Auburn University, AL; the University of Nevada at Las Vegas, NV; and the University of Colorado at Colorado Springs, CO. He has authored more than 100 research papers. He has also authored one book, *Compact Models for Integrated Circuit Design: Conventional Transistors and Beyond* (CRC Press, 2015); one book chapter on Technology Computer-Aided Design (TCAD), "Introduction to Technology Computer-Aided Design," in *Technology Computer Aided Design: Simulation for VLSI MOSFET* (C.K. Sarkar, ed., CRC Press, 2013); and holds 12 US patents. His research interests include nanoscale device and process architecture, TCAD, compact modeling, devices for renewable energy, and TCAD and R&D management.

Dr. Saha served as the 2016–2017 President of the Institute of Electrical and Electronics Engineers (IEEE) Electron Devices Society (EDS) and is currently serving as the Senior Past President of EDS, J.J. Ebers Award Committee Chair, and EDS Fellow Evaluation Committee Chair. He is a Fellow of IEEE and a Fellow of the Institution of Engineering and Technology (IET, UK), and a Distinguished Lecturer of IEEE EDS. Previously, he has served as the Junior Past President of EDS; EDS Awards Chair; EDS Fellow Evaluation Committee Member; EDS President-Elect; Vice President of EDS Publications; an elected member of the EDS Board of Governors; Editor-In-Chief of IEEE *QuestEDS*; Chair of EDS George Smith and Paul Rappaport Awards; Editor of Region 5&6 EDS Newsletter; Chair of EDS Compact Modeling Technical Committee; Chair of EDS North America West Subcommittee for Regions/Chapters; a member of the IEEE Conference Publications Committee; a member of the IEEE TAB Periodicals Committee; and the Treasurer, Vice Chair, and Chair of the Santa Clara Valley-San Francisco EDS chapter.

Dr. Saha served as the head guest editor for the IEEE Transactions on Electron Devices (T-ED) Special Issues (SIs) on *Advanced Compact Models and 45-nm Modeling Challenges* and *Compact Interconnect Models for Giga Scale Integration*; and as a guest editor for the T-ED SI on *Advanced Modeling of Power Devices and their Applications* and the IEEE JOURNAL OF ELECTRON DEVICES SOCIETY (J-EDS) SI on *Flexible Electronics from the Selected Extended Papers at 2018 IFETC*. He has also served as a member of the editorial board of the *World Journal of Condensed Matter Physics* (WJCMP), published by the Scientific Research Publishing (SCIRP).

1 Introduction

1.1 Fin FIELD-EFFECT TRANSISTORS

A FinFET or a "fin" field-effect transistor (FET) is a metal-oxide-semiconductor (MOS) device where the gate is placed on two sides of a thin vertical semiconductor body standing on a substrate. This novel device was invented by Yutaka Hayashi at the Electrotechnical Laboratory, Tsukaba, Japan in 1980 [1] and the manufacturing device technology was developed in the late 1990s by the research group led by Chenming Hu at the University of California Microfabrication Lab, Berkeley, CA [2]. The "Fin" in the acronym FinFET describes an ultrathin-body "fin" of a semiconductor material such as silicon as the channel of an FET device. In FinFET device architecture, the gate may be placed on two, three, four sides, or all around the ultrathin-body fin for improved device performance, and the devices could be fabricated on silicon or silicon-on-insulator (SOI) substrates [3,4]. The FinFETs offer significantly superior device performance including faster switching speed and higher current drive compared to the mainstream metal-oxide-semiconductor field-effect transistors (MOSFETs) [3–7]. Thus, considering the advantages of FinFETs over the conventional MOSFETs, Intel was the first to introduce triple-gate FinFET technology into mass production at the 22 nm node in the year 2011 [8]. This chapter presents a brief introduction and the history of the development of FinFET device technology.

Prior to the introduction of the FinFET structure, a brief overview of the mainstream MOSFET devices used for integrated circuit (IC) manufacturing technology, their limitations and scaling challenges, and the need for alternative devices such as FinFETs for IC manufacturing technology at the nanometer nodes is presented in Section 1.2.

1.2 OVERVIEW OF MOSFET DEVICES FOR INTEGRATED CIRCUIT MANUFACTURING

Over the past five decades, silicon-based microelectronic devices have revolutionized human society and continue to have an unprecedented impact on every aspect of modern society [7,9–11]. The basic building blocks of ICs are conductor, semiconductor, and insulator materials creating target electrical properties to design very large scale integrated (VLSI) circuits that are core elements of all modern electronic devices and systems. These electronic devices and systems enable communications, military, security, healthcare, energy saving, industrial automation, transport, autonomous vehicles, entertainment, infotainment, and digital society [10–12]. This microelectronics revolution started after the invention of bipolar junction transistors (BJTs) by W. Shockley, J. Bardeen, and W. Brattain at Bell Laboratories in 1947 [13,14], followed by the invention of first working ICs in September 1958 by J. Kilby of Texas Instruments [15,16]. Subsequently, in April 1959, Robert Noyce at Fairchild patented the practical monolithic IC technology for high volume manufacturing [17]. The inventions of BJTs and ICs paved the way for the realization of complete ICs with both active and passive elements on monolithic silicon substrates by the late 1950s. And, through the 1960s, the IC or microelectronics industry was dominated by BJT manufacturing technology.

In the late 1950s, semiconductor companies started the development of manufacturing technology for MOSFET device that was invented by J.E. Lilienfeld in 1926 [18]. In 1960, D. Kahng and M. M. Atalla fabricated working MOS transistors and demonstrated the first MOSFET amplifier enabling cost-effective integration of a large number of MOSFETs with interconnections on a single silicon chip [19,20]. In 1963, F. Wanlass invented the complementary MOS (CMOS) device configuration to build two-input acting on one output inverter circuit using both n-type and p-type MOSFET devices [21]. In 1965, Gordon Moore at Fairchild estimated that the number of transistors per chip would double about every two years which later transformed itself into a law known as "Moore's law" [22]. The underlying idea behind Moore's law was that the smaller devices improve almost every aspect of IC operation including a reduction in the cost and switching power consumption per transistor along with enhancement in memory capacity and speed.

The successful development of MOSFET device technology and the cost-effective integration of a large number of these devices in IC-chips started the progress of MOSFET technology from the early 1970s. And subsequently, the 1970s saw MOSFET technology begin to overtake BJT technology in terms of the functional complexity and level of integration. Moore's law became the guiding principle to scale down MOSFET device dimensions to manufacture lower cost, lower power, and higher computing speed IC-chips from each new generation of technology. To aid Moore's law for accurate estimation of device scaling, R. Dennard *et al.* at IBM quantified the scaling rules for IC process design in 1974 [23]. These scaling rules accelerated the global race to shrink physical dimensions of devices and manufacture more complex VLSI circuits enabling continuous increase in the integration density proposed by Moore's law to achieve higher speed and reduced power consumption of a digital MOS circuit at reduced cost. The high speed and low power consumption became the driving force for the microelectronics industry to achieve higher integration density of MOSFETs in every following generation of IC technology [24]. Thus, CMOS technology with its cost-effective technology solution became the pervasive technology for ICs from the 1980s [9].

Figure 1.1 shows an ideal conventional MOSFET device structure used in CMOS technology for manufacturing VLSI circuits. As shown in Figure 1.1, a MOSFET device is characterized by a channel length L_g, a channel width W, a gate oxide with thickness T_{ox}, substrate doping N_b, and source-drain with junction depth X_j. In VLSI circuits, NMOS (p-type body with $n+$ source-drain) and PMOS (n-type body with $p+$ source-drain) FET devices are integrated together and called CMOS transistors. In the context of device operation, a MOSFET is a four-terminal device with *gate*, *source*, *drain*, and *substrate or body* and the corresponding terminal voltages are V_g, V_s, V_d, and V_b, respectively, as shown in Figure 1.1 [9]. The device is symmetrical and cannot be distinguished without the applied bias. An applied gate bias V_g above

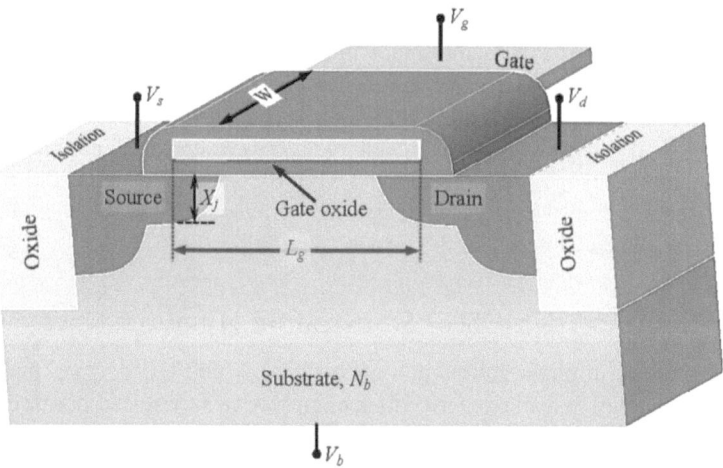

FIGURE 1.1 An ideal three-dimensional structure of an advanced four-terminal bulk-MOSFET device: here V_g, V_s, V_d, and V_b are the applied biases at the gate, source, drain, and body terminals, respectively; W and L_g are the channel width and channel length of the device, respectively; and X_j is the source-drain junction depth.

a certain threshold voltage (V_{th}) induces a conducting channel between the source and drain at the silicon/silicon dioxide (SiO_2) interface of the substrate and a drain bias V_d is used to generate current-voltage characteristics of the devices. The body terminal allows the modulation of the inversion layer from the gate as well as body to offer more flexibility of device performance in circuit operation. The basic operation of a MOSFET device is available in most books on semiconductor devices [9,25,26].

According to the MOSFET scaling rule [23], as L_g is scaled down, T_{ox} is scaled proportionately to maintain a strong capacitive coupling between the inversion channel and the gate terminal with respect to the other transistor terminals. This enables controlling of the short-channel effects (SCEs) including threshold voltage (V_{th}) roll-off, subthreshold swing (S) degradation, and drain-induced barrier lowering (DIBL) that act together to increase the off-state transistor leakage current (I_{off}) [9,27]. Also, scaling down L_g is accompanied by a corresponding increase in the body doping N_b and a decrease in source-drain junction depth X_j to reduce the sub-surface leakage paths below the channel inversion layer. Thus, the worldwide effort continued for the miniaturization of MOSFETs and CMOS technology to provide increasingly higher computing power, lower power consumption, and cheaper cost IC-chips at each new technology node [24,28].

With aggressive scaling of MOSFETs, the transistor dimensions soon reached the point at which the first-order assumptions about the physical effects and dopant distributions began to break down. For MOSFETs, the intrinsic device problem such as output conductance, velocity saturation, and subthreshold behavior all received substantial research interest and effort [9,29]. In order to continue device miniaturization following Moore's law, the US semiconductor industry association (SIA) formed the National Technology Roadmap for Semiconductors (NTRS) in 1994 and transition to the International Technology Roadmap for Semiconductors (ITRS) to include the global semiconductor companies in 2000 [30,31]. The ITRS defined the comprehensive guidelines for the future generations of technology by following Moore's law and Dennard's scaling rule. This scaling methodology has worked very well for several generations of VLSI technology. However, for 32 nm planar-CMOS technology and beyond, the conventional MOSFET scaling rules could no longer provide positive improvements in device performance set by Moore's law due to unsurmountable challenges [32–34] as described in Section 1.2.1.

1.2.1 CHALLENGES OF MOSFET SCALING AT THE NANOMETER NODE

The continuous miniaturization of conventional MOSFET devices and planar-CMOS technology to achieve performance improvement, reduced power consumption, and higher-density integration has become more challenging at the same rate of Moore's law due to several constraints imposed by the fundamental principles of device physics such as SCEs [32–37]. The major constraints to MOSFET device scaling in the sub-32 nm regime include degradation of leakage current and process variability-induced device parameter variation [9,26,27].

1.2.1.1 Leakage Current in Short Channel MOSFETs

As MOSFET devices are scaled down in the nanometer regime, the device performance becomes increasingly dependent on gate length, L_g [9,26,27]. As L_g decreases, the body doping concentration N_b is increased according to the scaling rule to suppress source-drain punchthrough and sub-surface leakage current (due to DIBL) [38]. This increase in N_b leads to a reduction in the carrier mobility due to the enhanced vertical electric field resulting in degradation in the current drive of the scaled devices [9]. The high vertical electric field at the source-drain junction due to high N_b, also increases the band-to-band tunneling and thereby off-state leakage current, I_{off}. Thus, scaling MOSFET gate length L_g in the decananometer regime degrades device characteristics and subthreshold swing as well as decreases V_{th}, causing V_{th} roll-off as shown in Figure 1.2 [9,35].

The subthreshold swing S degradation and V_{th} roll-off with decreasing L_g due to SCEs poses a severe challenge to control leakage current in devices beyond the 32 nm node [9,35]. This implies that the scaled MOSFETs cannot be turned off easily by lowering the gate voltage V_g due to SCEs even with reducing the gate-dielectric thickness in proportion to L_g following the scaling rule [9,27]. Thus, a major constraint for the continuous scaling of conventional bulk-MOSFETs is controlling the leakage current in the scaled devices [9,27,38] as shown in Figure 1.3. The leakage paths that are several nanometers below the silicon/gate-dielectric interface are primarily responsible for the observed leakage current in the scaled devices at the nanometer nodes [27]. This challenge to scaling L_g led to the enormous efforts on channel-profile engineering, shallow source-drain extensions (SDEs), and halo implants around SDEs [4,39–44] to continue scaling MOSFETs at the nanometer node. However, for the conventional MOSFETs below the 20 nm regime, the sub-surface leakage cannot be controlled by reducing the gate-dielectric thickness in the lowest possible value, even with zero thickness [27].

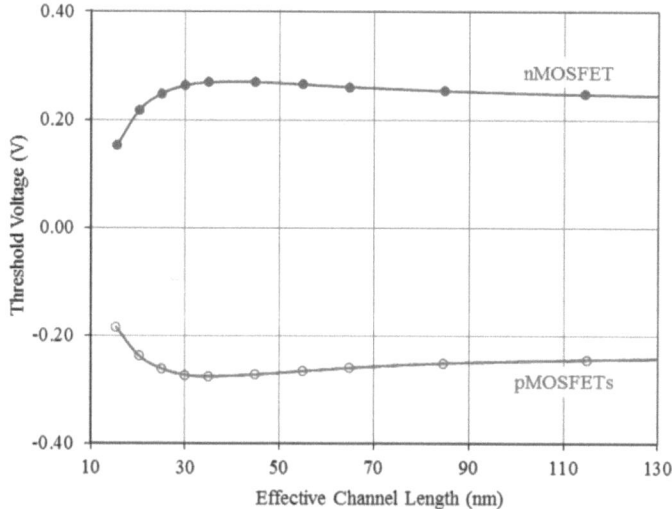

FIGURE 1.2   The threshold voltage V_{th} roll-off as a function of the effective channel length L_{eff} of nMOSFET and pMOSFET devices of a typical 40 nm CMOS technology with effective oxide thickness, T_{ox}(effective) = 1.5 nm. (V_{th} is defined as the gate bias, V_{gs} that is required to induce a conducting channel between the source and drain of MOSFETs enabling current flows in devices).

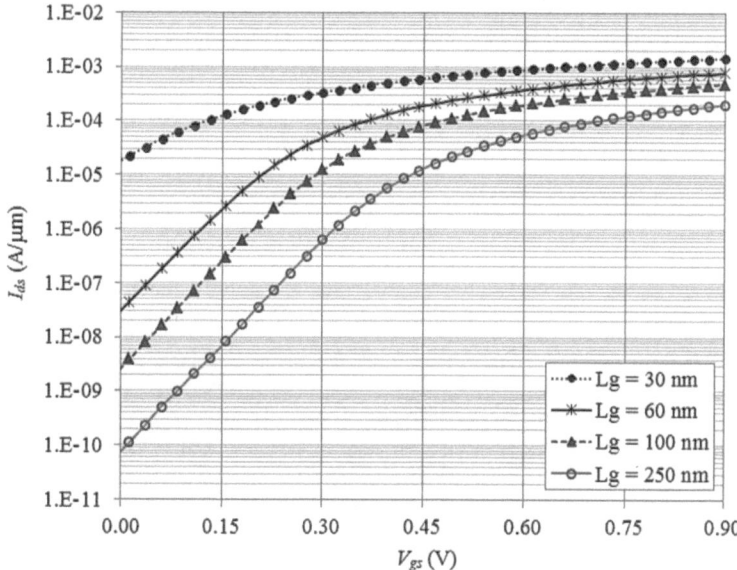

FIGURE 1.3   The drain current I_{ds} versus V_{gs} characteristics of the conventional nMOSFET devices of a typical 65 nm CMOS technology as a function of gate length; here the numerical simulation data are shown to illustrate the continuous increase in the off-state leakage current for the gate lengths below 60 nm.

1.2.1.2 Variability in MOSFETs

With the continued miniaturization of MOSFET devices [23,31,42–44], the performance variability of MOSFETs induced by process variability has become a critical
issue in the design of VLSI circuits using scaled CMOS technologies [45,46]. Process
variability in scaled CMOS technologies severely impacts the delay and power variability in VLSI devices, circuits, and systems, and this impact keeps increasing as
MOSFET devices and CMOS technologies continue to scale down [45–51]. Because
of SCEs, the device characteristics become increasingly sensitive to L_g variations,
and therefore, the process-induced variability imposes a serious challenge for continuous scaling of conventional MOSFETs [45–47]. As L_g decreases, N_b is increased
in keeping with the scaling rule to suppress source-drain punchthrough and sub-surface leakage current. The presence of heavy body doping N_b exacerbates the random
device-to-device variability in the form of random discrete doping (RDD) [46–48].
And, in scaled MOSFETs, the RDD is the major contributor to overall variability in
device performance due to V_{th} variation as shown in Figure 1.4 [9].

The effect of V_{th} variation causes variation in the performance of devices of the
conventional CMOS technology at the nanometer node as shown in Figure 1.5.
Figure 1.5 shows the distribution of ON-currents, IONN and IONP for the conventional 20 nm nMOSFETs and pMOSFETs, respectively. Here, IONN and IONP are
extracted at the supply bias of magnitude 1 V. The IONN *versus* IONP distribution
in Figure 1.5, clearly shows the impact of overall process variability on nanometer
scale device performance.

In order to overcome the major scaling challenges of the conventional planar
MOSFETs in the nanometer nodes as described in Section 1.2.1, it is important to
understand the physics of SCEs causing leakage current and the variability in device

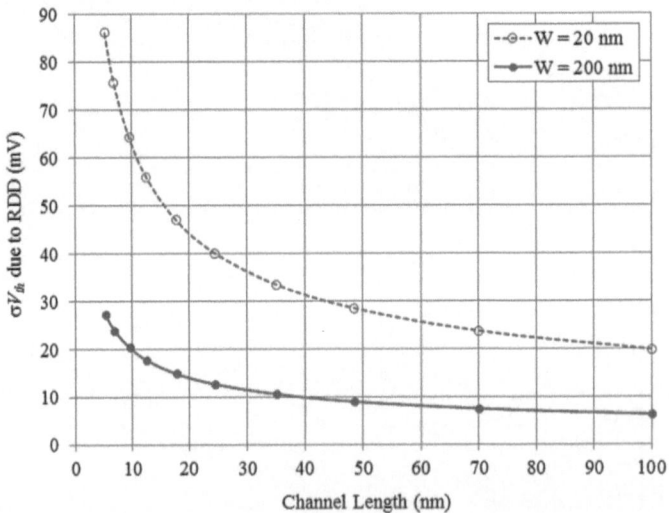

FIGURE 1.4 Threshold voltage variation of MOSFET devices for a typical 20 nm bulk-
CMOS technology as a function of device channel length for a channel width of 20 and
200 nm; here, σV_{th} is the variance in threshold voltage due to RDD; all data are taken from
the reference [9].

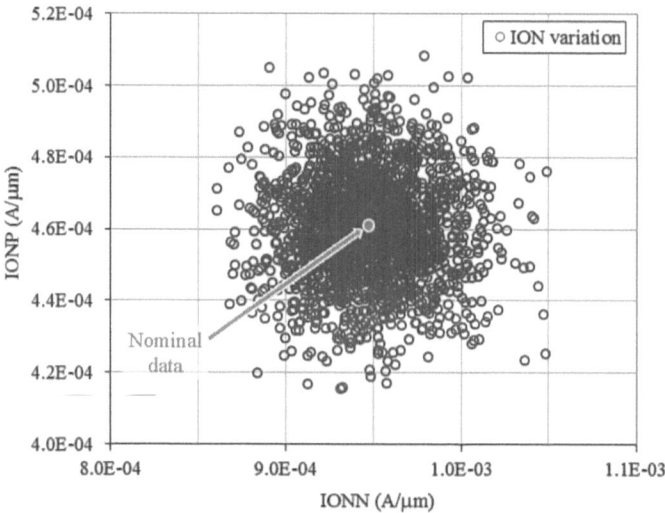

FIGURE 1.5 The distribution of ON-currents of *p*MOSFETs (IONP) and *n*MOSFETs (IONN) for a typical 20 nm CMOS technology showing the impact of process variation on device performance. The nominal values of IONP and IONN are also shown on the plot. All data are obtained by the industry standard circuit simulation tool using statistical device model for L_{eff} = 20 nm; W_{eff} = 1000 nm devices; and T_{ox}(effective) = 1.1 nm at gate bias, V_{gs} = 1 V = V_{ds} (drain bias).

performance as L_g decreases and discuss the alternative device structures to suppress SCEs and enable continuous scaling of devices to improve device performance according to Moore's law. In Section 1.2.2, the underlying physical principles of the above described scaling challenges for the conventional MOSFETs are briefly discussed.

1.2.2 PHYSICS OF MOSFET SCALING CHALLENGES

As shown in Figure 1.3, the I_{ds}–V_{gs} characteristics of the conventional MOSFETs of a planar-CMOS technology degrade in two major ways with scaling down L_g. First of all, S degrades and V_{th} decreases with decreasing L_g – that is, the device cannot be turned off easily by lowering V_{gs}. Secondly, S and V_{th} become increasingly sensitive to L_g variations – that is, the variation in device performance becomes more problematic as shown in Figures 1.4 and 1.5. These problems are referred to as the SCEs and the cause of SCEs in the conventional MOSFETs can be understood from Figure 1.6. Typically, a transistor is turned ON and OFF when V_g lowers and raises the potential of the channel, respectively; and thus, modulates the potential barrier between the channel and the source through the gate-to-channel capacitance, C_g (~C_{ox}, gate oxide capacitance). In an ideal transistor, the channel potential is controlled only by V_g and C_g. However, in real MOSFET devices, the channel potential is, also, subjected to the influence of V_d through the channel-to-drain coupling capacitance C_{dsc} as shown in Figure 1.6. For a device with large L_g, C_{dsc} is significantly smaller than C_g. Therefore, the influence of V_{ds} is insignificant in modulating the channel and V_{gs} solely controls

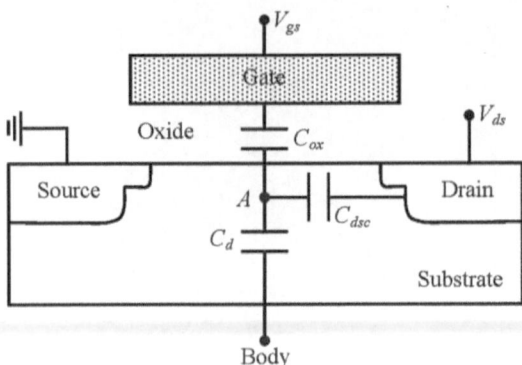

FIGURE 1.6 A conventional MOSFET device showing gate capacitance C_{ox}, bulk capacitance C_d, and the source-drain to channel coupling capacitances C_{dsc}; with scaling L_g, C_d increases due to the increase in C_{dsc} and controls the channel potential in the nanometer node devices; here, A represents the capacitance voltage divider node between C_{ox} and the effective capacitance of C_d and C_{dsc}.

the channel potential. As L_g decreases, the effective value of C_d increases [9,52,53] and V_{gs} loses its absolute control of the channel. In extreme cases, V_{gs} has less control than V_{ds} and the transistor can be turned on by V_{ds} only as shown in Figure 1.3 for a 30 nm MOSFET device. Before the extreme case (L_g = 30 nm) is reached, we observe the gradually deteriorating subthreshold device performance in Figure 1.3 with decreasing L_g from 250 nm and below.

The ideal MOSFET scaling rule to increase the gate control of the channel is to increase C_g by reducing the value of T_{ox} in proportion to L_g [23]. However, the conventional scaling rule is limited to improve V_{gs} control of the MOSFET channel and leakage current beyond the 20 nm regime even with the lowest possible T_{ox} due to fundamental physical constraint as shown in two-dimensional (2D) cross-section of an ideal planar MOSFET in Figure 1.7 [27]. It is seen from Figure 1.7 that the leakage paths far from the gate are worse than the surface leakage path since they are only weakly controlled by V_{gs}. As a result, the potential barriers along these weakly controlled paths can be easily lowered by V_{ds} through the large C_{dsc} in a small L_g device as shown in Figure 1.8 [9,27,38].

There are two possible options to overcome the above described scaling constraints in conventional MOSFETs at the nanometer node planar-CMOS technology. The first option is the introduction of new technologies and new materials into the conventional planar bulk-MOSFETs to allow further scaling and boost the performance of scaled transistors [54,55]. The second option being alternative transistor architectures such as ultrathin-body FETs and multiple-gate FETs which inherently have superior electrostatic control over the inversion channel [9,27] as discussed in Section 1.3.

1.3 ALTERNATIVE DEVICE CONCEPTS

In order to overcome the increasing challenges in continuous scaling of the conventional planar MOSFET devices, the major research and development efforts for the

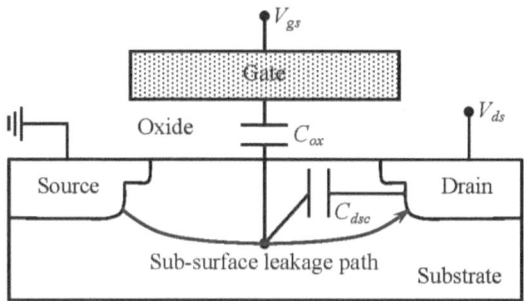

FIGURE 1.7 A conventional MOSFET device showing the sub-surface leakage path between the source and drain; the sub-surface leakage cannot be controlled by scaling gate oxide thickness in nanometer node devices due to C_{dsc} which becomes comparable or higher than C_{ox} to control the channel potential.

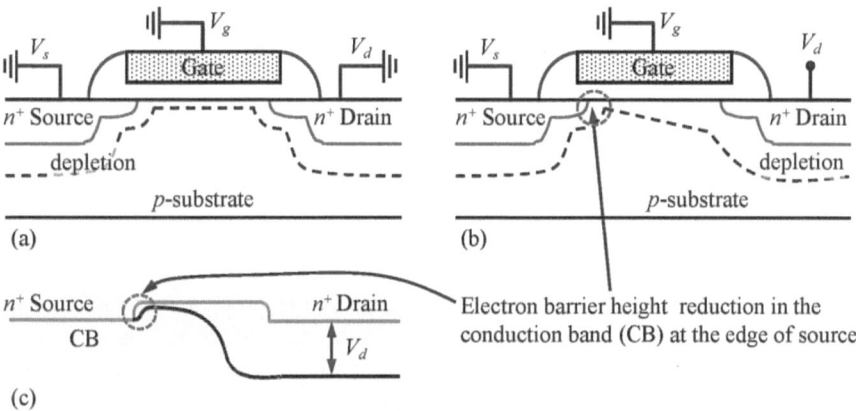

FIGURE 1.8 Lowering of source-to-channel potential in an n-channel device due to drain bias V_{ds}: (a) $V_{gs} = 0$ and $V_{ds} = 0$; (b) $V_{gs} = 0$ and $V_{ds} =$ supply voltage, V_{dd}; (c) plot of conduction band along the length of the device under zero-bias (top curve) and at drain-bias conditions (bottom curve).

last two decades have been to explore alternative device architectures [9,27,56–61] for VLSI manufacturing technology at the nanometer nodes. In Section 1.3.1, we briefly review the device options for continued scaling of planar-CMOS technology and thin-body devices as an alternative to conventional planar-CMOS technology.

1.3.1 UNDOPED OR LIGHTLY DOPED CHANNEL MOSFETs

1.3.1.1 Deeply Depleted Channel MOSFETs

Recently, advanced channel engineering has been developed to design nanoscale MOSFET devices with undoped or lightly doped channel to reduce the effect of RDD as shown in Figures 1.9 [62]. The channel is formed on an undoped epitaxial layer grown on silicon substrate followed by standard CMOS processing

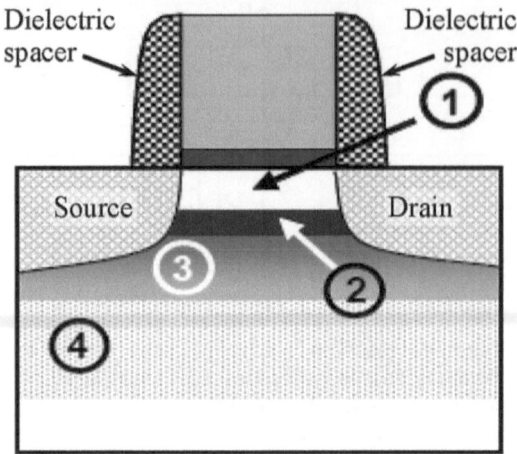

1. Undoped channel; **2.** V_{th}-control layer;
3. SCE control layer; **4.** Anti-punchthrough.

FIGURE 1.9 Deeply depleted channel MOSFET: In the schematic, region 1 is the undoped channel of the device, whereas, 2, 3, and 4 are the V_{th}-control, SCE suppression, and anti-punchthrough channel-type doped impurity layers, respectively.

steps [62,63]. This bulk-MOSFET structure is referred to as the deeply depleted channel (DDC) MOSFETs [62].

The DDC device shown in Figure 1.9 controls the process variability using undoped or lightly doped channel and suppresses the sub-surface leakage current by reducing C_{dsc} using heavily doped channel-type doping layers 2 and 3 deep into the channel [62].

1.3.1.2 Buried-Halo MOSFETs

In a planar-CMOS technology, a lightly doped channel is used in conjunction with a heavily doped channel-type implant, referred to as the "halo doping," around the source-drain to reduce sub-surface leakage [9]. The reported data show that the device architecture using two halo doping profiles, referred to as the "double-halo MOSFETs," reduces the leakage current as well as controls V_{th} variation in nanoscale devices [5,9,42–44]. Recently, a variability-tolerant double-halo MOSFET device architecture shown in Figure 1.10 has been invented to design undoped or lightly doped channel MOSFETs on undoped epitaxial layer to suppress SCEs and mitigate the risk of process variability in devices of planar-CMOS technology [64,65]. This new double-halo structure is referred to as the "buried-halo MOSFET" (BH-MOSFET). In BH-MOSFET architecture, well, optional V_{th}-adjust, and multiple halo implants are performed on bulk-silicon substrate prior to growing epitaxial-channel layer as shown in Figure 1.10, and the gate patterning and source-drain processing steps are performed after the epitaxy [64,65]. The reported data on V_{th}-variability clearly show a significant reduction in V_{th} variation due to RDD in nanoscale BH-MOSFETs compared to the conventional MOSFET devices [9,47,64].

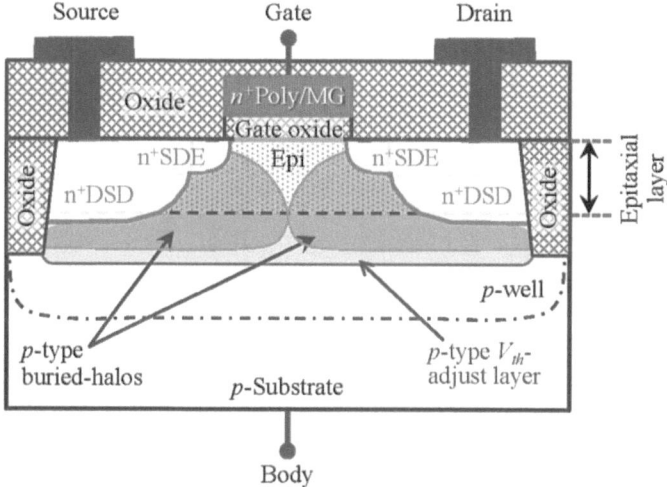

FIGURE 1.10 Buried-halo MOSFET device structure formed by up-diffusion of multiple halo implant profiles in the undoped epitaxial layer from silicon substrate; in the fabrication process, a heavily doped first halo is used around the source-drain extension (SDE) regions to suppress leakage paths near the silicon-surface closer to the gate and a lightly doped second halo is used around the deep source-drain (DSD) regions to suppress leakage paths far from the gate, whereas the undoped epitaxial channel reduces the process variability in devices.

Figure 1.11 shows the estimated V_{th} variability of the conventional and BH-MOSFET devices of a typical 20 nm planar-CMOS technology down to $L_g =$ 5 nm [9,47,64]. The data clearly show a significant reduction in V_{th} variation due to RDD in nanometer scale BH-MOSFETs compared to the conventional MOSFET devices.

1.3.2 Thin-Body Field-Effect Transistors

The novel device architecture to ensure greater gate control of the channel for FET devices is to use thin-body silicon as the channel. There are two ways to enhance the gate control of the body and reduce the drain control of the channel and C_{dsc} by (1) ultrathin body on an SOI substrate as described in Section 1.3.2.1 and (2) multiple gates around an ultrathin-body silicon channel as described in Section 1.3.2.2 [27].

1.3.2.1 Single-Gate Ultrathin-Body Field-Effect Transistors

The SCEs in a MOSFET can be significantly suppressed by using an ultrathin SOI substrate [66] to bring silicon closer to the gate. However, the improvement of SCEs in MOSFETs on SOI substrates depends on the technology parameters such as silicon film thickness t_{si}, gate-dielectric (silicon dioxide) thickness, and body doping concentration. The reported data show that the leakage current decreases with a decrease of t_{si} [67,68]. And, by reducing t_{si} to only around 7 to 14 nm, SCEs can be significantly suppressed by eliminating the worst leakage paths terminated in the buried silicon dioxide as shown in Figure 1.12 [67,68]. Furthermore, ultrathin-body FETs

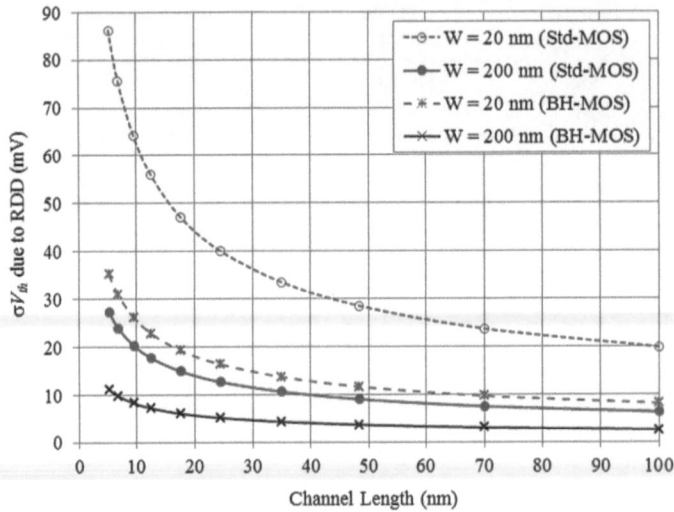

FIGURE 1.11 Buried-halo MOSFETs: Comparison of the simulated threshold voltage variation of the conventional (Std-MOS) and BH-MOSFET (BH-MOS) devices of a typical 20-nm bulk-CMOS technology as a function of the channel length for channel width 20 and 200 nm.

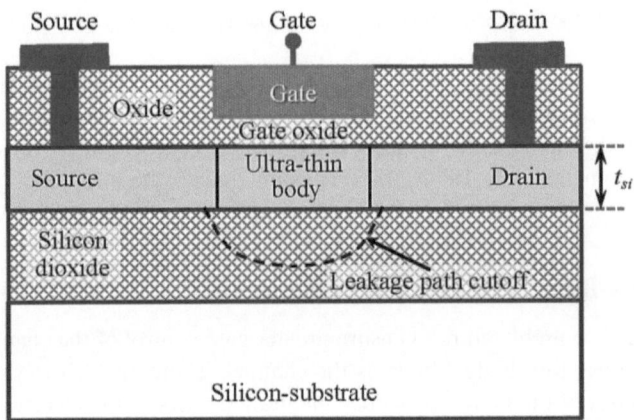

FIGURE 1.12 Ultrathin-body MOSFETs on SOI substrate: The structure shows that for an appropriate thickness of silicon, t_{si} body, the sub-surface leakage path is terminated in the buried oxide, thus suppressing the leakage current.

on undoped or lightly doped substrate reduces variability. The ultrathin-body FETs on SOI substrate, referred to as the UTB-SOI-MOSFETs, have emerged as one of the most promising devices for advanced VLSI circuits at the nanometer node [66–71].

1.3.2.2 Multiple-Gate Field-Effect Transistors

The SCEs can be more effectively controlled by using multiple gates, referred to as the multigate around the silicon body or channel of a MOSFET. Figure 1.13 shows a

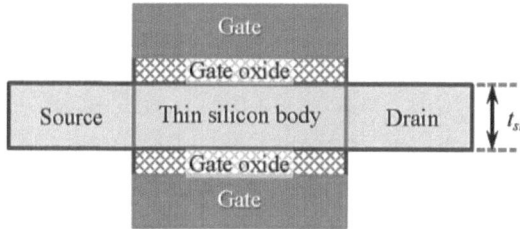

FIGURE 1.13 Double-gate MOSFETs: A thin silicon body with gates at the top and bottom of the silicon body to provide complete gate control of the body from the top and bottom, thus eliminating the sub-surface leakage paths. Here t_{si} is the thickness of the silicon body.

typical FET device with a thin silicon body and two gates; one above and the other below the body. For such a double-gate (DG) FET device with thin body shown in Figure 1.13, the potential leakage paths far from the top gate are closer to the bottom gate eliminating the potential sub-surface leakage far from the top gate; and the leakage paths far from the bottom gate are closer to the top gate, thus eliminating the leakage paths far from the bottom gate or sub-surface leakage. Thus, a DG-MOSFET offers stronger electrostatic control of the inversion channel by the gates above and below the silicon body. And, therefore, reduces the SCEs and makes the multigate FETs more scalable than the planar-CMOS devices on bulk substrate [72,73].

Thin-body architecture (Figure 1.13) also eliminates the requirement for heavy channel doping for suppressing SCEs contrary to the conventional scaling principle. Optional channel doping may be used to adjust V_{th} to the target specification instead of gate metal workfunction engineering [9,27]. The undoped or lightly doped thin body as the channel reduces RDD in multigate FETs and therefore, the variability in device performance can be eliminated. Furthermore, an undoped or lightly doped body reduces the average electric field in the channel which translates to an improvement in the carrier mobility, gate leakage currents, and device reliability. This, potentially, improves the bias-temperature stability (negative-bias temperature instability and positive-bias temperature instability) and the gate-dielectric tunneling leakage and wear out [27]. The combination of the light body doping and thin body offers steeper subthreshold swing and lower junction and body capacitances [27].

From the above discussions, we find that a thin-body DG-MOSFET structure shown in Figure 1.13 shows great potential to control SCEs and suppress leakage current by keeping the gate closer to the silicon body and reduce process variability. Then by rotating the DG-structure by 90 degrees to make it a vertical DG-MOSFET standing on a silicon or SOI substrate, it would be possible to achieve the lowest possible gate leakage current with self-aligned gates. Such a vertical DG-MOSFET with thin "fin"-like body structure is referred to as the "FinFET," as shown in Figure 1.14 and described in Section 1.4.

1.4 FinFET DEVICES FOR VLSI CIRCUITS AND SYSTEMS

A FinFET structure is one of the most feasible multigate configurations in terms of manufacturability. This multigate thin-body device configuration invented in 1980 [1]

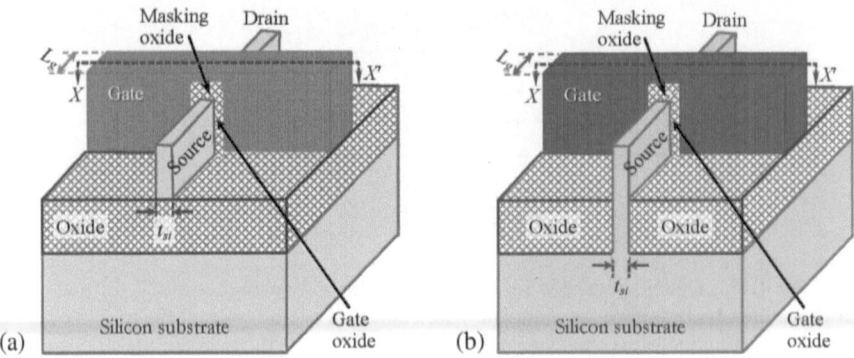

FIGURE 1.14 Ideal three-dimensional ultrathin-body DG-FinFET device structures: (a) on silicon-on-insulator (SOI) substrate and (b) on bulk-silicon substrate; here, L_g and t_{si} are the gate length and thin body silicon "fin" thickness, respectively; and XX' denotes the cutline to illustrate the 2D cross-sections of the 3D devices.

shows a larger immunity for SCEs compared to classical single-gate planar devices [9,27]. Figure 1.14 shows typical three-dimensional (3D) DG-FinFET structures where the gate controls the channel along the silicon sidewalls of the fin. In FinFETs, the silicon body can be controlled by either two gates, three gates, or four gates. The higher the number of gates the higher the electrostatic control of the channel, but there is a trade-off with the corresponding process complexity. FinFETs can be built on SOI substrate as shown in Figure 1.14(a) or on bulk-silicon substrate as shown in Figure 1.14(b). In DG-FinFET structures, the gate oxide is grown on the sidewalls of the fin body and a thick masking oxide is grown on the top of the fin as shown in Figure 1.14. Figures 1.15(a) and (b) show 2D cross-sections along the cutline $XX'$of 3D structures in Figures 1.14(a) and (b), respectively.

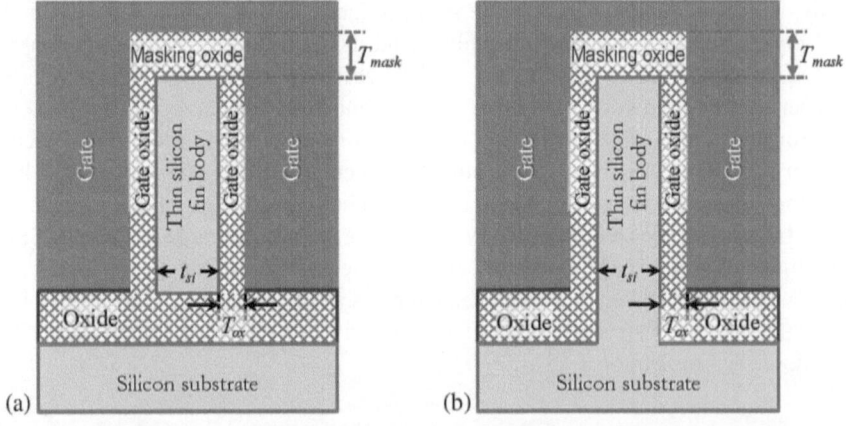

FIGURE 1.15 Two-dimensional cross-sections of the ideal DG-FinFET devices: (a) on SOI substrate and (b) on bulk-silicon substrate; here, the source-drain regions are normal to the surface; t_{si} is the thickness of the silicon body and T_{mask} is the thickness of the oxide layer on the top of the silicon body.

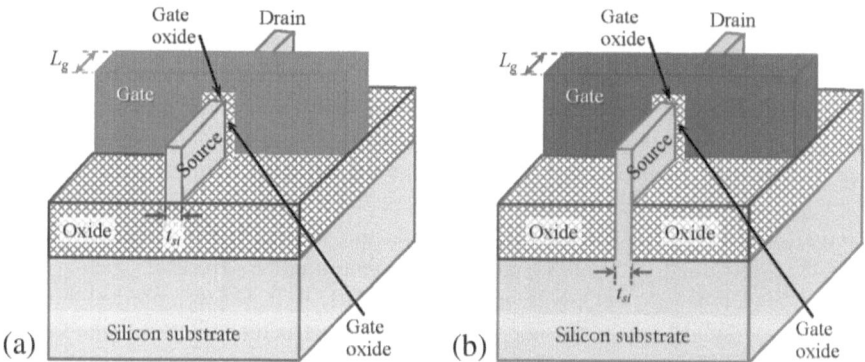

FIGURE 1.16   Ideal 3D triple-gate FinFET device structures: (a) on SOI substrate and (b) on bulk-silicon substrate; here, L_g and t_{si} are the gate length and fin thickness, respectively, and $T_{mask} = T_{ox}$.

The 2D structures shown in Figures 1.15(a) and (b) represent the vertical planes along the cutline XX' of the corresponding 3D structures shown in Figures 1.14(a) and (b), respectively. Here, the source-drain regions are perpendicular to the surface.

In triple-gate configuration of FinFETs, the same gate oxide thickness is grown on the sidewalls as well as on the top of the fin-channel as shown in Figure 1.16. Thus, in triple-gate FinFETs, the thickness and dielectric material of the masking oxide T_{mask} is the same as the sidewall gate oxide T_{ox}.

The multiple gates of the multigate FETs offer strong electrostatic control over the inversion channel and reduce the coupling between the source and drain in the subthreshold region [74,75]. First of all, the multigate FETs show great potential to mitigate the risk of process variability by using undoped channel. Secondly, a reduction of four orders of magnitude in the leakage current of multigate devices has been observed over the MOSFET devices of a comparable 32 nm planar-CMOS technology [8]. Thus, the multigate FETs are adopted for large-scale manufacturing of advanced VLSI technology at the nanometer nodes [8,74–77]. The ultrathin body enables the continuous scaling down of FETs by overcoming the major scaling constraints such as SCEs and RDD of the conventional bulk-MOSFETs discussed in Section 1.2.1.

Thus, the FinFETs are 3D vertical structures that stand on the planar substrate. Since the gate wraps around the channel, it provides sole control of the conducting channel from the source to drain. As a result, the off-state leakage current through the body is insignificant. The reported data show that the scaling of FinFETs gate length is related to the thickness of the channel [9,78].

Though the FinFET device technology was adopted for manufacturing in early 2011 [8,74–76], the research and development efforts of FinFET-type multigate device architecture started in the early 1980s as described in Section 1.5.

1.5   A BRIEF HISTORY OF FinFET DEVICES

As discussed in Section 1.4, a FinFET offers an excellent SCE immunity in scaled devices enabling continuous scaling of device dimensions towards their fundamental limit near the 3 nm regime [78]. Though FinFET device technology was introduced for manufacturing VLSI circuits in 2011, the research and development efforts on double-gate FETs started in the early 1980s [1]. The FinFET-type multigate FET device structure was originally invented by Yutaka Hayashi at the Electrotechnical Laboratory, Tsukuba, Japan in 1980 [1], as shown in Figure 1.17. The original patent was filed at the Japanese patent office on June 24, 1980, with an assigned application number of S55-85706 and was published prior to examination by *Public Notice Number*: S57-10973 on January 20, 1982, and the second *Public Notice Number*: H05-4822 on January 20, 1993, after patent examination. Hayashi's patent was granted on October 14, 1993, with patent No. JP,1791730,B. Hayashi noticed that the 2D cross-sectional view of the lateral or planar DG-structure resembles the Greek letter "Ξ" (xi) with channel as the *center bar* and two gates as *top and bottom bars* which corresponds to the English letter "X." Therefore, in the later reports, Hayashi named the lateral double-gate structure as the "XMOS" [79].

Hayashi's patent application on vertical channel multigate FETs has been in the public domain since January 1982. In 1987, K. Heida *et al.* reported a triple-gate vertical FET device with fully depleted silicon body and trench isolated sidewall gates on two sides of the silicon body [80]. Heida *et al.* showed that the sidewall gate electrodes increase the gate-controllability to the channel and fully depleted body of silicon improves switching operation of the device due to small body bias effect.

In 1989 and 1990, D. Hisamoto *et al.* reported a vertical channel ultrathin-body SOI transistor in triple-gate configuration called DELTA (depleted lean-channel transistor) with silicon-body thickness of 200 nm along with a simple process flow for device fabrication [81,82]. Using experimental and simulation data the authors' showed that the gate of the DELTA structure offers effective channel controllability, and its vertical ultrathin-body SOI structure provides superior device characteristics including suppression of SCEs, near-ideal subthreshold swing (62 mV/decade for

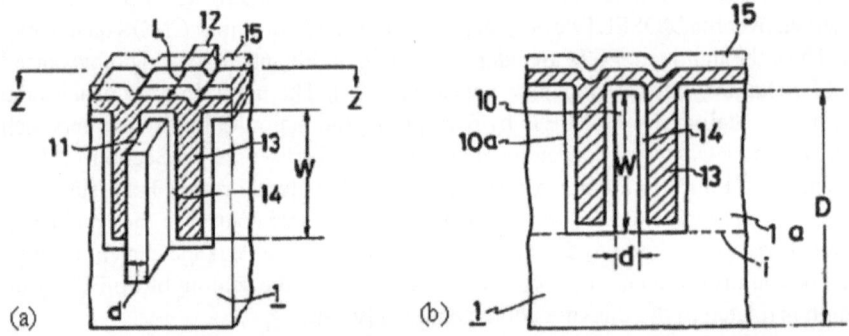

FIGURE 1.17 Multigate vertical device structure invented by Yutaka Hayashi in 1980: (a) 3D structure and (b) 2D cross-sectional view along the cutline ZZ shown in (a). In the figure, L, W, and d are the channel length, height of silicon body (region 10), and body thickness, respectively; regions 1 and 1a are the oxide and silicon, respectively, with i as the boundary between silicon and oxide; 10a is the silicon/silicon-dioxide interface; 11 and 12 are the source and drain, respectively; 13 and 14 are the gate electrode and gate oxide, respectively; and 15 is the passivation layer on the top of the device.

effective channel length L_{eff} = 0.57 μm, t_{si} = 0.15 μm, and T_{ox} = 8.5 nm), and high transconductance. In the 1991 report, Hisamoto *et al.* presented the detailed fabrication processes along with a simple mathematical formulation of DELTA device performance [83].

The real development of thin-body multiple-gate FETs structure with the new acronym FinFET started in the late 1990s at the University of California, Berkeley, CA. In 1996, the Berkeley research group led by Chenming Hu engaged in the development of FinFET device technology under the sponsored program on Advanced Microelectronics (AME) of the US government's Defense Advanced Research Projects Agency (DARPA). Under this DARPA AME sponsored program in 1998, Hisamoto *et al.* reported a folded-channel MOSFET with L_g down to 17 nm on SOI substrate using a spacer-based process [84]. This is the first report describing "fin" as the body of an *n*-channel MOSFET device. In 1999, Huang *et al.* reported the first *p*-channel FinFETs with SiGe gate, polysilicon-Ge source-drain, and nitride spacer [2]. In 2001, Choi *et al.* demonstrated sub-20 nm CMOS FinFET devices (t_{si} = 10 nm, T_{ox} = 2.1 nm) using spacer-defined lithography and Ge-raised source-drain technologies with excellent device characteristics and discussed the advantage of spacer-defined lithography in doubling the fin-density compared to the conventional lithography [85].

Subsequently, device researchers from industry and academia engaged in the study of FinFET devices and technology worldwide. In 2001, Kedzierski *et al.* demonstrated the feasibility of ultimate integration of FinFETs for ultra large scale integrated (ULSI) circuits using scalable lithography techniques, a combination of highly angled source-drain extension implants and selective silicon epitaxy to fabricate raised source-drain FinFETs with low parasitic series resistance [86]. They showed that for fully depleted symmetric DG-FinFETs with L_{eff} = 30 nm, t_{si} = 20 nm, and low extrinsic resistance R_{ext}, offer high ON currents of about 1.3 mA/μm and 0.85 mA/μm for *n*-type and *p*-type devices, respectively, at 1.5 V supply voltage with I_{off} ~200 nA/μm for both types of devices.

In 2002, Kedzierski *et al.* reported undoped channel FinFET devices with metal-gate using total gate silicidation [87]. The authors demonstrated that the metal-gate with dual-workfunction provides V_{th} controllability by metal workfunction engineering and the silicided-gate thin-body FinFETs with L_g = 100 nm, t_{si} = 25 nm, and T_{ox} = 1.6 nm offering higher mobility, lower gate-leakage, higher gate capacitance, and nearly ideal subthreshold swing. In the same year, Yu *et al.* reported complementary FinFET device characteristics demonstrating the scalability of FinFETs on SOI substrate down to 10 nm using a modified dual poly-Si gate CMOS fabrication process [88]. Also in 2002, Yang *et al.* demonstrated high performance, low leakage, and low active-power 25 nm FinFET-like multigate structure with a top gate, sidewall gates, and gate extensions under the silicon body resembling an Omega-shaped structure [89].

In 2003, Doyle *et al.* reported fully depleted high performance triple-gate FETs on SOI substrate [90,91]. The authors showed that the high performing non-planar CMOS-FET devices with L_g = 60 nm, t_{si} = 55 nm, fin height H_{fin} = 36 nm, and T_{ox} = 1.5 nm can be fabricated with targeted threshold voltages. In the same year, Park *et al.* reported fabrication processes to manufacture high performance 60 nm body-tied FinFETs or Omega-FETs with fin top thickness of 30 nm, fin bottom thickness of 61 nm, H_{fin} = 99 nm, and T_{ox} = 20 nm [92].

In 2004, Lee *et al.* demonstrated a highly manufacturable dynamic random access memories (DRAMs) with body-tied FinFET cell array on bulk-silicon substrate with dimensions t_{si} = 80 nm, H_{fin} = 100 nm, and L_g = 90 nm [93]. In the same year, Ha *et al.* demonstrated HfO_2 high-*k* gate-dielectric with effective oxide thickness down to 1.72 nm and molybdenum (MO) metal gate FinFET device technology with low gate leakage current, comparable carrier mobilities, and feasibility of V_{th} adjustment by workfunction engineering via nitrogen implantation [94]. In 2004, Yang *et al.* reported the scalability of FinFETs to 10 and 5 nm [95].

In 2005, Yang *et al.* studied the feasibility of undoped body multigate FETs and fully depleted single-gate SOI-FETs to assess SCEs and gate layout area using 3D numerical device simulation [6]. This study showed that the DG-FinFETs offer better scalability than the multigate FETs and single-gate fully depleted (FD) SOI-FETs (FDSOI-FETs) and are feasible for VLSI technology at the nanometer node. Similar simulation-based study has been reported comparing 20 nm FinFETs with the conventional high performance MOSFETs showing that FinFETs offer a better performance matrix than the bulk-MOSFETs [5]. And, the research and development efforts continue in industry, academia, and research laboratories.

The first reported FinFET circuit was a four-stage inverter chain using over 300 fins with L_g = 200 nm, t_{si} = 60 nm, and T_{ox} = 2.2 nm operating at 1.5 V by Rainey *et al.* in 2002 [96]. Nowak *et al.* reported the FinFET static random access memories (SRAMs) in 2002 [97] and FinFET ring oscillator in 2003 [98]. In 2004, Choi *et al.* demonstrated the manufacturability of triple-gate FETs based 20 MB SRAM array [99]. Subsequently, FinFETs have been used for various logic and memory applications including high speed digital ICs [100,101], analog ICs [102,103], SRAMs [104–106], flash memories [107–111], and DRAMs [93,112–114].

Intel was the first to introduce tri-gate FinFET technology into mass production for the 22 nm technologies and beyond [8], followed by the other major semiconductor manufacturing companies including Taiwan Semiconductor Manufacturing Company (TSMC), Samsung, and Globalfoundries for manufacturing ULSI circuits [74–76].

1.6 SUMMARY

This chapter presented an overview of the FinFET devices as an alternative to the mainstream MOSFETs for the manufacturing of VLSI circuits and systems at the 22 nm node and beyond. First of all, the emergence of MOSFETs and CMOS technology as the pervasive technology for IC manufacturing since the 1980s is discussed. It is shown that MOSFETs and CMOS technology could be continuously scaled down following Moore's law and Dennard's scaling rule to improve the speed and performance of ICs at reduced cost. However, it is shown that the scaling of MOSFETs according to Moore's law could not be continued for small-geometry devices in the nanometer regime due to the limitations of fundamental physical principles. Due to these limitations, the ON and OFF performance of the small geometry MOSFETs beyond the 22 nm node cannot be controlled to the specifications of the target circuits and systems. Also, the scalable alternative device structures for continued scaling in the nanometer regime to improve the performance of VLSI circuits and systems are discussed. It is shown that the FinFET, due to its ultrathin body and multigate configuration, can overcome the scaling limitations of the bulk-MOSFETs and can be used as the replacement for MOSFETs in the sub-22 nm regime technology nodes. Finally, a brief history of the emergence and development of FinFET device technology and ICs is discussed.

REFERENCES

1. Y. Hayashi, "MOS field effect transistor," JP Patent Application S55-85706, June 24, 1980.
2. X. Huang, W.-C. Lee, C. Kuo, *et al.*, "Sub 50-nm FinFET: PMOS." In: *IEEE International Electron Devices Meeting Technical Digest*, pp. 67–70, 1999.
3. J.-P. Colinge (ed.), *FinFETs and Other Multi-Gate Transistors*, Springer, New York, 2008.
4. J.P. Colinge, M.H. Gao, A. Romano-Rodriguez, H. Maes, and C. Clays, "Silicon-on-insulator 'Gate-all-around device.'" In: *IEEE Electron Devices Meeting Technical Digest*, pp. 595–598, 1990.
5. S. Saha, "Device characteristics of sub-20-nm silicon nanotransistors." In: *Proceedings of the SPIE Conference on Design and Process Integration for Microelectronic Manufacturing*, vol. 5042, pp. 172–180, 2003.
6. J.-W. Yang and J.G. Fossum, "On the feasibility of nanoscale triple-gate CMOS transistors," *IEEE Transactions on Electron Devices*, 52(6), pp. 1159–1164, 2005.
7. J.G. Fossum and V.P. Trivedi, *Fundamentals of Ultra-Thin-Body MOSFETs and FinFETs*, Cambridge University Press, Cambridge, UK, 2013.
8. J. Markoff, *Intel Increases Transistor Speed by Building Upward*, May 4, 2011. www.nytimes.com/2011/05/05/science/05chip.html.
9. S.K. Saha, *Compact Models for Integrated Circuit Design: Conventional Transistors and Beyond*, CRC Press, Taylor & Francis Group, Boca Raton, FL, 2015.
10. S.K. Saha, "Transitioning semiconductor companies enabling smart environments and integrated ecosystems," *Open Journal of Business & Management*, 6(2), pp. 428–437, 2018.
11. O. Vermesan and P. Friess (eds.), *Internet of Things – Converging Technologies for Smart Environments and Integrated Ecosystems*, River Publishers, Denmark, 2013.
12. S.K. Saha, "Emerging business trends in the microelectronics industry," *Open Journal of Business & Management*, 4(1), pp. 105–113, 2016.
13. W. Shockley, "Semiconductor amplifier," US Patent 2502488, April 4, 1950.
14. J. Bardeen and W.H. Brattain, "Three-electrode circuit element utilizing semiconductor materials," US Patent 2,524,035, October 3, 1950.
15. J.S. Kilby, "Miniaturized electronics circuits," US Patent 3138743, June 23, 1964.
16. J.S. Kilby, "Invention of the integrated circuit," *IEEE Transactions on Electron Devices*, 23(7), pp. 648–654, 1976.
17. R.N. Noyce, "Semiconductor device-and-lead structure," US Patent 2981877, April 26, 1961.
18. J.E. Lilienfeld, "Method and apparatus for controlling electric currents," US Patent 1,745,175, January 28, 1930.
19. D. Kang and M.M. Atalla, "Silicon-silicon dioxide field induced surface device." In: *IRE Solid Sate Device Research Conference*, Carnegie Institute of Technology, Pittsburgh, PA, 1960.

20. D. Kahng, "Electric field controlled semiconductor device," US Patent 3,102,230, August 27, 1963.
21. F.M. Wanlass, "Low stand-by power complementary field effect circuitry," US Patent 3,358,858, December 5, 1967.
22. G.E. Moore, "Cramming more components onto integrated circuits," *Electronics*, 38(8), pp. 114–117, 1965.
23. R.H. Dennard, F.H. Gaensslen, H.N. Yu., *et al.*, "Design of ion-implanted MOSFETs with very small physical dimensions," *IEEE Journal of Solid-State Circuits*, SC-9(5), pp. 256–268, 1974.
24. D.L. Critchlow, "MOSFET scaling - The driver of VLSI technology," *Proceedings of the IEEE*, 87(4), pp. 659–667, 1999.
25. Y. Taur and T.H. Ning, *Fundamentals of Modern VLSI Devices*, Cambridge University Press, Cambridge, 1998.
26. N. Arora, *MOSFET Models for VLSI Circuit Simulation: Theory and Practice*, Springer–Verlag, Wien, 1993.
27. Y.S. Chauhan, D.D. Lu, S. Venugopalan, *et al.*, *FinFET Modeling for IC Simulation and Design: Using the BSIM-CMG Standard*, Academic Press, San Diego, CA, 2015.
28. K.J. Kuhn, "Considerations for ultimate CMOS scaling," *IEEE Transactions on Electron Devices*, 59(7), pp. 1813–1828, 2012.
29. Y. Cheng and C. Hu, *MOSFET Modeling and BSIM3 User's Guide*, Kluwer Academic Publishers, Boston, MA / Dordrecht / London, 1999.
30. Semiconductor Industry Association, *The National Technology Roadmap for Semiconductors 1994*, http://www.rennes.supelec.fr/ren/perso/gtourneu/enseigne-ment/roadmap94.pdf
31. International Technology Roadmap for Semiconductors (ITRS), http://www.itrs2.net/
32. Y. Taur, D.A. Buchanan, W. Chen, *et al.*, "CMOS scaling into the nanometer regime," *Proceedings of the IEEE*, 85(4), pp. 486–504, 1997.
33. H. Iwai, "CMOS technology – Year 2010 and beyond," *IEEE Journal of Solid-State Circuits*, 34(3), pp. 357–366, 1999.
34. D.J. Frank, R.H. Dennard, E. Nowak, *et al.*, "Device scaling limits of Si MOSFETs and their application dependencies," *Proceedings of the IEEE*, 89(3), pp. 259–288, 2001.
35. G. Bertrand, S. Deleonibus, B. Previtali, *et al.*, "Toward the limits of conventional MOSFETs: Case of sub 30 nm NMOS devices," *Solid-State Electronics*, 48(4), pp. 505–509, 2004.
36. K.J. Kuhn, "Moore's law past 32nm: Future challenges in device scaling." In: *Proceedings of the 2009 13th International Workshop on Computational Electronics*, IWCE'09, pp. 1–6, 2009.
37. S.K. Saha, N.D. Arora, M.J. Deen, and M. Miura-Mattausch, "Advanced compact models and 45-nm modeling challenges," *IEEE Transactions on Electron Devices*, 53(9), pp. 1957–1960, 2006.
38. R.R. Troutman, "VLSI limitations from drain-induced barrier lowering," *IEEE Journal of Solid-State Circuits*, 14(2), pp. 383–391, 1979.
39. S. Saha, "Effects of inversion layer quantization on channel profile engineering for nMOSFETs with 0.1 μm channel lengths," *Solid-State Electronics*, 42(11), pp. 1985–1991, 1998.
40. S.K. Saha, "Method for forming channel-region doping profile for semiconductor device," US Patent 6,323,520, November 27, 2001.
41. S.K. Saha, "Drain profile engineering for MOSFET devices with channel lengths below 100 nm." In: *Proceedings of the SPIE Conference on Microelectronic Device Technology*, vol. 3881, pp. 195–204, 1999.
42. S.K. Saha, "Transistors having optimized source-drain structures and methods for making the same," US Patent 6,344,405, February 5, 2002.

43. S. Saha, "Scaling considerations for high performance 25 nm metal-oxide-semicon- ductor field-effect transistors," *Journal of Vacuum Science and. Technology B*, 19(6), pp. 2240–2246, 2001.

44. S. Saha, "Design considerations for 25 nm MOSFET devices," *Solid-State Electronics*, 45(10), pp. 1851–1857, 2001.

45. K.J. Kuhn, M.D. Giles, D. Becher, *et al.*, "Process technology variation," *IEEE Transactions on Electron Devices*, 58(8), pp. 2197–2208, 2011.

46. S.K. Saha, "Modeling process variability in scaled CMOS technology," *IEEE Design & Test of Computers*, 27(2), pp. 8–16, 2010.

47. S.K. Saha, "Compact MOSFET modeling for process variability-aware VLSI circuit design," *IEEE Access*, 2, pp. 104–115, 2014.

48. A. Asenov, "Random dopant induced threshold voltage lowering and fluctuations in sub-0.1 μm MOSFET's: A 3-D 'atomistic' simulation study," *IEEE Transactions on Electron Devices*, 45(12), pp. 2505–2513, 1998.

49. C.M. Mezzomo, A. Bajolet, A. Cathignol, R. Di Frenza, and G. Ghibaudo, "Characterization and modeling of transistor variability in advanced CMOS technolo- gies," *IEEE Transactions on Electron Devices*, 58(8), pp. 2235–2248, 2011.

50. K. Bernstein, D.J. Frank, A.E. Gattiker, *et al.*, "High-performance CMOS variability in the 65-nm regime and beyond," *IBM Journal of Research & Development*, 50(4/5), pp. 433–449, 2006.

51. S.K. Springer, S. Lee, N. Lu, *et al.*, "Modeling of variation in submicrometer CMOS ULSI technologies," *IEEE Transactions on. Electron Devices*, 53(9), pp. 2168–2178, 2006.

52. C.C. Hu, *Modern Semiconductor Devices for Integrated Circuits*, Prentice Hall, Upper Saddle River, NJ, 2010.

53. Z.H. Liu, C. Hu, J.-H. Huang, *et al.*, "Threshold voltage model for deep-submicrometer MOSFETs," *IEEE Transactions on Electron Devices*, 40(1), pp. 86–95, 1993.

54. M. Radosavljevic, B. Chu-Kung, S. Corcoran, *et al.*, "Advanced high-K gate dielectric for high-performance short-channel $In_{0.7}Ga_{0.3}As$ quantum well field effect transistors on silicon substrate for low power logic applications." In: *IEEE International Electron Devices Meeting Technical Digest*, pp. 319–322, 2009.

55. M. Caymax, G. Eneman, F. Bellenger, *et al.*, "Germanium for advanced CMOS anno 2009: A SWOT analysis." In: *IEEE International Electron Devices Meeting Technical Digest*, pp. 461–464, 2009.

56. F. Schwierz, "Graphene transistors," *Nature Nanotechnology*, 5(7), pp. 487–496, 2010.

57. B. Radisavljevic, A. Radenovic, J. Brivio, V. Giacometti, A. Kis, "Single-layer MoS2 transistors," *Nature Nanotechnology*, 6(3), pp. 147–150, 2011.

58. J. Kanghoon, L. Wei-Yip, P. Patel, *et al.*, "Si tunnel transistors with a novel silicided source and 46mV/dec swing." In: *Symposium on VLS Technology*, pp. 121–122, 2010.

59. A.I. Khan, D. Bhowmik, P. Yu, *et al.*, "Experimental evidence of ferroelectric negative capacitance in nanoscale heterostructures," *Applied Physics Letters*, 99(11), p. 113501- 1–113501-3, 2011.

60. M. Radosavljevic, G. Dewey, D. Basu, *et al.*, "Electrostatics improvement in 3-D tri- gate over ultra-thin body planar InGaAs quantum well field effect transistors with high-K gate dielectric and scaled gate-to-drain/gate-to-source separation." In: *IEEE International Electron Devices Meeting Technical Digest*, pp. 765–768, 2011.

61. K. Tomioka, M. Yoshimura, and T. Fukui, "Steep-slope tunnel field-effect transistors using III-V nanowire/si-heterojunction." In: *Symposium on VLS Technology*, pp. 47–48, 2012.

62. K. Fujita, Y. Tori, M. Hori, *et al.*, "Advanced channel engineering achieving aggressive reduction of VT variation for ultra-low-power applications." In: *IEEE International Electron Devices Meeting Technical Digest*, pp. 749–752, 2011.

63. J.D. Plummer, M.D. Deal, and P.B. Griffin, *Silicon VLSI Technology: Fundamentals, Practice and Modeling*, Prentice Hall, Upper Saddle River, NJ, 2000.

64. S.K. Saha, "Transistor structure and method with an epitaxial layer over multiple halo implants," US Patent 9,299,702, March 29, 2016.

65. S.K. Saha, "Transistor structure and fabrication methods with an epitaxial layer over multiple halo implants," US Patent No. 9,768,074, September 19, 2017.

66. S. Cristoloveanu and F. Balestra, "Introduction to SOI technology and transistors." In: *Physics and Operation of Silicon Devices and Integrated Circuits*, J. Gautier, (ed.), ISTE-Wiley, London, UK, New York, 2009.

67. Y.-K. Choi, K. Asano, N. Lindert, *et al.*, "Ultrathin-body SOI MOSFET for deep-subtenth micron era," *IEEE Electron Device Letters*, 21(5), pp. 254–255, 2000.

68. Q. Liu, A. Yagishita, N. Loubet, *et al.*, "Ultra-thin-body and BOX (UTBB) fully depleted (FD) device integration for 22nm node and beyond." In: *Symposium on VLSI Technology*, pp. 61–62, 2010.

69. V. Barral, T. Poiroux, F. Andrieu, *et al.*, "Strained FDSOI CMOS technology scalability down to 2.5nm film thickness and 18nm gate length with a TiN/HfO2 gate stack." In: *IEEE International Electron Devices Meeting Technical Digest*, pp. 61–64, 2011.

70. K. Cheng, A. Khakifirooz, P. Kulkarni, *et al.*, "Fully depleted extremely thin SOI technology fabricated by a novel integration scheme featuring implant-free, zero-silicon-loss, and faceted raised source/drain," *Symposium on VLSI Technology*, pp. 212–213, 2009.

71. O. Faynot, F. Andrieu, O. Weber, *et al.*, "Planar fully depleted SOI technology: A powerful architecture for the 20nm node and beyond." In: *IEEE International Electron Devices Meeting Technical Digest*, pp. 50–53, 2010.

72. H.-S.P. Wong, D.J. Frank, and P.M. Solomon, "Device design considerations for double-gate, ground plane and single-gate ultra-thin SOI MOSFETs at the 25nm channel length consideration." In: *IEEE International Electron Devices Meeting Technical Digest*, pp. 407–410, 1998.

73. K. Suzuki, T. Tanaka, Y. Tosaka, H. Horie, and Y. Arimoto, "Scaling theory for double-gate SOI MOSFETs," *IEEE Transactions on Electron Devices*, 40(12), pp. 2326–2329, 1993.

74. C. Auth, C. Allen, A. Blattner, *et al.*, "A 22-nm-high performance and low-power CMOS technology featuring fully-depleted tri-gate transistors, self-aligned contacts and high density MIM capacitors." In: *Symposium on VLS Technology*, pp. 131–132, 2012.

75. R. Merritt, *TSMC Taps ARM's V8 on Road to 16-nm FinFET*, October 16, 2012. www. eetimes.com/document.asp?doc_id=1262655.

76. D. McGrath, *Globalfoundries Looks to Leapfrog Fab Rival with New Process*, September 20, 2012. www.eetimes.com/document.asp?doc_id=1262552.

77. D. Hisamoto, W.-C. Lee, J. Kedzierski, *et al.*, "FinFET - A self-aligned double gate MOSFET scalable to 20 nm," *IEEE Transactions on Electron Devices*, 47(12), pp. 2320–2325, 2000.

78. P. Clarke, *KAIST Claims Record Size 3-nm FinFETs*, March 14, 2006. www.eetimes. com/document.asp?doc_id=1160025.

79. T. Sekigawa and Y. Hayashi, "Calculated threshold-voltage characteristics of an XMOS transistor having an additional bottom gate," *Solid-State Electronics*, 27(8–9), pp. 827–828, 1984.

80. K. Hieda, F. Horiguchi, H. Watanabe, *et al.*, "New effect of trench isolated transistor using side-wall gates." In: *IEEE International Electron Devices Meeting Technical Digest*, pp. 736–739, 1987.

81. D. Hisamoto, T. Kaga, Y. Kawamoto, and Takeda Eiji, "A fully depleted lean-channel transistor (DELTA) – a novel vertical ultra thin SOI MOSFET." In: *IEEE International Electron Devices Meeting Technical Digest*, pp. 833–836, 1989.

82. D. Hisamoto, T. Kaga, Y. Kawamoto, and Takeda Eiji, "A fully depleted lean-channel transistor (DELTA) – a novel vertical ultrathin SOI MOSFET," *IEEE Electron Device Letters*, 11(1), pp. 36–38, 1990.

83. D. Hisamoto, T. Kaga, and Takeda Eiji, "Impact of the vertical SOI 'DELTA' structure on planar device technology," *IEEE Transactions on Electron Devices*, 38(6), pp. 1419–1424, 1991.

84. D. Hisamoto, W.-C. Lee, J. Kedzierski, *et al.*, "A folded-channel MOSFET for deep-sub-tenth micron era." In: *IEEE International Electron Devices Meeting Technical Digest*, pp. 1032–1034, 1998.

85. Y.-K. Choi, N. Lindert, P. Xuan, *et al.*, "Sub-20nm CMOS FinFET technologies." In: *IEEE International Electron Devices Meeting Technical Digest*, pp. 421–424, 2001.

86. J. Kedzierski, D.M. Fried, E.J. Nowak, *et al.*, "High-performance symmetric-gate and CMOS-compatible Vt asymmetric-gate FinFET devices." In: *International Electron Devices Meeting Technical Digest*, pp. 437–440, 2001.

87. J. Kedzierski, D.M. Fried, E.J. Nowak, *et al.*, "Metal-gate FinFET and fully-depleted SO1 devices using total gate silicidation." In: *International Electron Devices Meeting Technical Digest*, pp. 247–250, 2002.

88. B. Yu, L. Chang, S. Ahmed, *et al.*, "FinFET scaling to 10nm gate length." In: *International Electron Devices Meeting Technical Digest*, pp. 251–254, 2002.

89. F.-L. Yang, H.-Y. Chen, F.-C. Chen, *et al.*, "25 nm CMOS Omega FETs." In: *International Electron Devices Meeting Technical Digest*, pp. 255–258, 2002.

90. B. Doyle, S. Datta, M. Doczy, *et al.*, "High performance fully-depleted tri-gate CMOS transistors," *IEEE Electron Device Letters*, 24(4), pp. 263–265, 2003.

91. B. Doyle, B. Boyanov, S. Datta, *et al.*, "Tri-gate fully-depleted CMOS transistors: Fabrication, design and layout." In: *Symposium on VLSI Technology*, pp. 133–134, 2003.

92. T. Park, S. Choi, D.H. Lee, *et al.*, "Fabrication of body-tied FinFETs (Omega MOSFETs) using Bulk Si Wafers." In: *Symposium on VLSI Technology*, pp. 135–136, 2003.

93. C.H. Lee, J.M. Yoon, C. Lee, *et al.*, "Novel body tied FinFET cell array transistor DRAM with negative word line operation for sub 60nm technology and beyond." In: *Symposium on VLSI Technology*, pp. 130–131, 2004.

94. D. Ha, H. Takeuchi, Y.-K. Choi, *et al.*, "Molybdenum-gate HfO_2 CMOS FinFET technology." In: *International Electron Devices Meeting Technical Digest*, pp. 643–646, 2004.

95. F.-L. Yang, D.-H. Lee, H.-Y. Chen, *et al.*, "5nm-gate nanowire FinFET." In: *Symposium on VLSI Technology*, pp. 196–197, 2004.

96. B.A. Rainey, D.M. Fried, M. Ieong, *et al.*, "Demonstration of FinFET CMOS circuits." In: *IEEE Device Research Conference Digest*, pp. 47–48, 2002.

97. E.J. Nowak, B.A. Rainey, D.M. Fried, *et al.*, "A functional FinFET-DGCMOS SRAM cell." In: *International Electron Devices Meeting Technical Digest*, pp. 411–414, 2002.

98. E.J. Nowak, T. Ludwig, I. Aller, *et al.*, "Scaling beyond the 65 nm node with FinFET-DGCMOS." In: *Proceedings of the IEEE Custom Integrated Circuits Conference*, pp. 339–342, 2003.

99. J.A. Choi, K. Lee, Y.S. Jin, *et al.*, "Large scale integration and reliability consideration of triple gate transistors." In: *International Electron Devices Meeting Technical Digest*, pp. 647–650, 2004.

100. W. Rosner, E. Landgraf, J. Kretz, *et al.*, "Nanoscale FinFETs for low power applications," *Solid-State Electronics*, 48(10–11), pp. 1819–1823, 2004.

101. K. von Arnim, E. Augendre, C. Pacha, *et al.*, "A low-power multi-gate FET CMOS technology with 13.9 ps inverter delay, large-scale integrated high performance digital circuits and SRAM." In: *Symposium on VLSI Technology*, pp. 106–107, 2007.

102. P. Wambacq, B. Verbruggen, K. Scheir, *et al.*, "Analog and RF circuits in 45 nm CMOS and below: Planar bulk versus FinFET." In: *Proceedings of the European Solid-State Device Research Conference Technical Digest*, pp. 53–56, 2006.

103. J.-P. Raskin, T.M. Chung, V. Kilchytska, D. Lederer, D. Flandre, "Analog/RF performance of multiple gate SOI devices: Wideband simulations and characterization," *IEEE Transactions on Electron Devices*, 53(5), pp. 1088–1095, 2006.

104. Z. Guo, S. Balasubramanian, R. Zlatanovici, *et al.*, "FinFET-based SRAM design." In: *Proceedings of the International Symposium on Low Power Electronics and Design Technical Digest*, pp. 2–7, 2005.

105. H. Kawauka, K. Okano, A. Kaneko, *et al.*, "Embedded bulk FinFET SRAM cell technology with planar FET peripheral circuit for hp 32 nm node and beyond." In: *Symposium on VLSI Technology*, pp. 86–87, 2006.

106. T. Park, H.J. Cho, J.D. Chae, *et al.*, "Characteristics of the full CMOS SRAM cell using body-tied TG MOSFETs (bulk FinFETs)," *IEEE Transactions on Electron Devices*, 53(3), pp. 481–487, 2006.

107. I.H. Cho, T.-S. Park, S.Y. Choi, *et al.*, "Body-tied double-gate SONOS flash (omega flash) memory device built on bulk Si wafer." In: *Proceedings of the Device Research Conference Technical Digest*, pp. 133–134, 2003.

108. T.-H. Hsu, H.-T. Lue, E.-K. Lai, *et al.*, "A high-speed BE-SONOS NAND flash utilizing the field enhancement effect of FinFET." In: *Symposium on VLSI Technology*, pp. 913–916, 2007.

109. C. Gerardi, S. Lombardo, G. Cina, *et al.*, "Highly manufacturable/low aspect ratio Si nano floating gate FinFET memories: High speed performance and improved reliability." In: *Proceedings of the Non-Volatile Semiconductor Memory Workshop Technical Digest*, pp. 44–45, 2007.

110. J.-R. Hwang, T.-L. Lee, H.-C. Ma, *et al.*, "20 nm gate bulk-FinFET SONOS flash." In: *IEEE Electron Devices Meeting Technical Digest*, pp. 154–157, 2005.

111. E.S. Cho, T.-Y. Kim, B.K. Cho, *et al.*, "Technology breakthrough of body-tied FinFET for sub 50 nm NOR flash memory." In: *Symposium on VLSI Technology*, pp. 110–111, 2006.

112. C. Lee, J.-M. Yoon, C.-H. Lee, *et al.*, "Enhanced data retention of damascene-finFET DRAM with local channel implantation and ⟨100⟩ fin surface orientation engineering." In: *IEEE Electron Devices Meeting Technical Digest*, pp. 61–64, 2004.

113. Y.-S. Kim, S.-H. Lee, S.-H. Shin, *et al.*, "Local-damascene-FinFET DRAM integration with p+ doped poly-silicon gate technology for sub-60 nm device generations." In: *IEEE Electron Devices Meeting Technical Digest*, pp. 315–318, 2005.

114. D.-H. Lee, S.-G. Lee, J.R. Yoo, *et al.*, "Improved cell performance for sub-50 nm DRAM with manufacturable bulk FinFET structure." In: *Symposium on VLSI Technology*, pp. 167–165, 2007.

2 Fundamentals of Semiconductor Physics

2.1 INTRODUCTION

The basic materials used for integrated circuit (IC) device fabrication are conductors, semiconductors, and insulators creating desired electrical properties to manufacture very large scale integrated (VLSI) circuits with target performance objective. Thus, the characteristics of an IC device depend on the properties of its constituent materials along with its geometrical and structural information. And, in atomic level, the IC device characteristics are modulated by the transport of current carrying fundamental constituents of materials, referred to as the *electrons* and *holes*. Again, the electronic properties of semiconductors primarily depend on the transport of the majority carrier electrons or holes. The semiconductors with majority carrier concentration as electrons are referred to as the *n*-type, whereas the semiconductors with majority carrier concentration as holes are referred to as the *p*-type. Thus, in order to understand the performance of IC transistors in general, and "fin" field-effect transistor (FinFET) devices in particular, it is essential to understand the basic physics of the *n*-type and *p*-type semiconductors along with the transport properties of electrons and holes in building IC devices. Though a number of published titles are available on the subject [1–16], the objective of this chapter is to present a brief overview of semiconductor physics, basics of *n*-type and *p*-type semiconductors, and the characteristics of an *n*-type and a *p*-type semiconductor in contact forming a *pn*-junction that are necessary to understand the theory and operations of FinFET devices in VLSI circuits and systems.

2.2 SEMICONDUCTOR PHYSICS

Crystalline silicon is widely used as the starting semiconducting material for manufacturing VLSI devices and system-on-chips (SoCs). Thus, unless otherwise specified, in this book, the semiconductor physics is described with reference to silicon material. Thin silicon wafers used in the IC fabrication processes are cut parallel to either the $\langle 111 \rangle$ or $\langle 100 \rangle$ crystal planes. However, the $\langle 100 \rangle$ material is most commonly used due to the fact that during IC fabrication processes, $\langle 100 \rangle$ wafers produce the lowest amount of charge at the silicon/silicon dioxide (Si/SiO_2) interface and offer higher carrier mobility [17,18]. Thus, it is of great interest to study how the electrons and holes are bonded in a silicon atom and understand their transport mechanism in a silicon crystal. And therefore, in this section, the energy band model and transport properties of electrons and holes are discussed.

2.2.1 Energy Band Model

In a silicon crystal, each atom has four valence electrons and four nearest neighboring atoms. Each atom shares its valence electrons with its four neighbors in a paired configuration called a *covalent bond*. It is predicted by quantum mechanics (QM) that the allowed energy levels of electrons in a solid are grouped into two bands, called the *valence band* (VB) and the *conduction band* (CB) as shown in Figure 2.1. These bands are separated by an energy range that the electrons in a solid cannot possess and is referred to as the *forbidden band* or *forbidden gap*. The VB is the highest energy band and its energy levels are mostly filled with electrons forming the *covalent bonds*. The CB is the next highest energy band with its energy levels nearly empty. The electrons which occupy the energy levels in the CB are called the *free electrons* or *conduction electrons*.

Typically, the energy is a complex function of momentum in a three-dimensional space and there are many allowed energy levels for a large number of electrons in silicon and therefore, the energy band diagram is also complex [19]. For the simplicity of representation, only the edge levels of each of the allowed energy bands are shown in the energy band diagram (Figure 2.1). In Figure 2.1, E_c and E_v are the bottom-edge of the CB and the top-edge of the VB, respectively, and E_g is the bandgap energy separating E_c and E_v. And, at any ambient temperature $T(K)$, E_g is given by

$$E_g = E_c - E_v. \tag{2.1}$$

When a valence electron is given sufficient energy ($\geq E_g$), it can break out of the chemical bonding state and excite into the CB to become a free electron leaving behind a vacancy, or *hole*, in the VB. A hole is associated with a positive charge since a net positive charge is associated with the atom from which the electron broke away. Note that both the electron and hole are generated simultaneously from a

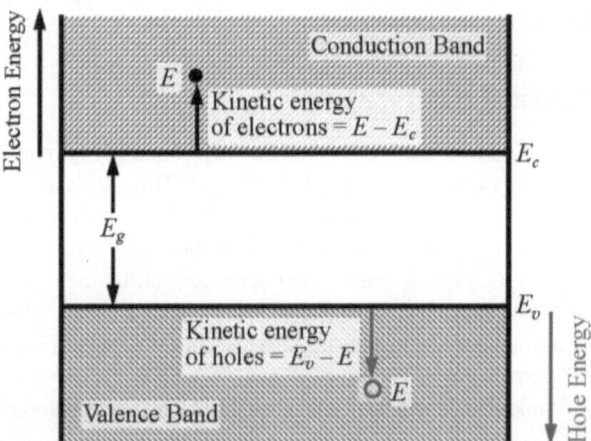

FIGURE 2.1 Energy band diagram of a semiconductor like silicon: E_c is the bottom-edge of the conduction band and E_v is the top-edge of the valance band which are separated by an energy gap $E_g = E_c - E_v$; in the figure, "•" represents the electrons and "O" represents the holes.

single event. The electrons move freely in the CB and holes move freely in the VB. In silicon, the bandgap is small (~1.12 eV), therefore, even at room temperature a small fraction of the valence electrons are excited into the CB generating electrons and holes. This allows limited conduction to take place from the motion of the electrons in the CB and holes in the VB. As shown in Figure 2.1, when an electron in the CB gains energy, it moves up to an energy $E > E_c$, while when a hole in the VB gains energy, it moves down to an energy $E < E_v$. Thus, the energy of the electrons in the CB increases upward while the energy of the holes in the VB increases downward.

The bandgap energy E_g for silicon at room temperature (300 °K) is ~1.12 eV. As the temperature increases, the value of E_g for most semiconductors decreases due to the increase in the crystal lattice spacing by thermal expansion. For silicon, the temperature coefficient of E_g at 300 °K temperature is: $\dfrac{dE_g}{dT} \cong -2.73 \times 10^{-4}$ eV K^{-1} [20]. The temperature dependence of E_g for silicon can be modeled by using polynomial functions valid for different range of temperatures [16,20]. However, in a circuit simulation tool like SPICE (Simulation Program with Integrated Circuit Emphasis) [21], the temperature dependence of E_g is modeled by [22]

$$E_g(T) = 1.160 - \frac{7.02 \times 10^{-4} T^2}{1108 + T} \tag{2.2}$$

where:

T is the temperature in Kelvin (K)
$E_g(T)$ is the energy gap in eV

2.2.2 CARRIER STATISTICS

The electrical properties of a semiconductor are determined by the number of carriers available for conduction. This number is determined from the density of states and the probability that these states are occupied by carriers. The probability that an available state with energy E is occupied by an electron under a thermal equilibrium condition is expressed by the *Fermi–Dirac* probability density function $f(E)$, also called the *Fermi function* [1–12] and is given by

$$f(E) = \frac{1}{1 + \exp\left(\dfrac{E - E_f}{kT}\right)} \tag{2.3}$$

where:

E_f is the *Fermi energy* or *Fermi level*
$k = 1.38 \times 10^{-23}$ J K^{-1} is the Boltzmann constant

From Equation 2.3, we find that at any $T > 0$, when $E = E_f$: $f(E) = \dfrac{1}{2}$ which means that the electron is equally likely to have an energy above E_f as below it. Thus, the Fermi level can be defined as the energy at which the probability of finding an electron, at any $T > 0$ °K, is exactly one half. Again, at absolute zero temperature

$(T = 0\ °K)$, $f(E) = 1$ for $E < E_f$ indicating that the probability of finding an electron below E_f is unity and above E_f is zero (that is, $f(E) = 0$ for $E > E_f$). In other words, at $T = 0\ °K$, all energy levels below E_f are filled and all energy levels above E_f are empty. At any finite temperature, some states above E_f are filled and some states below E_f become empty. As T increases above absolute zero, the function $f(E)$ changes as shown in Figure 2.2. Thus, the probability that the energy levels above E_f are filled increases with the increase of temperature. It is important to note that the Fermi function or Fermi energy applies only under the equilibrium condition.

Thus, Equation 2.3 describes the probability of an allowed energy state occupied by an electron with $E > E_f$. Then the probability of a state not occupied by an electron (with $E < E_f$) is given by

$$1 - f(E) = \frac{1}{1 + \exp\left(-\dfrac{E - E_f}{kT}\right)} \tag{2.4}$$

In other words, Equation 2.4 describes the probability function of the existence of a hole in silicon.

Again, we observe from Figure 2.2 that the probability distribution $f(E)$ makes a smooth transition from unity to zero as the energy increases across the Fermi level. The width of this transition is governed by the thermal energy kT. The value of thermal energy at room temperature is about 26 mV. Thus, for any energy at least several kT (~3kT) above E_f, the function $f(E)$ in Equations 2.3 and 2.4 can be approximated as

$$f(E) \cong \exp\left(-\frac{E - E_f}{kT}\right); \quad \text{for } E > E_f \tag{2.5}$$

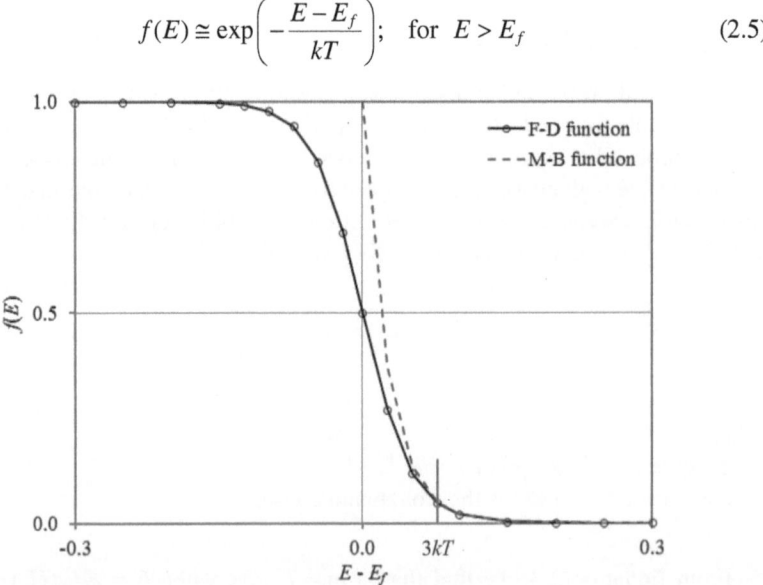

FIGURE 2.2 Fermi–Dirac (F–D) and Maxwell–Boltzmann (M–B) distribution functions in a semiconductor: the plots show that the F–D distribution can be approximated to an M–B distribution at any temperature T when $E - E_f > 3kT$.

and

$$1 - f(E) \cong \exp\left(-\frac{E_f - E}{kT}\right); \quad \text{for} \ \ E < E_f \tag{2.6}$$

Equations 2.5 and 2.6 are identical to Maxwell–Boltzmann (M–B) density function for classical gas particles [19]. For most device applications at room temperature, the function $f(E)$ given by Equation 2.5 is a good approximation as shown in Figure 2.2.

Fermi level can be considered to be the chemical potential for electrons and holes. Since the condition for any system in equilibrium is that the chemical potential must be constant throughout the system, it follows that the Femi level must be constant throughout a semiconductor in equilibrium.

2.2.3 Intrinsic Semiconductors

An *intrinsic semiconductor* is a perfect single crystal semiconductor with no impurities or lattice defects. In such materials, the VB is completely filled with electrons and the CB is completely empty. Therefore, in intrinsic semiconductors, there are no charge carriers at 0 °K. However, at higher temperatures, electron-hole pairs are generated as VB electrons are thermally excited across the bandgap to the CB. Thus, in intrinsic semiconductors, all the electrons in the CB are thermally excited from the VB. In other words, at a given temperature, the number of holes in the VB equals the number of electrons in the CB of an intrinsic semiconductor. Thus, if n and p are the concentrations of free electrons and holes, respectively, then

$$n = p = n_i \tag{2.7}$$

or,

$$np = n_i^2 \tag{2.8}$$

where:

n_i is called the *intrinsic carrier concentration* and is the free electron (or hole) concentration in an intrinsic semiconductor.

2.2.3.1 Intrinsic Carrier Concentration

We can derive an expression for the intrinsic carrier concentration in a semiconductor from the effective densities of carriers and probability distribution function. Thus, from Equation 2.5, we can show the expression for the concentration of electrons in the CB as

$$n \cong N_c \exp\left(-\frac{E_c - E_f}{kT}\right); \tag{2.9}$$

Similarly, from Equation 2.6, we can show the expression for the concentration of holes in the VB as

$$p \cong N_v \exp\left(-\frac{E_f - E_v}{kT}\right);$$

(2.10)

where:

N_c and N_v are the effective densities of states in the CB and VB, respectively.

The expressions for N_c and N_v are derived from QM considerations [2]. Both N_c and N_v are proportional to $T^{3/2}$. For an intrinsic semiconductor, $n = p = n_i$ and E_f is called the *intrinsic Fermi level* or the *intrinsic energy level*, E_i. Then (using $n = p = n_i$) we can write from Equations 2.9 and 2.10

$$N_c \exp\left(-\frac{E_c - E_f}{kT}\right)\Bigg|_{n=n_i, E_f = E_i} = N_v \exp\left(-\frac{E_f - E_v}{kT}\right)\Bigg|_{p=n_i, E_f = E_i}$$

(2.11)

Now, solving Equation 2.11 for E_f and using $E_f = E_i$, we get the expression for the *intrinsic energy* level as

$$E_i = E_f = \frac{E_c + E_v}{2} - \frac{kT}{2} \ln\left(\frac{N_c}{N_v}\right)$$

(2.12)

From Equation 2.12, it can be shown that the intrinsic Fermi level E_i, is only about 7.3 meV below the midgap at $T = 300$ °K. Since $kT \ll (E_c + E_v)$, Equation 2.12 can be simplified to

$$E_i = E_f \cong \frac{E_c + E_v}{2}$$

(2.13)

Thus, the intrinsic Fermi level in a semiconductor material is very close to the midpoint between the CB and VB. And, for all practical purposes, it can be assumed that E_i is in the middle of the energy gap. Thus, E_i is commonly referred to as the *midgap* energy level.

Now, in order to derive an expression for the intrinsic carrier concentration as a function of T, we multiply Equations 2.9 and 2.10, to get

or,

$$np = n_i^2(T) = N_c N_v \exp\left(-\frac{E_c - E_v}{kT}\right) = N_c N_v \exp\left(-\frac{E_g(T)}{kT}\right)$$

(2.14)

$$n_i(T) = CT^{3/2} \exp\left(-\frac{E_g(T)}{2kT}\right)$$

where:

C is a constant

$E_g(T)$ is the temperature-dependent bandgap energy defined in Equation 2.1

k is the Boltzmann constant (8.62×10^{-5} eV K^{-1})

The term kT has the dimension of energy and is called the *thermal energy* and is equal to 25.86 meV at $T = 300$ °K. Substituting the values for N_c and N_v [6,9], we can express Equation 2.14 as

$$n_i(T) = 3.9 \times 10^{16} T^{3/2} \exp\left(-\frac{E_g(T)}{2kT}\right). \qquad (2.15)$$

Now, if $E_g(T_{NOM})$ and $n_i(T_{NOM})$ represent E_g and n_i, respectively, at the nominal or reference temperature T_{NOM}, then we can show

$$n_i(T) = n_i(T_{NOM}) \cdot \left(\frac{T}{T_{NOM}}\right)^{3/2} \exp\left(-\frac{E_g(T)}{2kT} + \frac{E_g(T_{NOM})}{2kT_{NOM}}\right), \qquad (2.16)$$

where $E_g(T)$ is given by Equation 2.2. The above expression is used for calculation of n_i at any temperature T with $n_i = 1.45 \times 10^{10}$ cm^{-3} at $T = 300$ °K [9].

2.2.3.2   Effective Mass of Electrons and Holes

The electrons in the CB and holes in the VB move freely throughout the crystal like free particles, suffering only occasional scattering by impurities and defects present in the crystal. The free electrons experience Coulomb force due to the charged atomic cores of the host atoms on a regular lattice giving rise to a periodic potential energy. The effect of the periodic potential of the crystal lattice on the motion of electrons in the CB and holes in the VB is represented by the effective masses of the electrons (m_n^*) and holes (m_p^*), respectively. In practice, there are several types of mass used for a given material and carrier type [1–12]. The effective mass required to calculate the carrier (electron and hole) concentration is called the *density of states effective mass*, whereas that required to calculate carrier mobility is called the *conductivity effective mass*. These effective masses depend on temperature. There is a large variation in the reported values of m_n^* and m_p^* [20]. The commonly used values for the effective mass for electrons and holes at room temperature are given in Table 2.1 [9].

2.2.4   EXTRINSIC SEMICONDUCTORS

An *extrinsic semiconductor* is a semiconductor material with added elemental impurities called *dopants*. As we have discussed in Section 2.2.3, the intrinsic semiconductor at room temperature has an extremely low number of free-carrier concentration providing a very low conductivity. Thus, the added impurities introduce additional energy levels in the forbidden gap and can easily be ionized to add either electrons in the CB or holes in the VB, depending on the type of impurities and impurity levels in silicon as discussed below.

As we know, the silicon is a column-IV element in the Periodic Table of elements with four valence electrons per atom. There are two types of impurities in silicon that are electrically active: those from column-V such as arsenic (As), phosphorous (P), and antimony (Sb); and those from column-III such as boron (B). A column-V atom in a silicon lattice tends to have one extra electron loosely bound after forming

TABLE 2.1

Effective Mass Ratio for Silicon at 300 °K ($m_0 = 9.11 \times 10^{-31}$ kg is the Free Electron Mass)

Carriers	Density-of-states effective mass	Conductivity effective mass
Electrons (m_n^* / m_0)	1.08	0.26
Holes (m_p^* / m_0)	0.81	0.386

covalent bonds with silicon atoms as shown in Figure 2.3(a). In most cases, the thermal energy at room temperature is sufficient to ionize the impurity atom and free the extra electron to the CB. Such types of impurities (P, Sb, and As) are called the *donor* atoms, since they donate an electron to the crystal lattice and become positively charged. Thus, the P, Sb, and As doped silicon is called the *n*-type material, which contains excess electrons and its electrical conductivity is dominated by electrons in the CB. On the other hand, a column-III impurity atom in a silicon lattice tends to be deficient of one electron when forming covalent bonds with other silicon atoms as shown in Figure 2.3(b). Such an impurity (B) atom can also be ionized by accepting an electron from the VB, which leaves a freely moving hole that contributes to electrical conduction. These impurities (e.g., B) are called acceptors, since they accept electrons from the VB, and the doped silicon is called *p*-type that contains excess holes.

Thus, it is obvious from Figure 2.3 that the donor and acceptor atoms occupy substitutional lattice sites and the extra electrons or holes are very loosely bound, that is, can easily move to the CB or VB, respectively. In terms of energy band diagrams, the donors add allowed electron states in the bandgap close to the CB-edge as shown

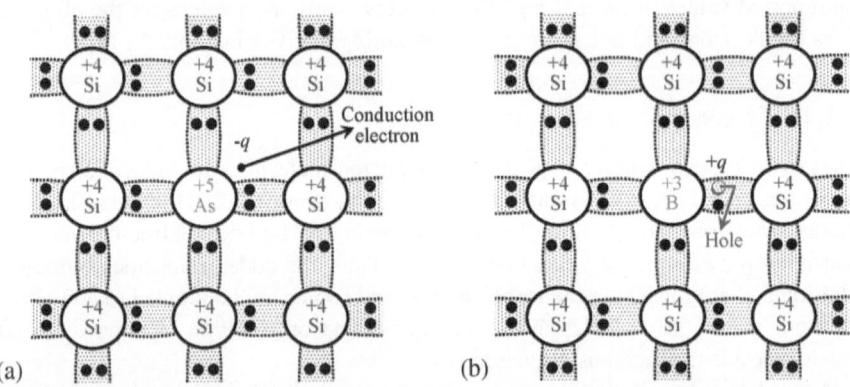

(a)   (b)

FIGURE 2.3 Extrinsic semiconductors forming covalent bonds: (a) an arsenic donor atom in silicon providing one electron for conduction in the CB; and (b) a boron acceptor atom in silicon creating a hole for conduction in the VB; in the figure, "•" represents electrons and "O" represents holes.

in Figure 2.4(a), whereas the acceptors add allowed states just above the VB-edge as shown in Figure 2.4(b). Also, Figure 2.4 shows the positions of the Fermi level due to donors (Figure 2.4(c)) and acceptors (Figure 2.4(d)). Donor levels contain positive charge when ionized (emptied). Acceptor levels contain negative charge when ionized (filled). A donor level E_d shown in Figure 2.4(a) is measured from the bottom of the CB, whereas an acceptor level E_a shown in Figure 2.4(b) is measured from the top of the VB. The ionization energies for the donors and acceptors are $(E_c - E_d)$ and $(E_a - E_v)$, respectively.

If silicon is doped such that $p = n$, then it is called *compensated* silicon. In practice, one type of impurity dominates over the other so that the semiconductor is either *n*-type or *p*-type. Again, a semiconductor is said to be *non-degenerate* if the Fermi level lies in the bandgap more than a few kT (~3kT) from either band-edge. Conversely, if the Fermi level is within a few kT (~3kT) of either band-edge, the semiconductor is said to be *degenerate*. In the *non-degenerate* case, the carrier concentration can be described by M–B statistics and Equations 2.5 and 2.6. However, for the case of degenerate doping where the dopant concentration is in excess of about 10^{18} cm^{-3} (heavy doping), the F-D distribution function given by Equations 2.3 and 2.4 must be used. Unless otherwise specified, we will assume the semiconductor to be non-degenerate.

2.2.4.1 Fermi Level in Extrinsic Semiconductor

In contrast to intrinsic semiconductors, the Fermi level in an extrinsic semiconductor is not located at the midgap. The Fermi level in an *n*-type silicon moves up towards the CB, consistent with the increase in the electron density described by Equation 2.9. On the other hand, the Fermi level in a *p*-type silicon moves towards the VB, consistent with the increase in the hole density described by Equation 2.10. These

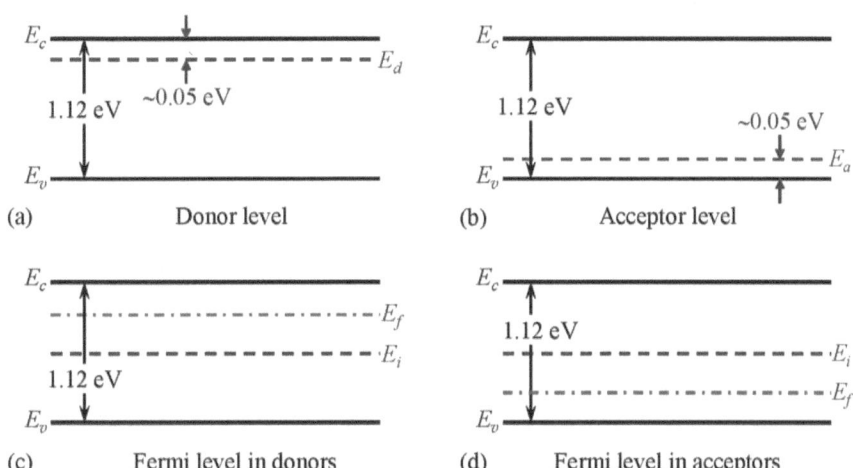

FIGURE 2.4 Energy band diagram representation in extrinsic semiconductors: (a) donor level, E_d and (b) acceptor level, E_a in silicon; (c) intrinsic energy level, E_i and Fermi level, E_f in an *n*-type semiconductor and (d) intrinsic energy level, E_i and Fermi level, E_f in a *p*-type semiconductor.

cases are depicted in Figure 2.4(c) and (d). The exact position of the Fermi level depends on both the ionization energy and concentration of dopants. For example, for an n-type material with a donor impurity concentration N_d, the charge neutrality condition in silicon requires that

$$n = N_d^+ + p, \qquad (2.17)$$

where:

N_d^+ is the density of ionized donors

Now, using Equation 2.4 we can write

$$N_d^+ = N_d \left[1 - f\left(E_d\right) \right] = N_d \left\{ 1 - \frac{1}{1 + (1/2)\exp\left[\left(E_d - E_f\right)/kT\right]} \right\} \qquad (2.18)$$

where:

$f(E_d)$ is the probability that a donor state is occupied by an electron in the normal state

E_d is the energy of the donor level

In Equation 2.18, the factor 1/2 in the denominator of $f(E_d)$ arises from the spin degeneracy (up or down) of the available electronic states associated with an ionized level [23]. Now, substituting Equations 2.9 and 2.10 for n and p, respectively, and Equation 2.18 for N_d^+ into Equation 2.17, we get

$$N_c \exp\left(-\frac{E_c - E_f}{kT} \right) = \frac{N_d}{1 + 2\exp\left[-\left(E_d - E_f\right)/kT \right]} + N_v \exp\left(-\frac{E_f - E_v}{kT} \right). \qquad (2.19)$$

Then Equation 2.19 can be solved for E_f. For an n-type semiconductor, $n \gg p$, therefore, the second term on the right-hand side of Equation 2.19 can be neglected. Now, if we assume $(E_d - E_f) \gg kT$, then $\exp\left(-\left(E_d - E_f\right)/kT\right) \ll 1$. Therefore, after simplification of Equation 2.19, we get,

$$E_c - E_f \cong kT \ln\left(\frac{N_c}{N_d} \right). \qquad (2.20)$$

In this case, the Fermi level is at least a few kT below E_d and essentially all the donor levels are ionized, that is, $n = N_d^+ = N_d$ for an n-type semiconductor. Then from Equation 2.8, the hole density in an n-type semiconductor is given by

$$p = \frac{n_i^2}{N_d}. \qquad (2.21)$$

Similarly, for a p-type silicon with a shallow acceptor concentration N_a, the Fermi level is given by

$$E_f - E_v = kT \ln\left(\frac{N_v}{N_a}\right). \tag{2.22}$$

In this case, the hole density is $p = N_a^- = N_a$, and the electron density is

$$n = \frac{n_i^2}{N_a} \tag{2.23}$$

Equations 2.20 and 2.22 can also be expressed in terms of E_f and E_i using Equations 2.9 and 2.10. Thus, from Equation 2.9, the intrinsic carrier concentration can be expressed as

$$n_i \cong N_c \exp\left(-\frac{E_c - E_i}{kT}\right) \tag{2.24}$$

Or,

$$E_c = E_i + kT \ln\left(\frac{N_c}{n_i}\right) \tag{2.25}$$

Then substituting for E_c from Equation 2.25 into Equation 2.20, we get for an n-type silicon

$$E_f - E_i = kT \ln\left(\frac{N_d}{n_i}\right). \tag{2.26}$$

Similarly, using Equation 2.10, we can express Equation 2.22 for a p-type silicon as

$$E_i - E_f = kT \ln\left(\frac{N_a}{n_i}\right). \tag{2.27}$$

Equations 2.26 and 2.27 are the measure of the Fermi level with reference to the midgap energy level for the n-type and p-type semiconductors, respectively.

2.2.4.2 Fermi Level in Degenerately Doped Semiconductor

For heavily doped silicon, the impurity concentration N_d or N_a can exceed the effective density of states N_c or N_v, so that $E_f \geq E_c$ and $E_f \leq E_v$ according to Equations 2.20 and 2.22. In other words, the Fermi level moves into the CB for $n+$ silicon, and into the VB for the $p+$ silicon. In addition, when the impurity concentration is higher than 10^{18} cm^{-3}, the donor (or acceptor) levels broaden into bands. This results in an effective decrease in the ionization energy until finally the impurity band merges with the CB (or VB) and the ionization energy becomes zero. Under these circumstances, the silicon is said to be *degenerate*. Strictly speaking, Fermi statistics should be used for the calculation of electron concentration when $(E_c - E_f) \leq kT$ [23]. For all practical purposes, it is a good approximation within a few kT to assume that the Fermi level of the degenerate $n+$ silicon is at the CB-edge, and that of the degenerate $p+$ silicon is at the VB-edge.

2.2.4.3 Electrostatic Potential in Semiconductor and Carrier Concentration

Conventionally, the electrostatic potential, ϕ, in a semiconductor is defined in terms of the intrinsic Fermi level (E_i) such that

$$\phi = -\frac{E_i}{q} \tag{2.28}$$

where:
$q = 1.6 \times 0^{-19}$ C is the electronic charge

The negative sign in Equation 2.28 is due to the fact that E_i is defined as the electron energy, whereas ϕ is defined for a positive charge.

Now, in an n-type non-degenerate semiconductor, the Fermi level E_f (or Fermi potential $\phi_f = -E_f/q$) lies above the intrinsic level E_i (or intrinsic potential $\phi_i = -E_i/q$) as shown in Figure 2.4(c). Then from Equation 2.26 we can write

$$N_d = n_i \exp\left(\frac{E_f - E_i}{kT}\right) = n_i \exp\left[\frac{q}{kT}\left(\phi_i - \phi_f\right)\right] \tag{2.29}$$

Similarly, in a p-type non-degenerate semiconductor, the Fermi level E_f (or Fermi potential ϕ_f) lies below the intrinsic level E_i (or intrinsic potential ϕ_i) as shown in Figure 2.4(d). Then from Equation 2.27 we can show

$$N_a = n_i \exp\left(\frac{E_i - E_f}{kT}\right) = n_i \exp\left[\frac{q}{kT}\left(\phi_f - \phi_i\right)\right] \tag{2.30}$$

At room temperature, the available thermal energy is sufficient to ionize nearly all acceptor and donor atoms due to their low ionization energies. Hence in non-degenerate silicon at room temperature, we can safely approximate the carrier concentrations as

$$n \approx N_d \quad (n\text{-type}) \tag{2.31}$$

$$p \approx N_a \quad (p\text{-type}) \tag{2.32}$$

In an n-type material, where $N_d \gg n_i$, electrons are *majority carriers* whose concentration is given by Equation 2.31. And, the hole concentration p_n (representing concentration of p in an n-type material) can be obtained using Equations 2.8 and 2.31 and is given by

$$p_n \cong \frac{n_i^2}{N_d} \tag{2.33}$$

Thus, the hole concentration p_n is much smaller than n_n ($\cong N_d$) in an n-type semiconductor. Therefore, the holes are *minority carriers* in an n-type semiconductor. Similarly, in a p-type semiconductor where $N_a \gg n_i$, holes are the majority carriers given by Equation 2.32, while the minority carrier electron concentration is given by

$$n_p \cong \frac{n_i^2}{N_a} \tag{2.34}$$

Since $n_p \ll p$, electrons are minority carriers in a p-type semiconductor. Consequently, we often use the terminology of majority and minority carriers.

From Equation 2.29, we can write for an n-type semiconductor

$$\phi_i - \phi_f = \frac{kT}{q} \ln\left(\frac{N_d}{n_i}\right) = v_{kT} \ln\left(\frac{N_d}{n_i}\right) \equiv -\phi_B \tag{2.35}$$

where:
$\phi_B \equiv (\phi_f - \phi_i)$ is called the *bulk potential* and is negative for n-type semiconductors
$v_{kT} = kT/q$ is called the thermal voltage

Similarly, from Equation 2.30, we can show for a p-type semiconductor

$$\phi_f - \phi_i = v_{kT} \ln\left(\frac{N_a}{n_i}\right) \equiv \phi_B \tag{2.36}$$

Thus, we can write a generalized expression for bulk potential in semiconductors as

$$\phi_B = \left(\phi_f - \phi_i\right) = \pm v_{kT} \ln\left(\frac{N_b}{n_i}\right) \tag{2.37}$$

where:
"+" sign is for p-type semiconductors with $N_b = N_a$
"−" sign is for n-type semiconductors with $N_b = N_d$

Note that the Fermi potential ϕ_f is not only a function of carrier concentration but also is dependent on temperature through n_i. From Equation 2.37, we observe that since n_i increases with temperature according to Equation 2.15, the magnitude of ϕ_B decreases and as n_i approaches N_b, ϕ_f approaches ϕ_i. Thus, with an increase in temperature, the Fermi level approaches the midgap position, that is, the intrinsic Fermi level. This implies that the semiconductor becomes intrinsic at high temperature. Thus, the doped or extrinsic silicon will become intrinsic if the temperature is high enough. The temperature at which this happens depends on the doping concentration. When the material becomes intrinsic, the device can no longer function and therefore, the intrinsic region is avoided in device operation [1].

The temperature coefficient of ϕ_f can be obtained by differentiating Equation 2.37 and is given by

$$\frac{d\phi_f}{dT} = \frac{1}{T}\left[\phi_f - \left(\frac{E_g}{2} + \frac{3}{2}v_{kT}\right)\right] \tag{2.38}$$

From Equation 2.38 we can show that $d\phi_f/dT \sim 1$ mV K^{-1}. If we set $\phi_i = 0$ as the reference potential, and substitute $n_i(T)$ from Equation 2.15 in Equation 2.37, then at any temperature T, ϕ_f can be written in terms of a reference temperature, T_{NOM} as

$$\phi_f(T) = \phi_f(T_{NOM})\cdot\left(\frac{T}{T_{NOM}}\right) - v_{kT}\left[\frac{3}{2}\ln\left(\frac{T}{T_{NOM}}\right) + \left(-\frac{E_g(T)}{2kT} + \frac{E_g(T_{NOM})}{2kT_{NOM}}\right)\right] \tag{2.39}$$

Equation 2.39 is used in circuit simulation tools for modeling the temperature dependence of ϕ_f.

2.2.4.4 Quasi-Fermi Level

Under thermal equilibrium conditions, the electron and hole concentrations are given by Equations 2.29 and 2.30 (using $n = N_d$ and $p = N_a$), respectively, maintaining the condition $pn = n_i^2$. However, when carriers are injected into the semiconductor or extracted out of the semiconductor, the equilibrium condition is disturbed. In non-equilibrium condition: (1) injection, $np > n_i^2$ or (2) extraction, $np < n_i^2$, we cannot use Equations 2.29 and 2.30. And, the carrier densities can no longer be described by a constant Fermi level through the system. Therefore, we define *quasi-Fermi* levels such that Equations 2.29 and 2.30 hold under non-equilibrium condition and are given by

$$n = n_i \exp\left(\frac{E_{fn} - E_i}{kT}\right) = n_i \exp\left[\frac{q}{kT}(\phi_i - \phi_n)\right] \tag{2.40}$$

$$p = n_i \exp\left(\frac{E_i - E_{fp}}{kT}\right) = n_i \exp\left[\frac{q}{kT}(\phi_p - \phi_i)\right] \tag{2.41}$$

where:
   E_{fn} and E_{fp} are the electron and hole quasi-Fermi levels, respectively
   $\phi_n \equiv (-E_{fn}/q)$ and $\phi_p \equiv (-E_{fp}/q)$ are the electron and hole quasi-Fermi potentials, respectively

It is to be noted that E_{fn} and E_{fp} *are the mathematical entities; their values are chosen so that the accurate carrier concentrations can be quantified in the non-equilibrium situations. Generally, $E_{fn} \neq E_{fp}$.*
   From Equations 2.40 and 2.41, we can show

$$pn = n_i^2 \exp\left(\frac{E_{fn} - E_{fp}}{kT}\right) \tag{2.42}$$

In equilibrium condition, $E_{fn} = E_{fp} = E_f$ and $\phi_n = \phi_p$ so that Equations 2.40 and 2.41 become the same as Equations 2.29 and 2.30 for $n = N_d$ and $p = N_a$, respectively. And, Equation 2.42 becomes $pn = n_i^2$.

2.2.5 CARRIER TRANSPORT IN SEMICONDUCTORS

In thermal equilibrium, the mobile (CB) electrons are in random thermal motion with an average velocity, $v_{th} \cong 1 \times 10^7$ cm sec^{-1} at 300 °K. However, due to the random thermal motion of electrons, no net current flows through the material. On the other hand, in the presence of an electric field E, electrons move opposite to the direction of E. This process is called *electron drift* and causes a net current flow through the material. Also, if there is a concentration gradient of carriers in the material, the carriers diffuse away from the higher concentration region to the lower concentration region producing a net current flow in the semiconductor. Thus, the carrier transport or current flow in a semiconductor is the result of two different mechanisms:

1. Drift of carriers (electrons and holes) caused by the presence of an electric field
2. Diffusion of carriers caused by the electron or hole concentration gradient in the semiconductor.

We will now consider both the drift and diffusion mechanisms of carriers in a semiconductor.

2.2.5.1 Drift of Carriers: Carrier Motion in Electric Field

The drift of carriers in a material depends on the crystal structure, level of impurities, and the strength of electric field that define the mobility of carriers, electrical conductivity of the material, and velocity saturation of carriers.

Carrier mobility: When an electric field is applied to a conducting medium containing free carriers, the carriers are accelerated in proportion to the force of the field. However, the accelerating carriers within a semiconductor will collide with various scattering centers including the atoms of the host lattice (*lattice scattering*), the impurity atoms (*impurity scattering*), and other carriers (*carrier-carrier scattering*). In the case of an electron, these different scattering mechanisms tend to redirect its momentum and, in many cases, tend to dissipate the energy gained from the electric field. Thus, under the influence of a uniform electric field, the process of energy gained from the field and energy loss due to the scattering balance each other and carriers attain a constant average velocity, called the *drift velocity* (v_d). At low electric fields, v_d is proportional to the electric field strength E and is given by

$$v_d = \mu E \tag{2.43}$$

where:

μ is the constant of proportionality and is called the mobility of the carriers in units of cm^2 V^{-1} sec^{-1}

The mobility is proportional to the time interval between collisions and inversely proportional to the effective mass of the carriers. The total mobility is determined by combining the mobilities for different scattering mechanisms such as mobility due to *lattice scattering* μ_L, mobility due to *ionized impurity* scattering μ_I, and so on. Assuming different scattering mechanisms are independent, we can write an expression for the total mobility using *Matthiessen's rule*

$$\frac{1}{\mu} = \frac{1}{\mu_L} + \frac{1}{\mu_I} + \cdots \tag{2.44}$$

The measurement data show that the electron mobility (μ_n) in an *n*-type silicon is about three times the hole mobility (μ_p) in a *p*-type silicon. This is due to the fact that the effective mass of electrons in the CB is much lighter than that of holes in the VB (Table 2.1).

The carrier mobility in bulk silicon is a function of the doping concentrations. Figure 2.5 shows the plots of electron and hole mobilities in silicon as a function of doping concentration at room temperature. It is observed from the plots that at low impurity levels, the mobilities are mainly limited by carrier collisions with the silicon lattice or acoustic phonons. As the doping concentration increases beyond 1×1^{15} per cm³, the mobilities decrease due to the increase in the collisions with the charged (ionized) impurity atoms through Coulomb interaction. At high temperatures, the mobility tends to be limited by lattice scattering and is proportional to $T^{-3/2}$, relatively insensitive to the doping concentration. At low temperatures, the mobility is higher; however, strongly depends on doping concentration as it becomes

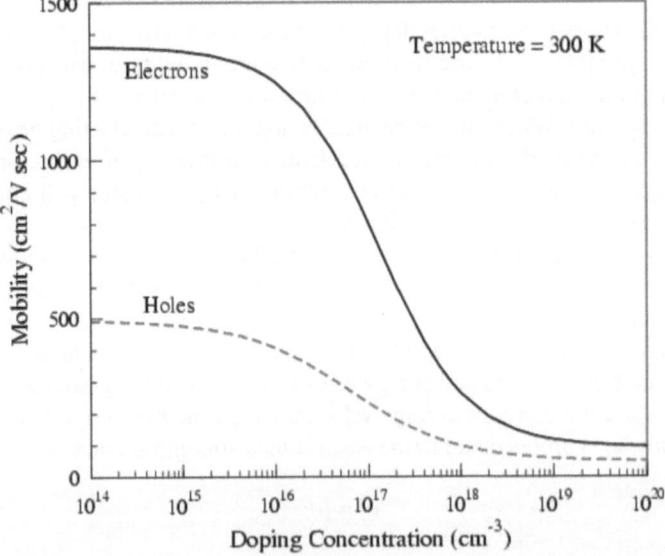

FIGURE 2.5 Electron and hole mobilities in bulk silicon at 300 °K as a function of doping concentration.

more limited by impurity scattering. The detailed temperature dependence of mobility can be found in [16,24]

The carrier mobility discussed above is the *bulk mobility* applicable to conduction in the silicon substrate far away from the surface. In the channel region of field-effect transistor (FET) devices, the current flow is governed by the *surface mobility*. The surface mobility is much lower than the bulk mobility due to additional scattering mechanism of carriers at the Si/gate-dielectric interface in the presence of high electric field normal to the channel [15].

Electrical conductivity: The drift of charge carriers under an applied electric field E results in a current, called the *drift current*. If in a homogeneous n-type silicon there are n number of electrons per unit volume and each electron, carrying a charge q, flow with a drift velocity v_d, then the *electron drift current* density is given by

$$J_{n,drift} = qnv_d = qn\mu_n E \qquad (2.45)$$

where:

$v_d = \mu_n E$

μ_n is the electron mobility

We know from Ohm's law that the resistivity ρ of a conducting material is defined as E/J_n; therefore, from Equation 2.45, the resistivity ρ_n due to electron current flow is given by

$$\rho_n = \frac{1}{qn\mu_n} \qquad (2.46)$$

Similarly, for a p-type silicon, the *hole drift current* density $J_{p,drift}$ and resistivity ρ_p are given by

$$J_{p,drift} = qpv_d = qp\mu_p E \qquad (2.47)$$

$$\rho_p = \frac{1}{qp\mu_p} \qquad (2.48)$$

where:

μ_p is hole mobility

If the silicon is doped with both donors and acceptors, then the total resistivity can be expressed as

$$\rho = \frac{1}{qn\mu_n + qp\mu_p} \qquad (2.49)$$

Thus, the resistivity of a semiconductor depends on the electron and hole concentrations and their corresponding mobilities. For a uniformly doped silicon substrate, the plots of the resistivity versus impurity concentration at 300 °K are shown in Figure 2.6. Since the electron mobility is higher than the hole mobility, the resistivity

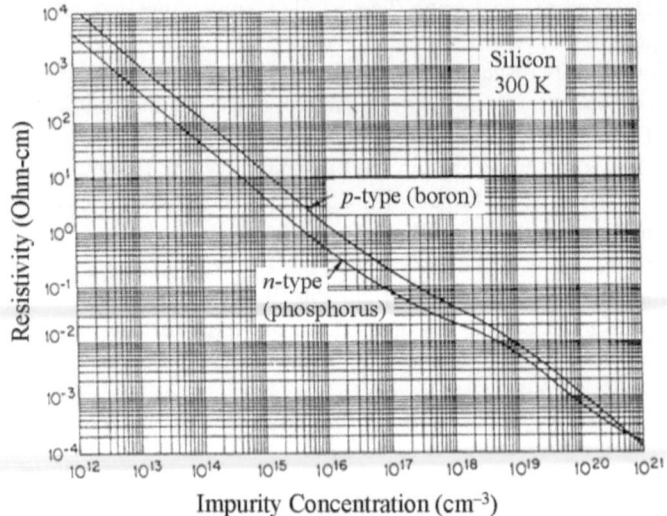

FIGURE 2.6 Impurity concentration versus resistivity of *n*-type and *p*-type silicon at 300 K [2].

[Equation 2.46] of an *n*-type doped silicon is lower than the resistivity [Equation 2.48] of a *p*-type doped silicon as shown in Figure 2.6.

Sheet Resistance: The resistance of a uniform conductor of length *L*, width *W*, and thickness *t* is given by

$$R = \rho \frac{L}{tW} \tag{2.50}$$

where:

ρ is the resistivity of the conductor in Ohm-centimeter

Typically, in an IC technology, the thickness *t* of a diffusion region is uniform and much less than both *L* and *W* of the region. Therefore, it is useful to define a new parameter ρ_{sh}, called the *sheet resistance*, which has the dimension of Ohm (Ω) and is given by

$$\rho_{sh} = \frac{\rho}{t} \tag{2.51}$$

Then Equation 2.50 becomes

$$R = \rho_{sh} \frac{L}{W} \tag{2.52}$$

From Equation 2.52, it is found that when $L = W$, the diffused layer becomes a square with $R = \rho_{sh}$. Thus, the total resistance of a diffusion line is simply ρ_{sh} times the number of squares in the path of current and is expressed in units of Ω per square ($\Omega/\square$). The process parameters that determine the sheet resistance of a layer are ρ and t of the layer [Equation 2.51]. Since the resistivity is a function of carrier concentration and mobility, both of which are functions of temperature, therefore, ρ_{sh} is temperature-dependent.

Velocity saturation: The mobility Equation 2.43 assumes a linear relationship between E versus v_d. However, this linear relationship is valid only for low electric field ($< 1 \times 10^4$ V cm^{-1}) and the carriers are at equilibrium with the lattice. At higher electric fields, the average carrier energy increases and they lose their energy by optical-phonon emission nearly as fast as they gain it from the field. This causes a decrease in μ from its low field value as the field increases until finally the drift velocity reaches a limiting value v_{sat}, referred to as the *saturation velocity*. This phenomenon is called the *velocity saturation*. For silicon, a typical value of $v_{sat} = 1.07 \times 10^7$ cm sec^{-1} for electrons and occurs at an electric field of about 2×10^4 V cm^{-1}. The corresponding values for holes are $v_{sat} \cong 8.34 \times 10^6$ cm sec^{-1} and $E \cong 5.0 \times 10^4$ V cm^{-1}.

It is observed that the measured value of drift velocity for electrons and holes in silicon is a function of the applied field E and can be approximated by an empirical relation [15,16,25]

$$v_d = v_{sat} \frac{E / E_{sat}}{\left[1 + \left(E / E_{sat}\right)^\beta\right]^{1/\beta}} \tag{2.53}$$

where:

E_{sat} is the critical electric field at which carrier velocity saturates

The parameters v_{sat}, E_{sat}, and β in Equation 2.53 are given in Table 2.2.

Figure 2.7 shows the calculated value of drift velocity for electrons and holes at 300 °K in silicon as a function of the applied field E obtained by Equation 2.53. At low fields, the carrier velocity increases linearly with the electric field indicating constant mobility. When the field exceeds about 2×10^4 V cm^{-1}, carriers begin to lose energy by scattering with optical phonons and their velocity saturates. As the field exceeds 100 KV cm^{-1}, carriers gain more energy from the field than they can lose by scattering. Consequently, their energy with respect to the bottom of the CB (for electrons) or top of the VB (for holes) begins to increase. The carriers are no longer at thermal equilibrium with the lattice. Since they acquire energy higher than the thermal energy (kT) they are called hot-carriers. It is these hot-carriers which are responsible for reducing the mobility at high fields. For a more heavily doped material, the low-field mobility is lower because of the impurity scattering. However, v_{sat}

TABLE 2.2

Parameters for Field Dependence of Drift Velocity for Silicon at 300 °K

Carriers	v_{sat} (cm sec^{-1})	E_{sat} (V cm^{-1})	β
Electrons	1.07×10^7	6.91×10^3	1.11
Holes	8.34×10^6	1.45×10^4	2.637

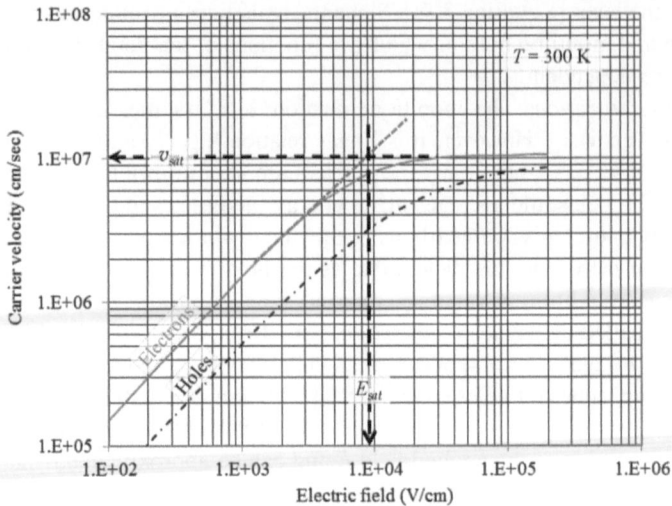

FIGURE 2.7   Drift velocities of electrons and holes in silicon at room temperature as a function of applied electric field showing velocity saturation at high electric fields.

remains the same, independent of impurity scattering. Also, v_{sat} is weakly dependent on temperature and decreases slightly as the temperature increases [16]. Figure 2.7 shows the carrier velocity as a function of electric field in silicon at 300 °K. It is observed from the plots that the carrier velocity increases linearly at low electric field, then the increase in the carrier velocity slows down with the increase in electric field, and finally above a certain critical electric field the carrier velocity saturates.

2.2.5.2   Diffusion of Carriers

In addition to the drift of electrons under the influence of an electric field, the carriers also diffuse if the carrier concentration is not uniform within a semiconductor. This leads to an additional component of current in proportion to the concentration gradient and is called the *diffusion current*. Thus, the diffusion is a gradient driven motion and occurs from the high concentration regions to low concentration regions as shown in Figure 2.8.

In order to calculate the diffusion current, let us consider the diffusion flux F due to concentration gradient dC/dx along the x-direction. Now, from Fick's first law [26]

$$F = -D\frac{dC}{dx} \tag{2.54}$$

where:
   D is the diffusion constant
   C is the carrier density

The negative sign on the right-hand side of Equation 2.54 is due to the fact that the carriers flow from the higher concentration to lower concentration in space, that

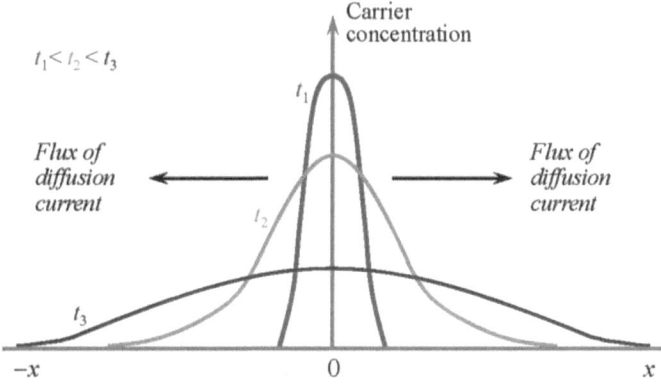

FIGURE 2.8 Diffusion of carriers from the high concentration region to low concentration regions due to concentration gradient shown over the time intervals $t_1 < t_2 < t_3$; t_1 is the initial time and the background concentration ≈ 0.

is, dC/dx is negative. If the carrier flow in a semiconductor material is electrons, then from Equation 2.54 the diffusion current flow due to the electron concentration gradient dn/dx is given by

$$J_{n,diff} = qD_n \frac{dn}{dx} \tag{2.55}$$

Similarly, the hole diffusion current due to hole concentration gradient dp/dx is given by

$$J_{p,diff} = -qD_p \frac{dp}{dx} \tag{2.56}$$

where:

D_n is the *diffusivity* or *diffusion constant* for electrons

D_p is the *diffusivity* or *diffusion constant* for holes

The negative sign in Equation 2.56 implies that the hole current flows in a direction opposite to the hole concentration gradient. And, D_n and D_p are related to the respective mobility by the relationship [9]

$$\frac{D_n}{\mu_n} = \frac{D_p}{\mu_p} = \frac{kT}{q} \equiv v_{kT} \tag{2.57}$$

Equation 2.57 is often referred to as Einstein's relation. For lightly doped silicon (e.g., $N_d \cong 1 \times 10^{15}$ cm^{-3}) at room temperature, $D_n = 38$ cm^2 sec^{-1} and $D_p = 13$ cm^2 sec^{-1}.

Non-uniformly doped semiconductors and built-in electric field: Let us consider an n-type material with non-uniformly doped N_d donor atoms as shown in Figure 2.9. Considering complete ionization of donor atoms, we have $n = N^+_d = N_d$.

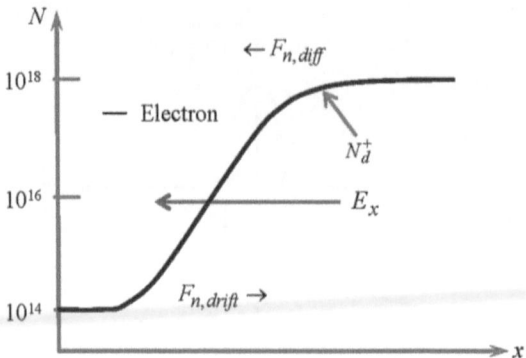

FIGURE 2.9 Drift and diffusion of carriers in a non-uniformly doped n-type semiconductor: $F_{n,diff}$ is the electron diffusion flux from the high concentration to low concentration regions, $F_{n,drift}$ is the drift flux of electrons due to the built-in electric field E_x set up by the ionized donors and diffused electrons in the semiconductor.

Due to the concentration gradient, the electrons diffuse from the high concentration region to the low concentration region. Then, from Equation 2.54, the diffusion flux of electrons is given by

$$F_{n,diff} = -D_n \frac{dn(x)}{dx} \qquad (2.58)$$

where:
the subscript n represents the parameters for electrons

As the electrons move (diffuse) away, they leave behind positively charged donor ions N_d^+ which try to pull electrons back causing drift flux of electrons from the low to high concentration regions. This drift of electrons from the low to high concentration regions sets up an electric field E_x from the high concentration to the low concentration regions as shown in Figure 2.9. Then, from Equation 2.45, the flux due to the drift of electrons is given by

$$F_{n,drift} = n(x)v_d = n\mu_n E_x \qquad (2.59)$$

An equilibrium is established when *diffusion = drift*. Here $n(x)$ is the number of electrons in the diffusion flux at any point x in the distribution and $\neq N_d(x)$. Therefore, a *built-in* electric field is established that prevents further diffusion of electrons. Then, from Equations 2.58 and 2.59, we get the expression for the *built-in electric field* for electrons in an n-type non-uniformly doped substrate as

$$E_x = -\frac{D_n}{\mu_n} \frac{1}{n} \frac{dn(x)}{dx} = -v_{kT} \frac{1}{n} \frac{dn(x)}{dx} \qquad (2.60)$$

Similarly, the built-in electric field for holes in a non-uniform p-type substrate is given by

$$E_x = \frac{D_p}{\mu_p} \frac{1}{p} \frac{dp(x)}{dx} = V_{kT} \frac{1}{p} \frac{dp(x)}{dx} \qquad (2.61)$$

In Equations 2.60 and 2.61 we have used Einstein's relation given in Equation 2.57. This built-in electric field favors the transport of the minority carriers if created by an external source.

2.2.6 GENERATION-RECOMBINATION OF CARRIER

In a semiconductor under thermal equilibrium, carriers possess an average thermal energy corresponding to the ambient temperature. This thermal energy excites some VB electrons to reach the CB. This upward transition of an electron from the VB to CB leaves behind a hole in the VB and an electron-hole pair is created. This process is called the *carrier generation* (G). On the other hand, when an electron makes a transition from the CB to VB, an electron-hole pair is *annihilated*. This reverse process is called the *carrier recombination* (R). Under thermal equilibrium, $G = R$ so that the carrier concentration remains the same and the condition $pn = n_i^2$ is maintained. The thermal G-R process is shown in Figure 2.10.

The equilibrium condition of a semiconductor is disturbed by optically or electrically introducing free carriers exceeding their thermal equilibrium value resulting in $pn > n_i^2$ or electrically removing carriers resulting in $pn < n_i^2$. The process of introducing carriers in excess of thermal equilibrium values is called the *carrier injection* and the additional carriers are called the *excess carriers*. In order to inject excess carriers optically, we shine light with energy $E = hv > E_g$ on an intrinsic semiconductor so that the valence electrons can be excited into the CB by the excess energy $\Delta E = (hv - E_g)$; where h and v are Planck's constant and frequency of light, respectively. In this band-to-band tunneling process, we get optically generated excess electrons (n_L) and holes (p_L) in the semiconductor as shown in Figure 2.10. Therefore, the total non-equilibrium values of carrier concentration are given by

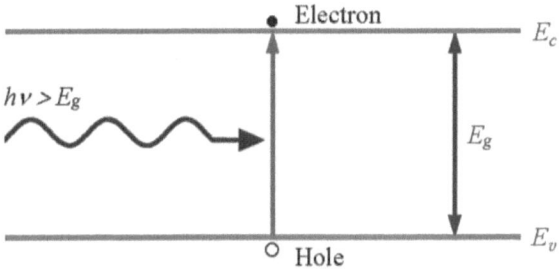

h = Planck's constant

v = frequency of incident light.

FIGURE 2.10 Band-to-band generation of electron-hole pairs under optical illumination with photon energy hv; where h and v are Planck's constant and the frequency of incident light, respectively; the symbol "•" represents electrons and "O" represents holes.

$$\left.\begin{array}{l} n = n_i + n_L \\[2mm] p = n_i + p_L \end{array}\right\} \quad \text{Injection of carriers by light} \qquad (2.62)$$

2.2.6.1 Injection Level

In the case of carrier injection into the semiconductor, we observe from Equation 2.62 that both n and p are greater than the intrinsic carrier concentration of the semiconductor and therefore, $pn > n_i^2$. If the injected carrier density is lower than the majority carrier density at equilibrium so that the latter remains essentially unchanged while the minority carrier density is equal to the excess carrier density, then the process is called *low level injection*. If the injected carrier density is comparable to or exceeds the equilibrium value of the majority carrier density, then it is called *high level* injection.

To illustrate the injection levels, we consider an n-type *extrinsic* semiconductor with $N_d = 1 \times 10^{15}$ cm^{-3}. Then from Section 2.2.4.1, the equilibrium majority carrier electron concentration is given by $n_{n0} = 1 \times 10^{15}$ cm^{-3}, whereas from Equation 2.21, the minority carrier hole concentration is given by $p_{n0} = 1 \times 10^{5}$ cm^{-3}. Here, n_{n0} and p_{n0} define the equilibrium concentrations of electrons and holes, respectively, in an n-type material. Now, we shine light on the sample so that 1×10^{13} cm^{-3} electron-hole pairs are generated in the material. Then using Equation 2.62, the total number of electrons $n_n \cong 1 \times 10^{15}$ cm^{-3} = n_{n0} and $p_n = 1 \times 10^{13}$ cm^{-3}. Thus, the majority carrier concentration n_n remains unchanged, whereas the minority carrier concentration p_n is increased significantly. This is an example of *low level injection*. On the other hand, if 1×10^{17} cm^{-3} electron-hole pairs are generated by incident light, then from Equation 2.62, we get $n_n \cong 1 \times 10^{17}$ cm^{-3} and $p_n = 1 \times 10^{17}$ cm^{-3} changing both the electron and hole concentrations in the semiconductor resulting in a *high level injection. The mathematics for high level injection is complex and therefore, we will consider only low level injection.*

2.2.6.2 Recombination Processes

The semiconductor material returns to equilibrium through recombination of injected minority carriers with the majority carriers in the case of carrier injection, or through generation of electron-hole pairs in the case of extraction of carriers.

The electron-hole recombination process occurs by transition of electrons from the CB to the VB. In a direct bandgap semiconductor like GaAs where the minimum of the CB aligns with the maximum of the VB, an electron in the CB can give up its energy to move down to occupy the empty state (hole) in the VB without a change in momentum as shown in Figure 2.11(a). Since the momentum (k) must be conserved in any energy level transition, an electron in GaAs can easily make direct transition from E_c to E_v across E_g. This is called the *direct* or *band-to-band recombination*. When the direct recombination happens, the energy given up by electron will be emitted as a photon which makes it useful for light-emitting diodes [Figure 2.11(b)].

Now, if we generate excess carriers (Δn, Δp) at a rate G_L due to the incident light, then for low level injection, we get $\Delta p = \Delta n = U\tau = G_L\tau$; where U is the net

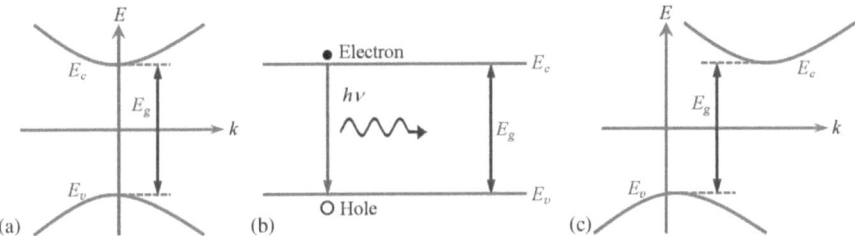

FIGURE 2.11   Bandgap in semiconductors: (a) direct bandgap; (b) band-to-band recombination in a direct bandgap semiconductor; and (c) indirect bandgap.

recombination rate and τ is the excess carrier lifetime. If p_o and n_o are the equilibrium concentrations of electrons and holes, respectively, and p and n are the respective total concentrations due to generation, then $\Delta p = p - p_o$ and $\Delta n = n - n_o$ and the net recombination rate due to direct recombination is given by

$$U = \frac{\Delta n}{\tau_n} = \frac{\Delta p}{\tau_p}$$ (2.63)

where:

τ_n and τ_p are the excess carrier electron and hole lifetime, respectively

It is to be noted that for band-to-band recombination, the excess carrier lifetime for an electron is equal to that of a hole since the single phenomenon annihilates an electron and a hole simultaneously.

For *indirect bandgap* semiconductors such as silicon and germanium (Figure 2.11(c)), the probability of direct recombination is very low. Physically, this means that the minimum energy gap between E_c and E_v does not occur at the same point in the momentum space as shown in Figure 2.11(c). In this case, for an electron to reach the VB, it must experience a change of momentum as well as energy to satisfy the conservation principle. This can be achieved by recombination processes through intermediate trapping levels, called the *indirect recombination* as shown in Figure 2.12.

In indirect bandgap semiconductors, the impurities that form electronic states deep in the energy gap assist recombination of electrons and holes. Here, the word *deep* indicates that the states are far away from the band-edges and near the center of the energy gap. These deep states are commonly referred to as the *recombination centers* or *traps*. Such recombination centers are usually unintentional impurities, which are not necessarily ionized at room temperature. These deep level impurities have concentrations far below the concentration of donor or acceptor impurities which have shallow energy levels. As an example, gold (Au) is a deep level impurity intentionally used in silicon to increase the recombination rate. This recombination via deep level impurities or traps is often referred to as the *indirect recombination* as shown in Figure 2.12. The *G-R* processes shown in Figure 2.12 consist of "1" electron capture by an empty center, "2" electron emission from an occupied center, "3" hole capture by an occupied center, and "4" hole emission by an empty center.

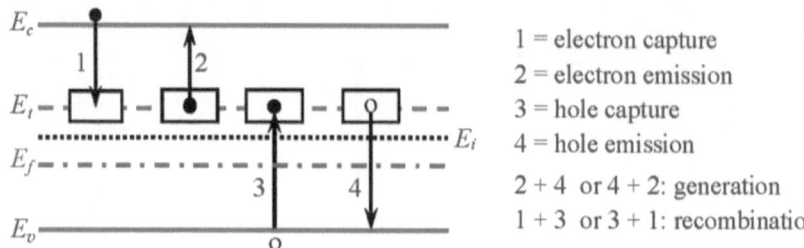

FIGURE 2.12 Generation and recombination in an indirect bandgap semiconductor; E_t is the trap level deep into the bandgap; 1, 2, 3, and 4 represent the generation and recombination processes; here, "•" represents electrons and "O" represents holes.

Now, let us consider the following example where an impurity like Au is introduced to provide a *trapping level* or a set of allowed states at energy E_t. The trap level E_t is assumed to act like an acceptor (it can also be neutral or negatively charged). Recombination is accomplished by trapping an electron and a hole. (The analysis can be easily extended to the case where the trap acts like a donor, i.e., positively charged or neutral charge states). The indirect recombination process was originally proposed by Shockley and Read [27] and independently suggested by Hall [28] and is therefore, often referred to as the *Shockley–Read–Hall* (SRH) recombination. By considering the transition processes shown in Figure 2.12, Shockley, Read, and Hall showed that for low level injection, the net recombination rate is given by

$$U = \frac{v_{th}\sigma N_t \left(pn - n_i^2\right)}{n + p + 2n_i \cosh\left[\dfrac{E_t - E_i}{kT}\right]} \tag{2.64}$$

where:

v_{th} = carrier thermal velocity ($\approx 1 \times 10^7$ cm sec^{-1})
σ = carrier capture cross-section ($\approx 10^{-15}$ cm^2)
N_t is the density of trap centers
$v_{th}\sigma N_t$ is the *capture probability* or capture cross-section

From Equation 2.64, we observe the following:

1. The "driving force" or the rate of recombination is proportional to $\left(pn - n_i^2\right)$, that is, the deviation from the equilibrium condition
2. $U = 0$ when $\left(np = n_i^2\right)$, that is, under the equilibrium condition
3. U is maximum when $E_t = E_i$, that is, trap levels near the midgap are the most efficient recombination centers

Now, for simplicity of understanding, let us consider the case when $E_t = E_i$. Then, from Equation 2.64, the net recombination rate is given by

$$U = \frac{v_{th}\sigma N_t \left(pn - n_i^2 \right)}{n + p + 2n_i}$$

(2.65)

Then for an *n-type semiconductor* with low level injection, $n \gg p + 2n_i$, then denoting $p = p_n$ as the total excess minority carrier concentration and $\left(p_{n0} = n_i^2 / n \right)$ as the equilibrium minority carrier concentration, we get after simplification of Equation 2.65

$$U = v_{th}\sigma N_t \left(p_n - p_{n0} \right) = \frac{\Delta p}{\tau_p}$$

(2.66)

where:

τ_p is the minority carrier hole lifetime in an *n*-type semiconductor and is given by

$$\tau_p = \frac{1}{v_{th}\sigma_p N_t}$$

(2.67)

In an *n*-type material, lots of electrons are available for capture. Therefore, Equation 2.66 shows that the minority carrier hole lifetime τ_p is the limiting factor in the recombination process in an *n*-type material.

Similarly, for a *p-type semiconductor,* we can show from Equation 2.65 that the net recombination rate for electrons is given by

$$U = \frac{\Delta n}{\tau_n}$$

(2.68)

where:

τ_n is the minority carrier electron lifetime in a *p*-type semiconductor and is given by

$$\tau_n = \frac{1}{v_{th}\sigma_n N_t}$$

(2.69)

Thus, for a *p*-type semiconductor, the minority carrier electron lifetime is the limiting factor in the recombination process.

The other recombination process in silicon that does not depend on deep level impurities and which sets an upper limit on lifetime is *Auger recombination.* In this process, the electrons and holes recombine without trap levels and the released energy (of the order of energy gap) is transferred to another majority carrier (a hole in a *p*-type and electron in an *n*-type silicon). Usually, Auger recombination is important when the carrier concentration is very high ($> 5 \times 10^{18}$ cm^{-3}) as a result of high doping concentration or high level injection.

2.2.7 BASIC SEMICONDUCTOR EQUATIONS

2.2.7.1 Poisson's Equation

Poisson's equation is a very general differential equation governing the operation of IC devices and is based on Maxwell's field equation that relates the charge density to the electric field potential. We know that the electric field E in a semiconductor is equal to the negative gradient of the electrostatic potential ϕ such that

$$E = -\frac{d\phi}{dx} \tag{2.70}$$

Mathematically, Poisson's equation (for silicon) is stated as

$$\frac{dE}{dx} = \frac{\rho(x)}{K_{si}\varepsilon_0} \tag{2.71}$$

or, using Equation 2.70,

$$\frac{d^2\phi}{dx^2} = -\frac{\rho(x)}{K_{si}\varepsilon_0} \tag{2.72}$$

where:
  $\rho(x)$ is the net charge density at any point x
  $\varepsilon_0 (= 8.854 \times 10^{-14} \text{ F cm}^{-1})$ is the permittivity of free space
  $K_{si} (= 11.8)$ is the relative permittivity of silicon

Now, if n and p are the free electron and hole concentrations, respectively, corresponding to N_d^+ and N_a^- ionized donor and acceptor concentrations, respectively, in silicon, then we can express Equation 2.72 as

$$\frac{d^2\phi}{dx^2} = -\frac{dE}{dx} = -\frac{q}{K_{si}\varepsilon_0}\left\{\left[p(x)-n(x)\right]+\left[N_d^+(x)-N_a^-(x)\right]\right\} \tag{2.73}$$

Assuming complete ionization of dopants, $N_d^+ = N_d$ and $N_a^- = N_a$, we can write Poisson's equation as

$$\frac{d^2\phi}{dx^2} = -\frac{q}{K_{si}\varepsilon_0}\left\{\left[p(x)-n(x)\right]+\left[N_d(x)-N_a(x)\right]\right\} \tag{2.74}$$

Equation 2.74 is a one-dimensional (1D) equation and can easily be extended to three-dimensional (3D) space. The 1D Poisson equation is adequate for describing most of the basic device operations. However, for small geometry advanced devices like FinFETs, the 2D (two-dimensional) or 3D Poisson's equation must be used.

Another form of Poisson's equation is Gauss' law, which is obtained by integrating Equation 2.71 and is given by

$$E = \frac{1}{K_{si}\varepsilon_0}\int \rho(x)dx = \frac{Q_s}{K_{si}\varepsilon_0} \tag{2.75}$$

It is to be noted that the semiconductor as a whole is charged neutral, that is, ρ must be zero. However, when the space charge neutrality does not apply, Poisson's equation must be used to describe the distribution of charge and electrostatic potentials in the semiconductor.

2.2.7.2 Transport Equations

In Section 2.2.5.1, we have shown that the electron current density $J_{n,drift}$ due to drift of electrons by an applied electric field is given by Equation 2.45. On the other hand, the electron diffusion current density $J_{n,diff}$ due to the concentration gradient in a semiconductor as described in Section 2.2.5.2 is given by Equation 2.55. Thus, when an electric field is present in addition to a concentration gradient, both the drift and diffusion currents will flow through the semiconductor. The total electron current density J_n at any point x is then simply equal to the sum of the drift and diffusion currents, that is, J_n ($= J_{n,drift} + J_{n,diff}$). Therefore, the total electron current in a semiconductor is given by

$$J_n = qn\mu_n E + qD_n \frac{dn}{dx} \tag{2.76}$$

Similarly, the total hole current density J_p ($= J_{p,drift} + J_{p,diff}$) is given by

$$J_p = qp\mu_p E - qD_p \frac{dp}{dx}, \tag{2.77}$$

so that the total current density $J = J_n + J_p$. The current Equations 2.76 and 2.77 are often referred to as the *transport* or *drift-diffusion equations* of current carriers.

Under thermal equilibrium, no current flows inside the semiconductor and therefore, $J_n = J_p = 0$. However, under the non-equilibrium conditions, J_n and J_p can be written in terms of quasi-Fermi potentials ϕ_n and ϕ_p for electric field, E in Equations 2.76 and 2.77, respectively, to get

$$J_n = -qn\mu_n \frac{d\phi_n}{dx}$$
$$J_p = -qp\mu_p \frac{d\phi_p}{dx} \tag{2.78}$$

All parameters have their usual meanings as defined earlier.

2.2.7.3 Continuity Equations

When carriers diffuse through a certain volume of a semiconductor, the current density leaving the volume may be smaller or larger depending upon the recombination or generation taking place inside the volume. Let us consider a small length Δx of a semiconductor shown in Figure 2.13 with cross-sectional area A in the xy plane.

From Figure 2.13, the hole current density entering the volume $A.\Delta x$ is $J_p(x)$, whereas that leaving is a $J_p(x + \Delta x)$. From the conservation of charge, the rate of change of hole concentration in the volume is the sum of: (a) net holes flowing out of the volume; and (b) net recombination rate. That is,

$$-\frac{\partial p}{\partial t}\Delta x = \left[\frac{1}{q}J_p(x+\Delta x)-\frac{1}{q}J_p(x)\right]+(G_p-R_p)\Delta x \tag{2.79}$$

The negative sign is due to the decrease of holes due to recombination; and G_p and R_p are the generation and recombination rates of holes in the volume, respectively. Then from Equation 2.79 we can show

$$-\frac{\partial p}{\partial t} = \frac{1}{q}\frac{\partial J_p}{\partial x}+(G_p-R_p) \tag{2.80}$$

Similarly, for electrons we can show

$$-\frac{\partial n}{\partial t} = -\frac{1}{q}\frac{\partial J_n}{\partial x}+(G_n-R_n) \tag{2.81}$$

where:

G_n and R_n are the generation and recombination rates of electrons, respectively

Equations 2.80 and 2.81 are called the *continuity equations* for holes and electrons, respectively, and describe the time-dependent relationship between current density, recombination and generation rates, and space. They are used for solving transient phenomena and diffusion with recombination-generation of carriers.

Equations 2.74, 2.78, 2.80, and 2.81 constitute a complete set of 1D equations to describe carrier, current, and field distributions in a semiconductor; however, they can easily be extended to 3D space. Given appropriate boundary conditions, we can solve them for any arbitrary device structure. Generally, we will be able to simplify them based on physical approximations.

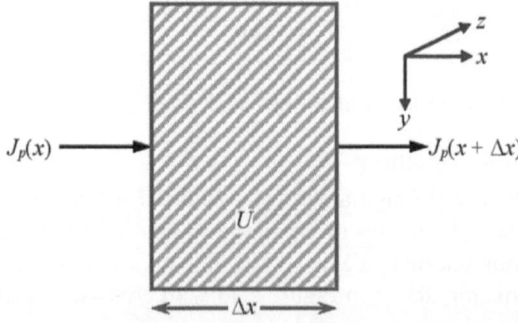

FIGURE 2.13 Current continuity in a semiconductor: $J_p(x)$ is the hole currents flowing into an elemental length Δx of the semiconductor and $J_p(x + \Delta x)$ is the net current flowing out after carrier generation-recombination processes inside the element; U is the net recombination rate.

2.3   THEORY OF *n*-TYPE AND *p*-TYPE SEMICONDUCTORS IN CONTACT

We have discussed the basic theory of intrinsic and *n*-type and *p*-type semiconductors in Sections 2.2.3 and 2.2.4, respectively. In this section, we will discuss the underlying physics of a semiconductor substrate in which *n*-type and *p*-type regions are adjacent to each other forming a junction called the *pn-junction*, or *pn-junction diode*, or simply *diode*. In reality, a silicon *pn*-junction is formed by counter doping a local region of a larger region of doped silicon as shown in Figure 2.14. The *pn*-junctions form the basis for all advanced semiconductor devices. Therefore, understanding their operation is basic to the understanding of FinFETs as well as most advanced IC devices.

2.3.1   Basic Features of *pn*-Junctions

A silicon *pn*-junction structure is an alternating type of *p*-type and *n*-type doped silicon layers. The *pn*-junctions can be fabricated in a variety of techniques on a silicon substrate using photo mask ⟶ Implant ⟶ Drive-in. A typical final impurity profile along the active region can be simplified as an *erfc* or *Gaussian* as shown in Figure 2.14(b) and (c).

The 2D cross-section of a typical *pn*-junction in Figure 2.14(a) shows the basic structure with an *n*-type region doped on a *p*-type substrate. The line *A* in Figure 2.14(a) shows a vertical cutline along the intrinsic or active region of the *pn*-junction. The 1D doping profile along the cutline *A* of the active device is shown in Figure 2.14(b) and (c). The metallurgical junction depth X_j is indicated as the point where the net impurity concentrations of donors and acceptors are equal or

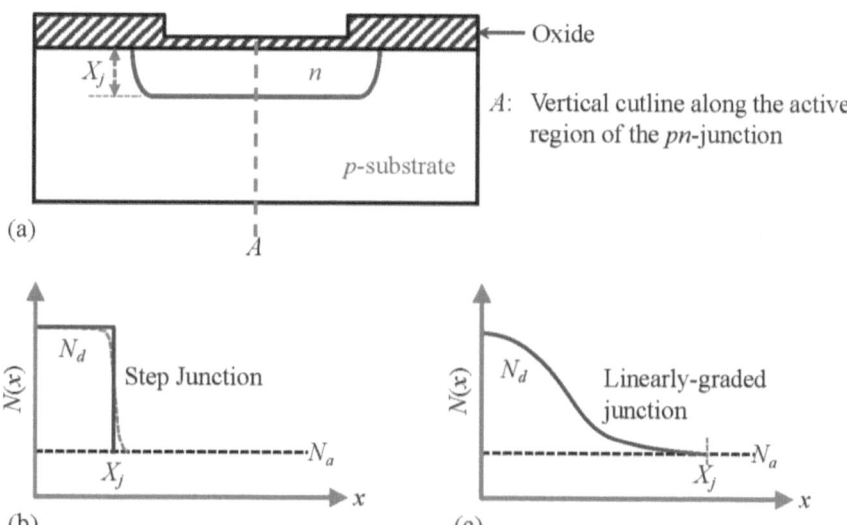

FIGURE 2.14   A typical *pn*-junction: (a) 2D cross-section showing the cut line along the active region of the structure to obtain 1D doping profile; (b) 1D doping profile of an abrupt or step junction; and (c) 1D doping profile of a graded junction.

compensated. For mathematical analysis of a *pn*-junction, the actual impurity profile is approximated by a *step* or *abrupt* (high-low) doping profile for shallow junctions as shown in Figure 2.14(b) or a *linearly graded* profile (deep junctions) shown in Figure 2.14(c). A step doping profile is characterized by constant *p*-type dopant of concentration N_a that changes with position in a stepwise fashion to a constant *n*-type dopant of concentration N_d and vice versa.

From the 1D impurity profiles in Figure 2.14(b) and (c), we find that there is a large carrier concentration gradient at the junction resulting in carrier diffusion. Holes from the *p*-side diffuse into the *n*-side leaving behind negatively charged acceptor ions $\left(N_a^-\right)$ and electrons from the *n*-side diffuse into the *p*-side leaving behind positively charged donor ions $\left(N_d^+\right)$. Consequently, a space charge region is formed (negative charge on the *p*-side and positive charge on the *n*-side) creating thereby an electric field *E*, and hence, a potential difference as shown in Figure 2.15. The direction of the field (*n*-region to *p*-region) is such that it opposes further diffusion of carriers so that, in thermal equilibrium, the net flow of carriers is zero, that is, an electric field is set up which tends to pull *electrons* and *holes* back to the original positions. The internal potential difference between the two sides of the junction is called the *built-in potential* or *barrier height* ϕ_{bi}. The space charge region on two sides of the metallurgical junction is often called the *depletion region*, because the region is depleted of the free carriers, or *transition region*, or *space charge region*.

Figure 2.16(a) shows the energy-band diagram of isolated *p*-type and *n*-type silicon materials. As discussed in Section 2.2.4.1, the Fermi level for a *p*-type silicon lies close to its VB and that for an *n*-type silicon lies close to its CB. Also, the Fermi level of a semiconductor is flat, that is, spatially constant, when there is no current flow in it [1]. Therefore, as the *p*-type region and the *n*-type region are brought together to form a *pn*-junction, the Fermi level must remain flat across the entire structure if there is no current flow in and across the junction. This causes the energy band bending as shown in Figure 2.16(b). Thus, the *built-in potential* ϕ_{bi} is the potential difference between the energy bands on the *p*-side and *n*-side of the *pn*-junction as shown in Figure 2.16(b).

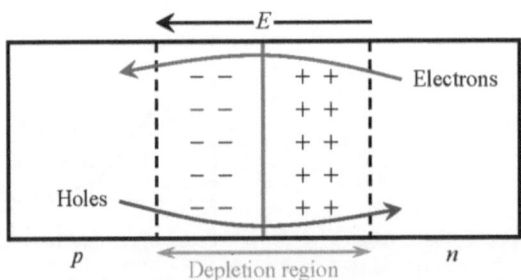

FIGURE 2.15 Formation of built-in electric field *E* due to the space charges left behind by mobile carriers after diffusion from the high to low concentration regions on both sides of the *pn*-junction.

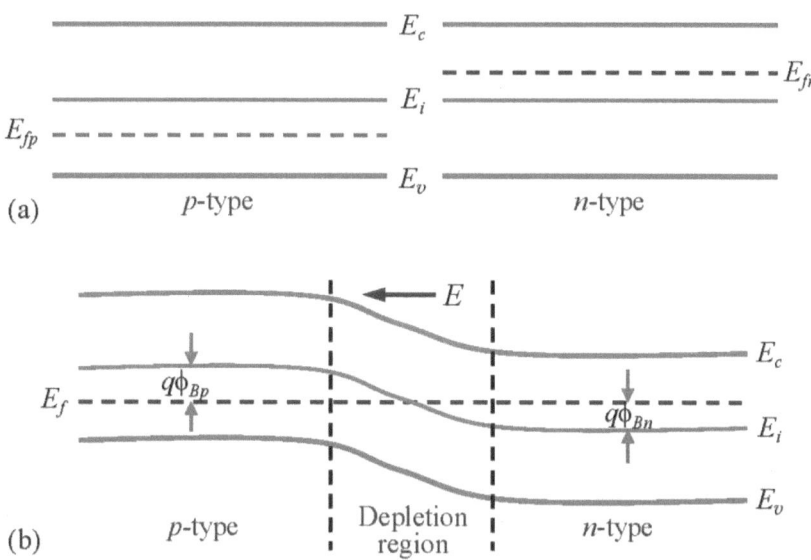

FIGURE 2.16 Energy band diagram of a *pn*-junction at equilibrium: (a) isolated *n*- and *p*-regions; (b) *p* and *n* regions are in contact to form a *pn*-junction. In the figure, ϕ_{Bp} and ϕ_{Bn} are the bulk potential of the *p*-type and *n*-type silicon, respectively.

2.3.2 BUILT-IN POTENTIAL

In a *pn*-junction at equilibrium, the diffusion of carriers is balanced by the drift of carriers by the *built-in* electric field. For clarity of description of both the *p*-side and *n*-side of a *pn*-junction simultaneously, we will add a subscript *p* to the symbols associated with the parameters on the *p*-side, and subscript *n* to the corresponding symbols associated with the parameters on the *n*-side. For example, E_{fp} and E_{fn} denote the Fermi level, on the *p*-side and *n*-side, respectively. Similarly, n_p and p_p denote the electron concentration and hole concentration, respectively, on the *p*-side and n_n and p_n denote the electron concentration and hole concentration, respectively, on the *n*-side. Thus, p_p and n_n specify the majority carrier concentrations, while n_p and p_n specify the minority carrier concentrations.

Now, let us consider the *pn*-junction at thermal equilibrium. If both *n*-side and *p*-side are non-degenerately doped to concentration of N_d and N_a, respectively, then the potentials with respect to the separation between its Fermi level, which is flat across the junction and corresponding intrinsic energy levels are given by Equations 2.29 and 2.30, namely

$$\phi_{in} - \phi_{fn} = \frac{kT}{q} \ln\left(\frac{N_d}{n_i}\right) = \frac{kT}{q} \ln\left(\frac{n_{no}}{n_i}\right) \equiv \phi_n$$

$$\phi_{fp} - \phi_{ip} = -\frac{kT}{q} \ln\left(\frac{N_a}{n_i}\right) = -\frac{kT}{q} \ln\left(\frac{p_{po}}{n_i}\right) \equiv \phi_p$$

(2.82)

where:

n_{n0} and p_{p0} represent the equilibrium concentrations in the n-type and p-type semiconductors, respectively

ϕ_n and ϕ_n are the potentials at the neutral edges of the depletion of the n-type and p-type regions, respectively

Since at equilibrium, E_f is a constant across the pn-junction, that is, $\phi_{fp} = \phi_{fn}$, therefore, the built-in potential across a pn-junction is the difference between the intrinsic energy levels shown in Figure 2.16(b). Then from the expressions in Equation 2.82, the built-in potential across a pn-junction is given by

$$\phi_{bi} = \phi_n - \phi_p = \frac{kT}{q} \ln\left(\frac{n_{n0}p_{p0}}{n_i^2} \right) \tag{2.83}$$

From Equation 2.8, the pn-product $p_{no}n_{n0} = n_i^2 = p_{p0}n_{p0}$; therefore, Equation (2.83) can also be written as

$$\phi_{bi} = \frac{kT}{q} \ln\left(\frac{n_{n0}p_{p0}}{n_i^2} \right) = v_{kT} \ln\left(\frac{N_a N_d}{n_i^2} \right) \tag{2.84}$$

or,

$$\phi_{bi} = v_{kT} \ln\left(\frac{n_{n0}}{n_{p0}} \right) = v_{kT} \ln\left(\frac{p_{p0}}{p_{n0}} \right) \tag{2.85}$$

Thus, at thermal equilibrium ϕ_{bi} given by Equation 2.84 or 2.85 exists across a pn-junction without an applied bias to counteract the carrier diffusion across the junction. The typical value of ϕ_{bi} is in-between 0.5–0.9 V for silicon pn-junctions and is strongly dependent on temperature due to dependence on n_i. Furthermore, ϕ_{bi} across a pn-junction increases as N_d and/or N_a increase as seen from Equation 2.84.

2.3.3 STEP JUNCTIONS

The analysis of a pn-junction is much simpler if the junction is assumed to be abrupt, that is, the doped impurities are assumed to change abruptly from p-type on one side to n-type on the other side of the junction. The abrupt junction approximation is reasonable for modern VLSI devices, where the use of ion implantation for doping the junctions, followed by low-thermal cycle diffusion and/or annealing, result in fairly abrupt junctions. Besides, the abrupt-junction approximation often leads to closed-form solutions for easier understanding of device physics.

2.3.3.1 Electrostatics

The analysis of an abrupt junction becomes even simpler using depletion approximation in which a pn-junction is approximated by three regions as illustrated in

Figure 2.17. In Figure 2.17, both the bulk p-region, $x < -x_p$, and bulk n-region, $x > x_n$, are assumed to be charge neutral, while the transition region, $-x_p < x < x_n$, is assumed to be depleted of mobile electrons and holes. The width W_d of the depletion region can be obtained by solving Poisson's Equation 2.74 as repeated below:

$$\frac{d^2\phi}{dx^2} = -\frac{q}{K_{si}\varepsilon_0}\left[\left(p(x)-n(x)\right)+\left(N_d(x)-N_a(x)\right)\right] \qquad (2.86)$$

Now, let us assume that the free carrier concentrations, n and p, are negligibly small compared to the fixed ionized impurities $N_d^+ \cong N_d$ and $N_a^- \cong N_a$ over the entire region defined by the depletion width bounded by $-x_p$ and x_n, that is, $N_d \gg n_n$ or p_n and $N_a \gg p_p$ or n_p as shown in Figure 2.17. This assumption is often referred to as the *depletion approximation*. It is often used during the development of analytical device models.

For simplicity of mathematical formulations, we will assume that all the donor and acceptor atoms within the depletion region are ionized and that the junction is abrupt without compensation, that is, there are no donor impurities on the p-side and no acceptor impurities on the n-side. With these assumptions, Equation 2.86 becomes

$$\frac{d^2\phi}{dx^2} = \frac{qN_a(x)}{K_{si}\varepsilon_0} \quad \text{for } -x_p < x < 0 \qquad (2.87)$$

and

$$\frac{d^2\phi}{dx^2} = -\frac{qN_d(x)}{K_{si}\varepsilon_0} \quad \text{for } 0 < x < x_n \qquad (2.88)$$

Then integrating Equation 2.87 from $x = -x_p$ to a point $x = -x$ and Equation 2.88 from a point $x = x > 0$ to $x = x_n$ with the boundary condition $d\phi/dx = 0$ at $x = -x_p$ and $x = x_n$, we can show the electric field distribution in the depletion region. Thus, assuming a step pn-junction so that N_a and N_d are uniform in the p- and n-regions, respectively,

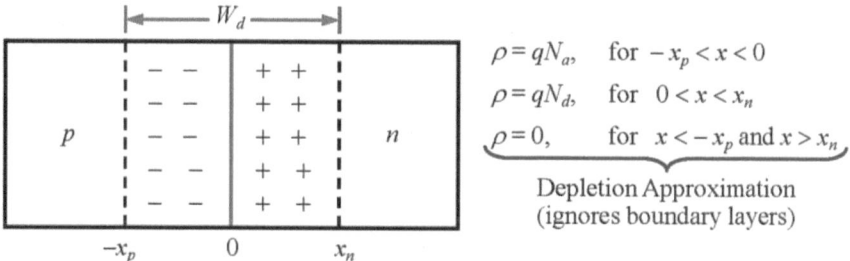

$\rho = qN_a,$ for $-x_p < x < 0$
$\rho = qN_d,$ for $0 < x < x_n$
$\rho = 0,$ for $x < -x_p$ and $x > x_n$

Depletion Approximation
(ignores boundary layers)

FIGURE 2.17 The charge condition in a pn-junction under the *depletion approximation* in three different regions: equilibrium depletion region is bounded by $-x_p$ and x_n on the p-region and n-region, respectively, and p and n neutral regions; the depletion region is assumed to be free of mobile carriers with $\rho = 0$.

and using depletion approximation, we obtain after the first integration of Equation 2.87 and 2.88 under the boundary conditions $d\phi/dx = -E(x) = 0$ at $x = -x_p$ and $x = x_n$

$$E(x) = -\frac{qN_a}{K_{si}\varepsilon_0}\left(x_p - x\right), \quad \text{for } -x_p < x < 0 \tag{2.89}$$

$$E(x) = -\frac{qN_d}{K_{si}\varepsilon_0}\left(x_n - x\right), \quad \text{for } 0 < x < x_n \tag{2.90}$$

Since the electric field must be continuous at $x = 0$, we get from Equations 2.89 and 2.90, the maximum electric field E_{max} as

$$E_{max} = -\frac{qN_a}{K_{si}\varepsilon_0}x_p = -\frac{qN_d}{K_{si}\varepsilon_0}x_n \tag{2.91}$$

or

$$qN_a x_p = qN_d x_n \tag{2.92}$$

where:
x_p is the width of the depletion region on the p-side of the pn-junction
x_n is the width of the depletion region on the n-side of the pn-junction

Equation 2.92 expresses the distribution of charge on either side of the junction and shows that the negative charge on the p-side exactly equals the positive charge on the n-side. Equation 2.92 also shows that the width of the depletion regions on each side of the junction varies inversely with the doping concentration; the higher the doping concentration, the narrower the depletion region. Furthermore, Equations 2.89 and 2.90 show that E varies linearly between 0 and E_{max} as shown in Figure 2.18.

Let ϕ_m be the total potential drop across the pn-junction, that is, $\phi_m = \left[\phi(x_n) - \phi(x_p)\right]$. Then from Equation 2.70 the total potential drop across the pn-junction from $x = -x_p$ to $x = x_n$. is given by

$$\phi_m = \int_{-x_p}^{x_n} d\phi(x) = -\int_{-x_p}^{x_n} E(x)dx = -\frac{E_{max}\left(x_n + x_p\right)}{2} \tag{2.93}$$

$$= -\frac{E_{max}}{2}W_d$$

where:
$W_d = (x_n + x_p)$ is the total width of the depletion layer
$E_{max}(x_n + x_p)/2$ is the area under the $E(x)$ versus x plot

Thus, it is observed from Equation 2.93 that ϕ_m is equal to the area under the $E(x)$ versus x plot in Figure 2.18. Then eliminating E_{max} from Equations 2.91 and 2.93, we can show that

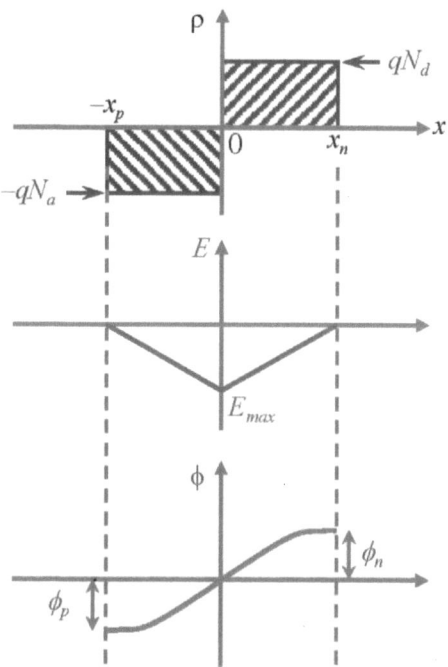

FIGURE 2.18 Depletion approximation of a pn-junction: the equilibrium distribution of charge ρ; electric field E; and electrostatic potential ϕ within the depletion region.

$$W_d = \sqrt{\frac{2\varepsilon_0 K_{si}(N_a + N_d)}{qN_aN_d}}\phi_m \tag{2.94}$$

In Equation 2.94, $\phi_m = \phi_{bi}$ is the potential difference between the p and n sides of a pn-junction at equilibrium condition without an external bias. Thus, we can estimate the W_d from Equation 2.94.

In order to derive expressions to estimate x_p and x_n, we integrate Equations 2.89 and 2.90 by replacing $E = -d\phi/dx$ [Equation 2.70]. Then we integrate Equation 2.89 from $x = -x_p$ to $x = 0$ using the boundary condition: $\phi(-x_p) = -\phi_p$ and $\phi(0) = \phi_p(0)$; and Equation 2.90 from $x = 0$ to $x = x_n$ using the boundary condition: $\phi(0) = \phi_n(0)$ and $\phi(x_n) = \phi_n$. Then, considering $\phi_p(0) = \phi_n(0)$ since the electrostatic potential must be the same at the p/n-interface and $\phi_{bi} = (\phi_n - \phi_p) \equiv \phi_m$, we can show after simplification

$$x_p = \sqrt{\frac{2K_{si}\varepsilon_0}{q}\frac{N_d}{N_a(N_a + N_d)}\phi_{bi}} \tag{2.95}$$

And

$$x_n = \sqrt{\frac{2K_{si}\varepsilon_0}{q}\frac{N_a}{N_d(N_a + N_d)}\phi_{bi}} \tag{2.96}$$

So that the total depletion width $W_d (= x_p + x_n)$ becomes

$$W_d = \sqrt{\frac{2K_{si}\varepsilon_0}{q}\left(\frac{1}{N_a} + \frac{1}{N_d}\right)\phi_{bi}} \qquad (2.97)$$

Equation 2.97 shows that W_d strongly depends on the doping on the lightly doped side and particularly, W_d is inversely proportional to the square root of the doping concentration on the lightly doped side. Also, W_d given by Equation 2.97 is the thermal equilibrium value without any external voltage applied to the pn-junction.

Now, we know from Equations 2.91 and 2.92 that the depletion charge per unit area on either side of the depletion region is given by

$$Q_{dep} = qN_a x_p = qN_d x_n = E_{max}K_{si}\varepsilon_0 \qquad (2.98)$$

Then substituting for E_{max} and W_d from Equations 2.93 and 2.94, respectively, in Equation 2.98, the depletion layer capacitance per unit area can be shown as

$$C_d = \frac{d|Q_{dep}|}{d\phi_m} = \frac{K_{si}\varepsilon_0}{W_d} \qquad (2.99)$$

Equation 2.99 shows that the depletion capacitance of a pn-junction is equivalent to a parallel plate capacitor of separation W_d and permittivity of silicon K_{si}. Physically, this is due to the fact that only the mobile charge at the edges of the depletion layer, not the space charge within the depletion region, responds to changes in the applied voltage.

2.3.4 *pn*-JUNCTIONS UNDER EXTERNAL BIAS

We discussed in Section 2.3.1 that in an equilibrium condition without an external bias, the Fermi level is spatially constant across a pn-junction. However, an externally applied voltage, V_d across the pn-junction has the effect of shifting the Fermi level of the bulk neutral n-region relative to that of the bulk neutral p-region. That is, the total potential drop is the sum of the built-in potential and the externally applied potential, namely

$$\phi_m = \phi_{bi} \pm V_d \qquad (2.100)$$

Where the "+" sign is for the case where the junction is reverse biased and $\phi_m > \phi_{bi}$, and the "−" sign is for the case where the junction is forward biased and $\phi_m < \phi_{bi}$. Thus, when the pn-junction is in a non-equilibrium condition, with voltage V_d applied to it, then as stated earlier, the potential barrier height becomes $(\phi_{bi} - V_d)$, so that the depletion width [Equation 2.97] as a function of voltage becomes

$$W_d = \sqrt{\frac{2K_{si}\varepsilon_0}{q}\left(\frac{1}{N_a} + \frac{1}{N_d}\right)\cdot(\phi_{bi} - V_d)} \qquad (2.101)$$

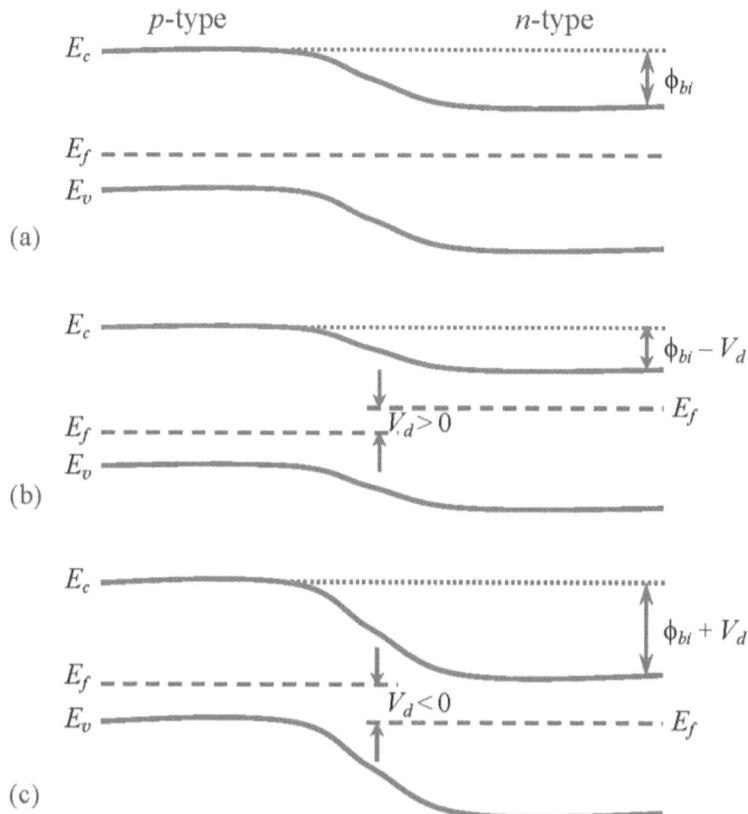

FIGURE 2.19 The energy band diagram of a typical *pn*-junction under different biasing conditions: (a) equilibrium; (b) forward bias; (c) reverse bias.

This shows that a forward bias V_d ($\equiv V_f$) will result in a decrease in the depletion width due to the decrease in the barrier height, while a reverse bias $-V_d$ ($\equiv V_r$) will result in an increase in the depletion width due to a higher barrier height as shown in Figure 2.19.

Now, using Equation 2.95 for x_p or 2.96 for x_n in Equation 2.91, the maximum electric field, E_{max} in the depletion region becomes

$$E_{max} = \sqrt{\frac{2q}{K_{si}\varepsilon_0}\frac{N_aN_d}{(N_a+N_d)}(\phi_{bi}-V_d)} \tag{2.102}$$

Equation 2.102 shows that the higher the reverse voltage (e.g., $-V_d$), the higher the electric field across the *pn*-junction.

2.3.4.1 One-Sided Step Junction

If the impurity concentration on one side of a *pn*-junction is much higher than the other side, the junction is called a *one-sided step junction*. In this case, the depletion region extends almost entirely into the lighter doped side. For example, in the case of an n^+p junction ($N_d \gg N_a$), we find from Equations 2.95 and 2.96 that $x_n \ll x_p$ and the depletion width W_d is almost entirely in the p-side. Thus, from Equation 2.101, we can show that the general expression for W_d for a one-sided step junction is

$$W_d = \sqrt{\frac{2K_{si}\varepsilon_0}{qN_b} \cdot \left(\phi_{bi} \pm V_d\right)}, \qquad (2.103)$$

where:

$N_b = N_a$ for n^+p junction
$N_b = N_d$ for p^+n junction

A more accurate result for the depletion width can be obtained by considering majority carrier distribution tails or spillover (electrons in the n-side and holes in the p-side by Debye length L_d) as shown by the dashed lines in Figure 2.20. Each spillover contributes a correction factor v_{kT} to ϕ_{bi}. The depletion width is still given by Equation 2.103 except that ϕ_{bi} is replaced by ($\phi_{bi} - 2v_{kT}$) so that a more accurate expression for W_d of a one-sided step junction becomes

$$W_d = \sqrt{\frac{2K_{si}\varepsilon_0}{qN_b} \cdot \left(\phi_{bi} - 2v_{kT} \pm V_d\right)} \qquad (2.104)$$

Equation 2.104 is used for accurate estimation of W_d in a one-side step pn-junction. However, for the biasing range used in the modern VLSI circuits, Equation 2.103 is accurate within about 3%.

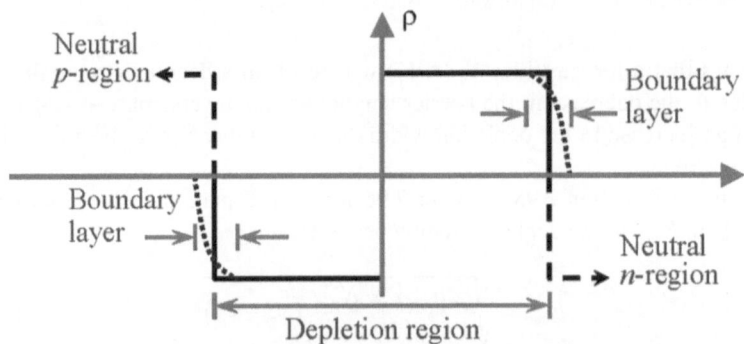

$\rho \cong 0$ outside the depletion region; $\rho \cong |N_a - N_d|$ within the depletion region; boundary layer spread $\approx 3L_d$

FIGURE 2.20 Majority carrier spill over (broken lines) outside the depletion region forming a boundary layer of width about $3L_d$ at the boundary of the neutral bulk region; L_d is the Debye length defining the abruptness of the junction.

2.3.5 CARRIER TRANSPORT ACROSS *pn*-JUNCTIONS

In considering $I - V$ characteristics of a *pn*-junction, it is much more convenient to work with the quasi-Fermi potentials, instead of the intrinsic potential. In Section 2.2.7.2, the electron and hole current densities J_n and J_p, respectively, for doped semiconductors are expressed in terms of quasi-Fermi potentials [Equations 2.78] as

$$J_n = -qn\mu_n \frac{d\phi_n}{dx}$$

$$J_p = -qp\mu_p \frac{d\phi_p}{dx}$$

(2.105)

where:
  ϕ_n and ϕ_p are the quasi-Fermi potentials for electrons and holes defined in Equations 2.40 and 2.41, respectively, and are given by

$$\phi_n = \phi_i - v_{kT} \ln\left(\frac{n}{n_i}\right)$$

$$\phi_p = \phi_i + v_{kT} \ln\left(\frac{p}{n_i}\right)$$

(2.106)

2.3.5.1 Relationship between Minority Carrier Density and Junction Potential

Under the forward bias V_d, the barrier to majority carrier flow is reduced and the electrons are injected from the *n*-region to the *p*-region and holes are injected from the *p*-region to the *n*-region. The electrons going from the *n*-region to the *p*-region become the minority carriers in the *p*-region. Similarly, *holes* going from the *p*-region to the *n*-region become the minority carriers in the *n*-region. Therefore, the minority carrier behavior is of fundamental importance to understand the behavior of a *pn*-junction. The minority carriers injected across the barrier will tend to recombine if given sufficient time. They will also tend to diffuse away from the region of the junction.

In order to calculate *pn*-junction current in thermal equilibrium, let us consider n_{no} and p_{po} as the equilibrium majority carrier concentrations in the neutral *n*- and *p*-regions, respectively; and n_{po} and p_{no} are the equilibrium minority carrier electron and hole concentrations in the neutral *p*- and *n*-regions, respectively, as shown in Figure 2.21. Then, from our discussions on carrier concentration in Section 2.2.4.3, we get in the *neutral n*-region

$$n_{no} \cong N_d; \quad p_{no} \cong \frac{n_i^2}{N_d}$$

(2.107)

and in the *neutral p*-region

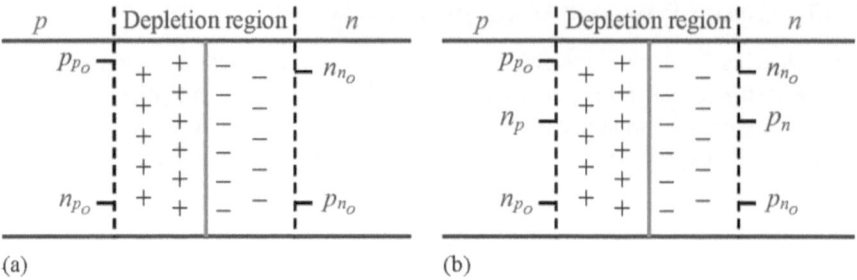

(a)   (b)

FIGURE 2.21 Carrier concentrations at the edge of the depletion region for low level injection: (a) *pn*-junction at equilibrium where p_{p0} and n_{n0} are the equilibrium majority carrier hole and electron concentrations in the *p*-type and *n*-type regions, respectively, whereas n_{p0} and p_{n0} are the equilibrium minority carrier electron and hole concentrations in the *p*-type and *n*-type regions, respectively; and (b) *pn*-junction after minority carrier n_p and p_n injection in the bulk *p*-region and *n*-region, respectively.

$$p_{p0} \cong N_a; \quad n_{p0} \cong \frac{n_i^2}{N_a}. \tag{2.108}$$

Now, from Equation 2.85, the equilibrium carrier concentrations in a *pn*-junction are given by the expressions

$$\phi_{bi} = \begin{cases} v_{kT} \ln\left(\dfrac{n_{n0}}{n_{p0}}\right); & \text{in terms of electron concentrations} \\[4mm] v_{kT} \ln\left(\dfrac{p_{p0}}{p_{n0}}\right); & \text{in terms of hole concentrations} \end{cases} \tag{2.109}$$

Then, from Equation 2.109, we can write the majority carrier expressions for a *pn*-junction at equilibrium

$$n_{n0} = n_{p0} \exp\left(\frac{\phi_{bi}}{v_{kT}}\right)$$

$$p_{p0} = p_{n0} \exp\left(\frac{\phi_{bi}}{v_{kT}}\right) \tag{2.110}$$

Now, under the applied bias V_d, we replace ϕ_{bi} by $(\phi_{bi} \pm V_d)$; therefore, from Equation 2.110, the non-equilibrium carrier concentrations are given by

$$n_n = n_p \exp\left(\frac{\phi_{bi} - V_d}{v_{kT}}\right);$$

$$p_p = p_n \exp\left(\frac{\phi_{bi} - V_d}{v_{kT}}\right), \tag{2.111}$$

where:

n_p is the non-equilibrium minority carrier electron concentration at the edge of the depletion region in the neutral p-region as shown in Figure 2.21(b)

p_n is the non-equilibrium minority carrier hole concentration at the edge of the depletion region in the neutral n-region as shown Figure 2.21(b)

Now, let us further assume a *low level injection*, that is, the injected carrier densities are lower than the background concentrations ($n_p \ll p_{p0}; p_n \ll n_{n0}$) so that $n_n = n_{no}$ and $p_p = p_{po}$. Then, from Equations 2.110 and 2.111, we get:

$$n_p = n_{p0} \exp\left(\frac{V_d}{v_{kT}}\right);$$

$$p_n = p_{n0} \exp\left(\frac{V_d}{v_{kT}}\right).$$

(2.112)

In Equation 2.112, n_p and p_n are the injected minority carrier concentrations at the edge of the depletion region in the p- and n-regions, respectively. Each expression in Equation 2.112 defines the minority carrier densities at the *edge* of the space charge region under an applied bias and is the most important boundary condition governing a pn-junction. Each relates the minority carrier concentration at the boundary of the depletion layer to its thermal-equilibrium value and to the applied voltage across the junction. Equation 2.112 applies to both a forward biased ($V_d > 0$) junction resulting in $n_p \gg n_{p0}$ at $x = -x_p$ and $p_n \gg p_{n0}$ at $x = x_n$, and to a reverse biased ($V_d < 0$) junction resulting in $n_p \ll n_{p0}$ at $x = -x_p$ and $p_n \ll p_{n0}$ at $x = x_n$. The expressions in Equation 2.112 can also be expressed as

$$n_p = \frac{n_i^2}{p_{p0}} \exp\left(\frac{V_d}{v_{kT}}\right)$$

$$p_n = \frac{n_i^2}{n_{n0}} \exp\left(\frac{V_d}{v_{kT}}\right)$$

(2.113)

Again, for a low level injection in the p-region: $p_{p0} = p$ and $n_p = n$; similarly, in the n-region: $n_{n0} = n$ and $p_n = p$ therefore, we get from Equation 2.113

$$pn = n_i^2 \exp\left(\frac{V_d}{v_{kT}}\right)$$

(2.114)

Equation 2.114 defines the pn-product of carriers at the depletion edge under the applied voltage V_d as shown in Figure 2.21. Thus, the applied bias in a pn-junction sets up the following processes as shown in Figure 2.22:

- The injected carriers in the neutral n- and p-regions momentarily set up electric fields with the respective majority carriers in each region
- Each momentary field draws in majority carriers in the respective region

- These majority carriers neutralize the injected carriers and re-establish the charge neutrality
- While this process continues, the injected minority carriers diffuse into the n- and p-regions, that is, the recombination process takes place over some distance

The distribution of carriers in each region of the pn-junction is shown in Figure 2.23. The majority carrier concentrations shown by broken lines remain unchanged, whereas the minority carrier concentration decays exponentially and approaches to the equilibrium concentration in each side of the junction.

The injected excess carriers set up a momentary electric field E in the regions of excess carrier concentration. Then the current due to this drift electric field in the n-region is $J_{n,drift} = q\mu_n nE$ [Equation 2.45] for majority carrier *electrons* and $J_{p,drift} = q\mu_p pE$ [Equation 2.47] for minority carrier *holes*. Since in the n-region, $n \gg p$, therefore, the hole drift current is negligible in the neutral n-region. Similarly, the electron drift

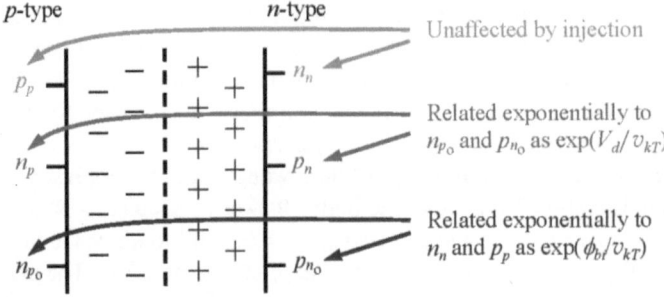

FIGURE 2.22 Carriers in a pn-junction under applied bias showing the corresponding dependence on built-in potential and applied bias; assume low level injection.

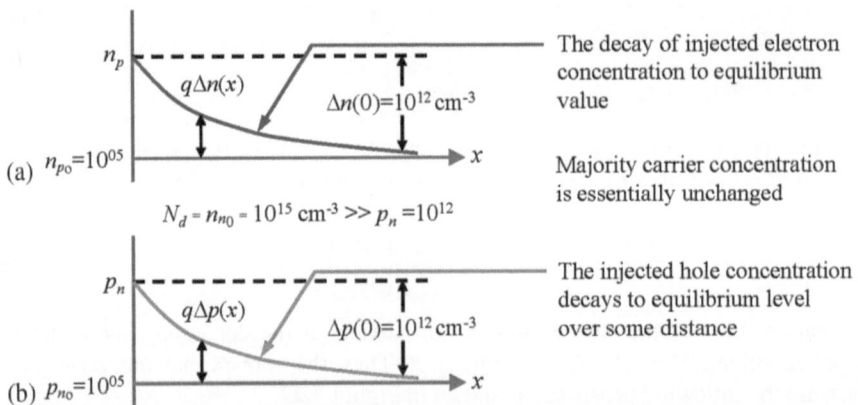

FIGURE 2.23 The carrier profile in a pn-junction with applied bias for low level injection: (a) decay of the injected minority carrier electrons in the neutral p-region; (b) decay of the injected minority carrier holes in the neutral n-region; injected carrier concentration, 1×10^{12} cm^{-3} is much lower than the majority carrier concentration, 1×10^{15} cm^{-3}.

current is negligible in the neutral p-region. Thus, the injected minority carriers move primarily by diffusion while the majority carriers are pulled to the junction by drift. Since the injected minority carriers control the current flow in a pn-junction, the current flow in pn-junctions can be considered as the diffusion current only. *Thus, we see that the minority carriers really control the behavior of pn-junctions.*

2.3.6 *pn*-Junctions *I* – *V* Characteristics

We discussed in Section 2.3.1 that the diffusion of carriers caused by the electron and hole concentration gradient across a pn-junction in equilibrium is balanced by the built-in electric field in the depletion region. As a result, the drift component of the current due to the built-in electric field is exactly balanced out by the diffusion component of the current across the junction and the net current flow in the pn-junction is zero. However, when an external voltage is applied, these current components are no longer balanced and a net current flows in the pn-junction. If the carriers are generated by light or some other external means, the thermal equilibrium is disturbed, and the current can also flow in a pn-junction. In this section, the current flow in a pn-junction due to an external applied voltage is described.

Let us consider a forward biased pn-junction (applied positive bias at the p-region and negative bias at the n-region). Then the electrons are injected from the n-side into the p-side, and holes are injected from the p-side to the n-side. If the generation and recombination in the depletion region are negligible, then the hole current leaving the p-side is the same as the hole current entering the n-side. Similarly, the electron current leaving the n-side is equal to the electron current entering the p-side. To determine the total current flowing in the pn-junction, we need to determine either the hole current entering the p-side or the electron current entering the n-side of the pn-junction.

The starting point for describing $I - V$ characteristics of a pn-junction is the continuity equations. From Equation 2.81, the electron continuity equation is given by

$$-\frac{\partial n}{\partial t} = -\frac{1}{q}\frac{\partial J_n}{\partial x} + (G_n - R_n) \qquad (2.115)$$

where:
R_n and G_n are the recombination and generation rates of electrons, respectively

Equation 2.115 can be rewritten as

$$\frac{\partial n}{\partial t} = \frac{1}{q}\frac{\partial J_n}{\partial x} - \frac{n - n_0}{\tau_n}, \qquad (2.116)$$

where:
τ_n is the electron lifetime defined in terms of the excess electron concentration n over the thermal equilibrium value n_0 in Equations 2.63 and 2.69 and is given by

$$\tau_n \equiv \frac{n - n_0}{R_n - G_n} \qquad (2.117)$$

Now, substituting for J_n from Equation 2.76 in Equation 2.116, we get

$$\frac{\partial n}{\partial t} = n\mu_n \frac{\partial E}{\partial x} + \mu_n E \frac{\partial n}{\partial x} + D_n \frac{\partial^2 n}{\partial x^2} - \frac{n - n_0}{\tau_n} \qquad (2.118)$$

Equation 2.118 is the generalized formulation that is solved under the appropriate boundary conditions to derive an expression for the electron current flow across a *pn*-junction under an applied bias.

Now, in order to calculate the current in a *pn*-junction, we assume that the injected minority carriers move away from the depletion region by diffusion only, that is, *diffusion approximation*. In addition, we make the following assumptions to calculate the *pn*-junction current:

1. The step junction profile is applicable
2. The depletion approximation is valid
3. Low level injection is maintained in the bulk
4. No generation-recombination occurs in the depletion region
5. There is no voltage drop in the bulk region so that V_d is sustained entirely across the depletion region
6. The width of the bulk *p*- and *n*-regions outside the depletion region is much longer than the minority carrier diffusion length for holes and electrons L_p and L_n, respectively, (long-base diode)

With the above simplifying assumptions, the current through an ideal *pn*-junction can be shown as

$$I_d = I_s \left[\exp\left(\frac{V_d}{v_{kT}}\right) - 1 \right] \qquad (2.119)$$

The parameter I_s in Equation 2.119 is called the reverse saturation current and is given by

$$I_s = \begin{cases} qA_d n_i^2 \left[\dfrac{D_p}{N_d L_p} + \dfrac{D_n}{N_a L_a} \right]; & W_n > L_p \quad \text{and} \quad W_p > L_n \\[4mm] qA_d n_i^2 \left[\dfrac{D_p}{N_d W_n} + \dfrac{D_n}{N_a W_p} \right]; & W_n < L_p \quad \text{and} \quad W_p < L_n \end{cases} \qquad (2.120)$$

where:
A_d is the active area of the *pn*-junction
W_n and W_p are the width of the neutral *n*- and *p*-regions, respectively
D_n and D_p are the minority carrier electron and hole diffusion constants, respectively
L_n and L_p are the minority carrier electron and hole diffusion lengths, respectively

Real *pn*-junctions may represent intermediate cases, that is, $W_n > L_p$ and $W_p < L_n$ and vice versa. In either case, the lightly doped side of the junction largely determines the diode current I_d in Equation 2.119. Figure 2.24 shows a typical $I - V$ characteristic of a *pn*-junction.

2.3.6.1 Temperature Dependence of pn-Junction Leakage Current

From Equation 2.120, we observe that the temperature dependence of the electron and hole diffusion currents is dominated by the temperature dependence of the parameter n_i^2, which is proportional to $\exp(-E_g/kT)$ as shown in Equation 2.14, where E_g is the bandgap energy. Then substituting for $n_i(T)$ from Equation 2.14 into Equation 2.120, we can show the temperature dependence of I_s with reference to the reference temperature, T_{NOM} as

$$I_s(T) = I_s(T_{NOM})\left(\frac{T}{T_{NOM}}\right)^3 \exp\left(\frac{E_g(T_{NOM})}{kT_{NOM}} - \frac{E_g(T)}{kT}\right)$$

$$= I_s(T_{NOM})\left(\frac{T}{T_{NOM}}\right)^{XTI} \exp\left(\frac{E_g(T_{NOM})}{kT_{NOM}} - \frac{E_g(T)}{kT}\right) \quad (2.121)$$

where:
$XTI = 3$ is used as a fitting parameter to account for any deviation of *pn*-junction performance from the ideal formulation

For the simplicity of *pn*-junction modeling for circuit analysis, Equation 2.121 is expressed as

$$I_s(T) = I_s(T_{NOM})\cdot\exp\left[\frac{\dfrac{E_g(T_{NOM})}{kT_{NOM}} - \dfrac{E_g(T)}{kT} + XTI \ln\left(\dfrac{T}{T_{NOM}}\right)}{NJ}\right] \quad (2.122)$$

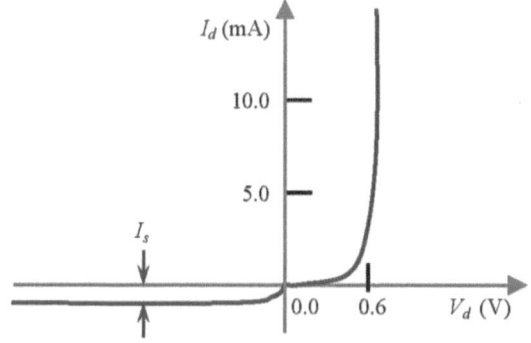

FIGURE 2.24 Current-voltage characteristics of a typical *pn*-junction: I_s is the reverse saturation current; an applied voltage of about 0.6 V is required to overcome the built-in potential for device conduction.

where:

XTI and NJ are the *temperature exponent coefficient* fitting parameters to optimize the performance of *pn*-junctions at any T with reference to T_{NOM}.

2.3.6.2 Limitations of *pn*-Junction Current Equation

The ideal expression for *pn*-junction current given by Equation 2.119 accurately describes the characteristics of a *pn*-junction over a certain range of applied voltage. However, the ideal Equation 2.119 becomes inaccurate over a significant range of device operations both in the forward and reverse biased modes.

The current voltage characteristics of a *forward biased* silicon *pn*-junction diode are shown in Figure 2.25 where the ideal diode current is shown by the broken line. Two different regions of non-ideal behaviors are shown in this plot which is due to real device effects. In deriving Equation 2.119, we have assumed that all the minority carriers cross the depletion region (*assumption 4*). However, in real devices, some of these carriers recombine through trapping centers and must be accounted for. Then, using the SRH theory of generation and recombination, it can be shown that the space charge recombination current I_{rec} is

$$I_{rec} = \frac{qA_d n_i W}{2\tau_{rec}} \exp\left(\frac{V_d}{2v_{kT}}\right) \tag{2.123}$$

In Equation 2.123, τ_{rec} is the lifetime associated with the recombination of excess carriers in the depletion region. Here, τ_{rec} is analogous to, but usually greater than, τ_n and τ_p (Section 2.2.6.2) in the neutral regions and is approximately equal to $2\sqrt{\tau_p \tau_n}$.

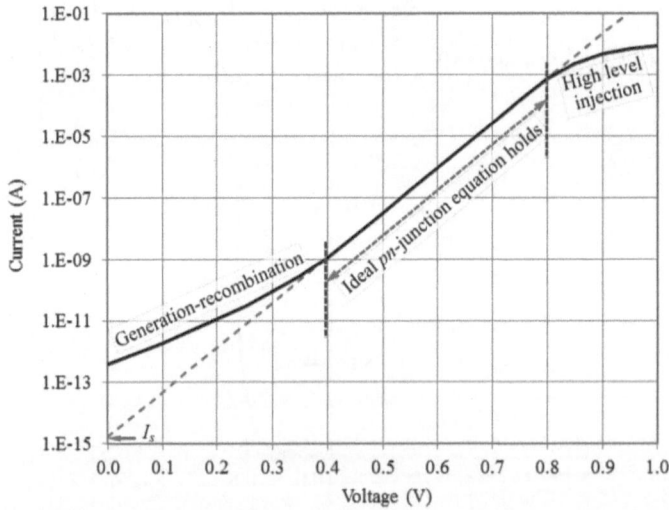

FIGURE 2.25 Forward characteristics of a real *pn*-junction: plot shows the deviation of ideal current equation at the low and high current levels due to generation-recombination and high level injections, respectively.

Thus, the total pn-junction saturation current I_s is the sum of the currents given in Equations 2.120 and 2.123. The result is a larger total current than that predicted by the ideal Equation 2.119, particularly in the low current level. In general, until V_d reaches a value of about 0.4 V, the neutral region diffusion current will be less than I_{rec}. Thus, I_{rec} dominates in the silicon diode at very small current levels and is negligibly small at higher current levels.

At high current levels, the injected minority carrier density is comparable to the majority carrier concentration (high level injection), and therefore, *assumption 3* is invalid. For high level injection, majority carrier concentration increases significantly above its equilibrium value giving rise to an electric field. Thus, in such cases both drift and diffusion components of currents must be considered. The presence of the electric field results in a voltage drop across this region, and thus, reduces the effective applied voltage across the junction resulting in a lower current than expected. It can be shown that under high level injection the diode current I_d is given by

$$I_d = \frac{qA_d n_i D_p}{W} \exp\left(\frac{V_d}{2v_{kT}}\right) \quad \text{(high-level injection)} \tag{2.124}$$

Equation 2.124 indicates that high level current depends on $1/2v_{kT}$ rather than $1/v_{kT}$ as shown in Figure 2.25. Thus, depending on the magnitude of the applied forward voltage, the current through a pn-junction can be represented by an empirical expression

$$I_d = I_s \left[\exp\left(\frac{V_d}{n_E v_{kT}}\right) - 1 \right] \tag{2.125}$$

where:
  n_E is called the *ideality factor* and is a measure of the deviation of the real and
      ideal $I - V$ plots

In Equation 2.125, $n_E = 2$ when recombination current dominates or there is high level injection and $n_E = 1$ when diffusion current dominates

In the case of a *reverse biased pn*-junction, Figure 2.26 shows the current through the pn-junction where I_s is the current due to an ideal pn-junction Equation 2.119. Clearly, the current in a real pn-junction does not saturate at $-I_s$ as predicted by Equation 2.119. This is because when the pn-junction is reverse biased, generation of electron-hole pairs in the depletion region takes place, which was neglected in the ideal pn-junction equation. In fact, the generation current dominates because carrier concentrations are smaller than their thermal equilibrium values. Again, using SRH theory, it can be shown that the generation current I_{gen} is

$$I_{gen} = \frac{qA_d n_i W_d}{2\tau_{gen}} \tag{2.126}$$

where τ_{gen} is the generation lifetime of the carriers in the depletion region and is approximately equal to $2\tau_p$ when $\tau_p = \tau_{gen}$. Note that while I_s is proportional to n_i^2,

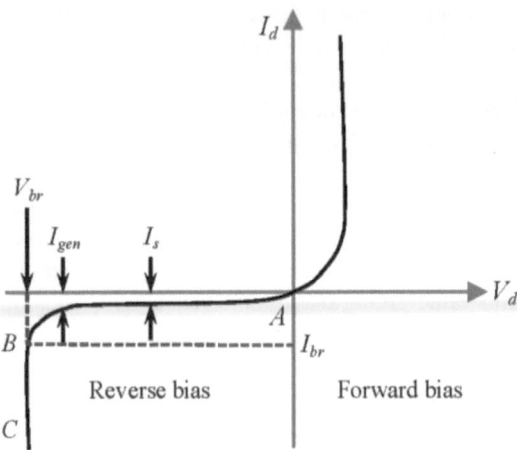

FIGURE 2.26 Reverse characteristics of a real *pn*-junction: V_{br} and I_{br} are the breakdown voltage and current, respectively; I_s is the ideal reverse saturation current; and I_{gen} is the generation current in the depletion region.

I_{gen} is proportional to n_i only. Thus, I_{gen} will dominate when n_i is small as is the case at room and low temperatures. Furthermore, since the space charge width W_d increases as the square root of the reverse bias [Equation 2.103], the generation current increases with reverse bias voltage as shown in Figure 2.26. Thus, taking I_{gen} into account, the total reverse current I_r becomes $I_r \equiv -I_d = -(I_s + I_{gen})$. This value of I_r agrees very well with the measured value of reverse current as well as provides proper voltage dependence of the reverse current in properly constructed planar silicon *pn*-junctions.

In real *pn*-junctions, there is a third component of leakage current, called the surface leakage current I_{sl}. This current can be treated as a special case of I_{gen} modeled at the surface where a high concentration of dislocations at the oxide-silicon interface, often referred to as fast surface states, provide additional generation centers over those present in the bulk. It is very much process-dependent and is responsible for large variation in the leakage current. Both process and electrically induced defects at the surface generally increase the generation rate by an order of magnitude compared to the bulk recombination-generation rate. In that case, I_{sl} dominates over the other components of I_r and is thus responsible for higher leakage current for a *pn*-junction compared to that predicted by the sum of I_{gen} and I_s. Leakage current is highly temperature-dependent due to the presence of n_i term. Also, note that the generation limited leakage current is proportional to n_i while diffusion limited leakage current is proportional to n_i^2.

2.3.6.3 Bulk Resistance

At high current levels, the bulk resistance and metal-silicon contact resistance can produce a significant voltage drop (invalidating *assumption 5*) resulting in a smaller voltage across the *pn*-junction and thus a lower current. For theoretical analysis,

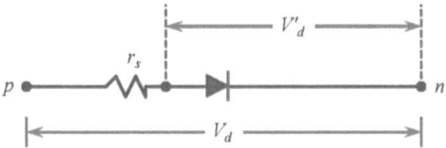

FIGURE 2.27 Circuit schematic of a *pn*-junction at high current level: r_s represents the overall resistance of the device due to the contact and neutral bulk regions, V_d is the applied bias, and V_d' is the voltage drop across the *pn*-junction.

the bulk resistance and contact resistance are usually combined into one resistor called series resistance r_s. Thus, if V_d is the applied voltage to the *pn*-junction terminals and V_d' is the voltage across the junction resulting in the current I_d as shown in Figure 2.27, we have

$$V_d = V_d' + r_s I_d \qquad (2.127)$$

Under the ideal condition when $r_s = 0$, then $V_d = V_d'$ and I_d is given by Equations 2.119 or 2.125. However, due to r_s at high current level, the expression for $I - V$ of a *pn*-junction becomes

$$I_d = I_s \left[\exp\left(\frac{V_d - I_d \cdot r_s}{n_E v_{kT}} \right) - 1 \right] \qquad (2.128)$$

Re-arranging Equation 2.128, we can show

$$V_d = n_E v_{kT} \ln\left(1 + \frac{I_d}{I_s} \right) + I_d \cdot r_s \qquad (2.129)$$

Clearly, when I_d is large, the terminal voltage, V_d increases linearly with I_d since the voltage drop, $I_d \times r_s$ in the neutral regions of a *pn*-junction, increases faster than the logarithmic term Equation 2.129.

2.3.6.4 Junction Breakdown Voltage

It is observed from Equation 2.126 that the *reverse* (or leakage) current of a *pn*-junction depends on W_d which in turn depends on the reverse bias $V_d = V_r$ [Equation 2.101]. Also, we notice from Equation 2.102 that the electric field in the depletion region increases with the increase in V_r. When the field reaches a certain critical value E_c corresponding to the reverse voltage $V_r = V_{br}$, called the *breakdown voltage*, a small increase in the reverse bias causes a very large increase in the current as shown in Figure 2.26 (region BC). This phenomenon is often referred to as the *breakdown condition* and is a most important consideration in device design. The breakdown occurs when the carriers moving through the depletion region acquire sufficient energy creating new electron-hole pairs by impact ionization [24,29,30]. The newly generated electron-hole pairs can also acquire sufficient energy from the field to create additional electron-hole pairs. Since electrons and holes travel in opposite directions, the carriers can multiply a few times in the depletion region before they reach the contact electrodes. This multiplication process results in an avalanche effect. The resulting breakdown voltage V_{br} is called the *avalanche break-down* voltage and is obtained from Equation 2.102 using $V_d = -V_r$ so that

$$E_{max} = \sqrt{\frac{2q}{K_{si}\varepsilon_0} \frac{N_a N_d}{(N_a + N_d)}(\phi_{bi} + V_r)} \tag{2.130}$$

At the breakdown condition: $E_{max} = E_c$ is the critical electric field at which the break-down occurs and $V_r = V_{br}$; since $V_{br} \gg \phi_{bi}$, we can safely neglect ϕ_{bi} in Equation 2.130 to obtain the expression for breakdown voltage for a *pn*-junction

$$V_{br} = \frac{K_{si}\varepsilon_0 E_c^2}{2q}\left(\frac{1}{N_a} + \frac{1}{N_d}\right) \tag{2.131}$$

Equation 2.131 shows that any increase in the doping, either of the *n*- or *p-region*, results in a decrease in the breakdown voltage, V_{br}. Also, it shows that V_{br} is con-trolled by the concentration N_b of the doped region and is proportional to $1/N_b$. In a particular *pn*-junction, V_{br} generally varies as $N_b^{-2/3}$ [14]. For a moderately doped silicon (1×10^{14} to 1×10^{16} cm^{-3}), the value of $E_c \sim 4 \times 10^5$ V cm^{-1} and to a first approximation V_{br} is independent of doping [31].

If the *pn*-junction is heavily doped ($N_b > 1 \times 10^{18}$ cm^{-3}) on both sides, the depletion layer is very narrow. Carriers cannot gain enough energy within the depletion region, so that the avalanche breakdown is not possible. However, in the depletion region, the electric field is high and the value of E_{max} approaches to 1×10^6 V cm^{-1}. In such a heavily doped p^+n^+ junction under reverse bias, electrons at the VB of the p^+side *tun-nel* through the forbidden gap into the CB of the n^+side. This tunneling process can be approximated by a particle penetrating a triangular potential barrier, with a height higher than its energy by the semiconductor bandgap E_g [15,32,33]. This *tunneling* process contributes to the current resulting in breakdown of the junction. This mecha-nism of breakdown is called the *Zener breakdown*. However, in the source-drain *pn*-junction of a FinFET device, the avalanche breakdown dominates [15].

2.3.7 *pn*-JUNCTION DYNAMIC BEHAVIOR

The *pn*-junctions are often subjected to varying voltages. In such dynamic opera-tions, charges in the *pn*-junction vary resulting in an additional current not predicted by the static current [Equation 2.125] due to stored charges. There are two types of stored charge in a *pn*-junction: (1) the charge Q_{dep} due to the depletion or space charge region on each side of the junction; and (2) the charge Q_{dif} due to minority carrier injection. Note that these injected (excess) mobile carriers generate current I_d and also represent a stored charge Q_{dif} in a *pn*-junction. The latter is given by the area between the curve representing p_n (or n_p) and the steady state level p_{no} (or n_{po}) as shown in Figure 2.23. These two types of stored charges result in two types of capacitances, the junction capacitance C_j due to Q_{dep} and the diffusion capacitance C_{dif} due to Q_{dif} as discussed in the following Sections 2.3.7.1 and 2.3.7.2, respectively.

2.3.7.1 Junction Capacitance

In a *pn*-junction, a small change in the applied voltage causes an incremental change in the depletion charge Q_{dep} due to the corresponding change in the depletion width. If the applied voltage is returned to its original value, the carriers flow in such a direction that the previous increment of charge is neutralized. The response of the *pn*-junction to the incremental voltage thus results in an effective capacitance C_j referred to as the *transition capacitance, junction capacitance,* or *depletion layer capacitance.* We know that the capacitance per unit area is defined as the incremental charge dQ_{dep} per unit area induced by an applied voltage dV_d. Therefore, we can write

$$C_j = \frac{dQ_{dep}}{dV_d} \tag{2.132}$$

Considering, $Q_{dep} = qN_a x_p = qN_d x_n$ from Equation 2.92, we can show

$$C_j = qN_a \frac{dx_p}{dV_d} = N_d \frac{dx_n}{dV_d} \tag{2.133}$$

Then using Equation 2.95 or 2.96, the *pn*-junction capacitance per unit area can be shown as

$$C_j = \sqrt{\frac{qK_{si}\varepsilon_0}{2(\phi_{bi} - V_d)}\left(\frac{N_a N_d}{N_a + N_d}\right)} \tag{2.134}$$

Equation 2.134 is the expression for the diode capacitance for a step profile in terms of the physical parameters of the *pn*-junction. However, Equation 2.134 is valid for $V_d < \phi_{bi}$, that is, for reverse bias only. Comparing Equations 2.134 and 2.101, we can show that the expression for C_j is given by

$$C_j = \frac{K_{si}\varepsilon_0}{W_d} \tag{2.135}$$

Equation 2.135 implies that the junction capacitance is equivalent to that of a parallel plate capacitor [Equation 2.99] with silicon as the dielectric separated by a distance W_d, the depletion width. Though the derivation of Equation 2.134 is based on a step profile, it can be shown that *the relationship is valid for any arbitrary doping profile.*

It should be pointed out that although the *pn*-junction capacitance can be calculated using the parallel plate capacitor formula, there are differences between the two types of capacitors. While true parallel plate capacitance is independent of the applied voltage, *pn*-junction capacitance given by Equation 2.134 becomes voltage-dependent through W_d [Equation 2.101]. Therefore, the total charge in a *pn*-junction cannot be obtained by simply multiplying the capacitance by the applied voltage, although a small variation in the charge can still be obtained by multiplying a small variation in the voltage by the instantaneous capacitance value. Another difference is that in a *pn*-junction, the dipoles in the transition region have their positive charge in the n-side depletion region and negative charge in the p-side depletion region, while in a parallel plate capacitor the separation between the charges in the dipoles is much less and they are distributed homogenously throughout the dielectric.

For a one-sided step junction, e.g., n^+p diode with $N_d \gg N_a$, Equation 2.134 can be simplified as

$$C_j = \sqrt{\frac{qK_{si}\varepsilon_0 N_a}{2(\phi_{bi} - V_d)}} \tag{2.136}$$

For the simplicity of pn-junction capacitance estimation, it is convenient to express C_j in terms of its bias independent parameter C_{j0} at equilibrium, that is, estimation at $V_d = 0$. Then from Equation 2.134, we can show

$$C_{j0} = \sqrt{\frac{qK_{si}\varepsilon_0}{2\phi_{bi}}\left(\frac{N_a N_d}{N_a + N_d}\right)} \tag{2.137}$$

Then substituting Equation 2.137 into Equation 2.134, the junction capacitance for a pn-junction is given by

$$C_j = \frac{C_{j0}}{\sqrt{1 - (V_d / \phi_{bi})}} \tag{2.138}$$

In IC pn-junctions, the doping profile is neither abrupt nor linearly graded as assumed in the derivation for C_j and therefore, to calculate the capacitance for real devices, we replace the one-half power in the denominator of Equation 2.138 with m_j, called the junction *grading coefficient*, to obtain the following generalized expression for C_j of a pn-junction

$$C_j = \frac{C_{j0}}{\left[1 - (V_d / \phi_{bi})\right]^{m_j}} \tag{2.139}$$

For IC pn-junctions, the value of m_j ranges between 0.2 and 0.6. Figure 2.28 shows the plot of the junction capacitance C_j as a function of the applied bias V_d obtained by Equation 2.139. It is observed from Figure 2.28 that the capacitance C_j decreases as the reverse biased $|V_d|$ increases ($V_d < 0$). However, when the pn-junction is forward bias ($V_d > 0$), the capacitance C_j increases and becomes infinite at $V_d = \phi_{bi}$ as observed in Figure 2.28 (Curve 1). This is because Equation 2.139 is no longer applicable due to depletion approximation becoming invalid. For simplicity of computer-aided design (CAD) of VLSI circuits, the forward biased pn-junction capacitance derived in Equation 2.139 is simplified by a series expansion of the denominator and is given by [15,34]

$$\left(1 - \frac{V_d}{\phi_{bi}}\right)^{-m_j} = 1 + m_j \frac{V_d}{\phi_{bi}} + \cdots \tag{2.140}$$

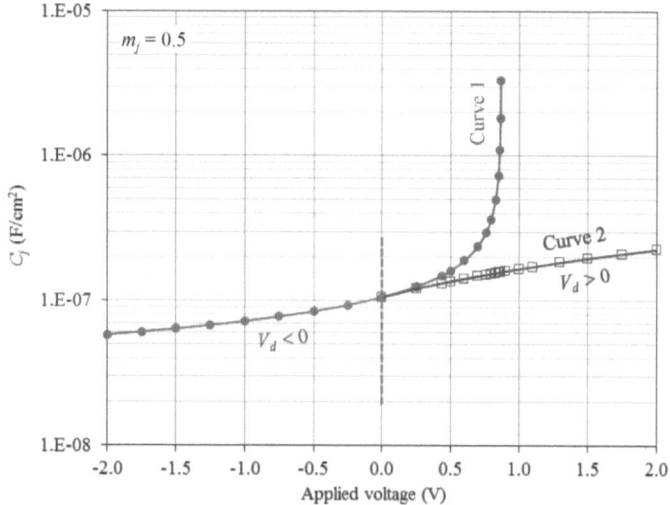

FIGURE 2.28 Plot of junction capacitance of a *pn*-junction obtained by using the expressions in Equation 2.141: Curve 1 is obtained for reverse bias and $V_d < \phi_{bi}$ and Curve 2 is obtained for forward bias using the analytical expression to ensure convergence in circuit simulation during forward biasing of the *pn*-junction.

Then, neglecting the higher order terms in Equation 2.140, we can express C_j of a *pn*-junction for the entire range of applied bias, V_d as

$$C_j = \begin{cases} C_{j0}\left(1 - \dfrac{V_d}{\phi_{bi}}\right)^{-m_j}; & V_d < 0 \\[3mm] C_{j0}\left(1 + m_j \dfrac{V_d}{\phi_{bi}}\right); & V_d > 0. \end{cases} \tag{2.141}$$

2.3.7.2 Diffusion Capacitance

The diffusion capacitance C_{dif} of a *pn*-junction is due to the variation of the stored charge density Q_{dif} associated with the excess minority carrier injection in the bulk region with incremental change in the applied forward bias. C_{dif} is called the diffusion capacitance, because the minority carriers move across the bulk region by diffusion. Since Q_{dif} is proportional to the current I_d, for an n^+p-junction, we can write

$$Q_{dif} = \frac{1}{A_d}\tau_p I_d \tag{2.142}$$

where:

τ_p is the minority carrier hole lifetime

For a short base *pn*-junction, τ_p is replaced by the carrier transit time τ_t of the *pn*-junction. However, in the case of a long base *pn*-junction, the transit time is the excess minority carrier lifetime. Differentiating Equation 2.142, we can show

$$C_{dif} = \frac{dQ_{dif}}{dV_d} = \frac{\tau_p I_s}{A_d v_{kT}} \exp\left(\frac{V_d}{v_{kT}}\right) \qquad (2.143)$$

In deriving C_{dif} per unit area in Equation 2.143, we have used Equation 2.119 for I_d. A more accurate derivation shows that the value of C_{dif} is half of the value obtained by Equation 2.143.

It should be noted that under a forward bias $V_d (= V_f)$, the value of C_{dif} increases faster with increasing voltage due to the exponential dependence on V_d [Equation 2.143] compared to C_j [Equation 2.141]. However, under a reverse bias, C_j decreases slowly with increasing magnitude of $V_d (= -V_r)$, compared to C_{dif}. Therefore, C_j is the dominant capacitance for pn-junctions at reverse bias and low for forward bias, whereas diffusion capacitance C_{dif} is dominant for forward bias pn-junctions.

2.3.7.3 Small-Signal Conductance

In the large signal model discussed in Sections 2.3.7.1 and 2.3.7.2, we did not place any restrictions on the allowed voltage variation. However, in some circuit situations, voltage variations are sufficiently small so that the resulting small current variations can be expressed using linear relationships. This is referred to as the *small-signal* behavior of a pn-junction. Examples of linear relations are the capacitances C_j and C_{dif} in Equations 2.141 and 2.143, respectively, as they represent an overall nonlinear charge storage effect in terms of linear circuit elements (capacitors), although we did not label them as such.

For small variations about the operating point, which is set by the DC condition, the nonlinear junction current can be linearized so that the incremental current in a pn-junction is proportional to the incrementally applied bias. This linear relationship is used to calculate the small-signal conductance g_d as

$$g_d = \frac{dI_d}{dV_d} \qquad (2.144)$$

Using Equation 2.119 for I_d, we can show

$$g_d = \frac{I_s}{v_{kT}} \exp\left(\frac{V_d}{v_{kT}}\right) = \frac{1}{v_{kT}}\left(I_d + I_s\right) \qquad (2.145)$$

Equation 2.145 clearly shows that g_d is proportional to the slope of the DC characteristics at the operating point. When the diode is forward biased, I_d is much larger than I_s and therefore, g_d is proportional to I_d. However, when the diode is reverse biased $I_d = -I_s$ and therefore, from Equation 2.145, g_d becomes zero. In real pn-junctions, $g_d \neq 0$ in the reverse bias condition due to the fact that the generation current I_{gen} [Equation 2.126] is a dominant conduction mechanism.

2.3.8 *pn*-Junction Equivalent Circuit

Figure 2.29 shows a typical small-signal equivalent circuit of a pn-junction. In Figure 2.29, r_s represents the series resistance due to ohmic drop across the neutral

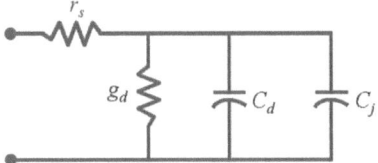

FIGURE 2.29 An equivalent circuit for a *pn*-junction showing the relevant circuit elements: r_s is the series resistance of the neutral *n*- and *p*-regions; C_j is the junction capacitance; C_d is the diffusion capacitance; and g_d is the small-signal conductance.

n- and *p*-regions, C_j is the junction capacitance, C_d is the diffusion capacitance due to the minority carrier diffusion through the neutral regions and g_d is the small-signal conductance of the *pn*-junction.

2.4 SUMMARY

This chapter presented a brief overview of the basic semiconductor physics and theory of extrinsic semiconductors forming *pn*-junctions for the basic understanding of the FinFET device and technology. First of all, the basic properties of intrinsic semiconductor materials including bond and band structures, intrinsic carrier concentration, and energy levels were discussed. Then the behavior of the extrinsic semiconductors, carrier statistics of electrons and holes, carrier transport, and fundamental semiconductor equations were presented. After the discussion of *p*-type and *n*-type semiconductors, a brief overview of the operational principles of *n*-type and *p*-type semiconductors forming *pn*-junctions were described. Then the electrostatic and dynamic characteristics of *pn*-junctions were discussed. Finally, a basic equivalent circuit model of a typical *pn*-junction was presented.

Physical Constants

Constants	Symbol	Magnitude	Units
Electronic charge	q	1.602×10^{-19}	C
Free-electron mass	m	9.11×10^{-28}	g
Boltzmann's constant	k	1.38×10^{-23}	J K^{-1}
		8.62×10^{-5}	eV K^{-1}
Planck's constant	h	6.25×10^{-34}	J.s
Permittivity of free space	ε_0	8.854×10^{-14}	F cm^{-1}
Thermal voltage at 300°K	$kT/q = v_{kT}$	0.02586	V
Thermal energy at 300°K	kT	0.02586	eV

REFERENCES

1. B.G. Streetman and S.K. Banerjee, *Solid State Electronic Devices*, 7th edition, Prentice Hall, Englewood Cliffs, NJ, 2014.
2. S.M. Sze and K.K. Ng, *Physics of Semiconductor Devices*, John Wiley & Sons, New York, 2007.
3. Y. Taur and T.H. Ning, *Fundamentals of Modern VLSI Devices*, Cambridge University Press, Cambridge, 1998.
4. M. Shur, *Physics of Semiconductor Devices*, Prentice Hall, Englewood Cliffs, NJ, 1990.
5. M. Zambuto, *Semiconductor Devices*, McGraw-Hill, New York, 1989.
6. S. Wang, *Fundamentals of Semiconductor Theory and Devices*, Prentice Hall, Englewood Cliffs, NJ, 1989.
7. E.S. Yang, *Microelectronic Devices*, McGraw-Hill, New York, 1988.
8. R.F. Pierret, *Advanced Semiconductor Fundamentals*, Addison-Wesley, Reading, MA, 1987.
9. R.S. Muller and T.I. Kamins, *Device Electronics for Integrated Circuits*, John Wiley & Sons, New York, 1986.
10. R.M. Warner, Jr. and B.L. Grung, *Transistors–Fundamentals for the Integrated Circuit Engineer*, John Wiley & Sons, New York, 1983.
11. R.A. Smith, *Semiconductors*, 2nd edition, Cambridge University Press, London, 1978.
12. A.S. Grove, *Physics and Technology of Semiconductor Devices*, John Wiley & Sons, New York, 1967.
13. G.W. Neudeck, *The PN Junction Diode*, Addison-Wesley, Reading, MA, 1989.
14. D.J. Roulston, *Bipolar Semiconductor Devices*, McGraw-Hill, New York, 1990.
15. S.K. Saha, *Compact Models for Integrated Circuit Design: Conventional Transistors and Beyond*, CRC Press, Taylor & Francis Group, Boca Raton, 2015.
16. N. Arora, *MOSFET Models for VLSI Circuit Simulation: Theory and Practice*, Springer–Verlag, Vienna, 1993.
17. P. Balk, P.G. Burkhardt, and L.V. Gregor, "Orientation dependence of built-in surface charge on thermally oxidized silicon," *IEEE Proceedings*, 53(12), pp. 2133–2134, 1965.
18. M. Aoki, K. Yano, T. Masuhara, S. Ikeda, and S. Meguro, "Optimum crystallographic orientation of submicron CMOS devices." In: *International Electron Devices Meeting Technical Digest*, pp. 577–580, 1985.
19. S.O. Kasap, *Principles of Electronic Materials and Devices*, 4th edition, McGraw Hill Education, New York, 2018.
20. M.A. Green, "Intrinsic concentration, effective density of states, and effective mass in silicon," *Journal of Applied Physics*, 67(6), pp. 2944–2954, 1990.
21. L. Nagel and D. Pederson, "Simulation program with integrated circuit emphasis," University of California, Berkeley, Electronics Research Laboratory Memorandum No. UCB/ERL M352, 1973.
22. Y.P. Varshni, "Temperature dependence of energy gap in semiconductors," *Physica*, 34(1), pp. 149–154, 1967.
23. S.K. Ghandhi, *The Theory and Practice of Microelectronics*, Wiley, New York, 1984.
24. N.D. Arora, J.R. Hauser, and D.J. Roulston, "Electron and hole mobilities in silicon as a function of concentration and temperature," *IEEE Transactions on Electron Devices*, 29(2), pp. 292–295, 1982.
25. C. Jacoboni, C. Canali, G. Ottaviani, and A.A. Quaranta, "A review of some charge transport properties of silicon," *Solid-State Electronics*, 20(2), pp. 77–89, 1977.
26. A. Fick, "Ueber Diffusion," *Annalen der Physik und Chemie*, 170(1), pp. 59–86, 1855.
27. W. Shockley and W.T. Read, "Statistics of the recombination of holes and electrons," *Physical Review*, 87(5), pp. 835–842, 1952.

28. R.N. Hall, "Electron-hole recombination in germanium," *Physical Review*, 87(2), pp. 387–387, July 1952.
29. S. Saha, C.S. Yeh, and B. Gadepally, "Impact ionization rate of electrons for accurate simulation of substrate current in submicron devices," *Solid-State Electronics*, 36(10), pp. 1429–1432, 1993.
30. S. Saha, "Extraction of substrate current model parameters from device simulation," *Solid-State Electronics*, 37(10), pp. 1786–1788, 1994.
31. G.W. Neudeck, *The PN Junction Diode*, vol. II, 2nd edition, Modular Series on Solid-State Devices, Addison-Wesley, Reading, MA, 1987.
32. C. Zener, "A theory of electrical breakdown of solid dielectrics," *Proceedings of the Royal Society of London*, A145(8555), pp. 523–529, 1934.
33. E.O. Kane, "Zener tunneling in semiconductors," *Journal of Physics and Chemistry of Solids*, 12(2), pp. 181–188, 1960.
34. N. Paydavosi, T.H. Morshed, D.L. Lu, *et al.*, *BSIM4v4.8.0 MOSFET Model User's Manual*, University of California, Berkeley, CA, 2013.

3 Multiple-Gate Metal-Oxide-Semiconductor (MOS) System

3.1 INTRODUCTION

The metal-oxide-semiconductor (MOS) structure, commonly referred to as the MOS capacitor system or MOS capacitor, is a two-terminal device with one electrode connected to the metal and the other electrode connected to the semiconductor forming a voltage-dependent capacitor. A typical MOS capacitor could be a single-gate (SG) or a multiple-gate system. A multiple-gate structure, hereafter, referred to as the "multigate" MOS capacitor, consists of a semiconductor body surrounded by gate stacks. The gates can be connected together or can be electrically isolated. Depending on the gate dielectric thickness, the structure can be double-gate (DG), triple-gate or tri-gate (TG), or gate-all-around (GAA) MOS system [1]. The simplest multigate structure consists of the configuration, *metal-oxide-semiconductor-oxide-metal*, that is, two MOS capacitors in parallel with a top gate, a bottom gate, and a common semiconductor body. An MOS capacitor is a very useful device both for evaluating the fabrication processes of fin field-effect transistor (FinFET) integrated circuits (ICs) and predicting the transistor performance. Therefore, MOS capacitor structures are included in the test chip for IC process and device characterization of an IC fabrication technology.

The MOS capacitor has been the subject of numerous investigations and the detailed description of the early development can be found in the literature [2]. The primary objective of this chapter is to build the foundation for the basic understanding of FinFET device operation and mathematical formulations described in Chapters 5 and 10 of this book. In order to achieve this objective, we first discuss the behavior of multigate MOS capacitors and then develop the charge-voltage (Q–V) relationships which will be used for detailed formulation of FinFET device performance and models for computer-aided design (CAD) of very large scale integrated (VLSI) circuits. Unless otherwise specified, we will assume that the substrate of MOS capacitor is undoped (however, not intrinsic) or lightly doped silicon.

3.2 MULTIGATE MOS CAPACITORS AT EQUILIBRIUM

Typical multigate MOS capacitor structures are shown in Figure 3.1. Figure 3.1(a) shows an ideal multigate MOS capacitor structure with a silicon body of thickness t_{si} surrounded by gate stack consisting of gate oxide and gate metal; whereas Figure 3.1(b) shows an ideal two-dimensional (2D) structure of a DG-MOS capacitor

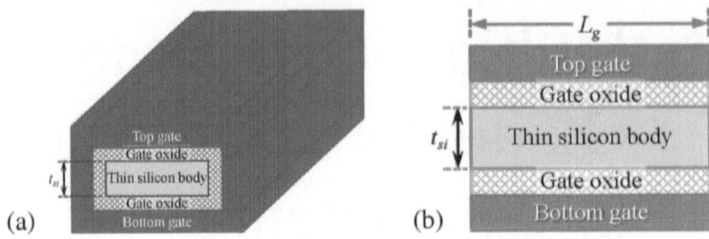

(a)   (b)

FIGURE 3.1 Typical multigate MOS capacitor structures: (a) an ideal three-dimensional MOS capacitor system with a silicon body surrounded by gate stack including gate oxide and metal gate electrode; and (b) an ideal two-dimensional MOS capacitor system with gate stack at the top and bottom of the silicon body; t_{si} is the thickness of silicon body.

with a silicon body, a top gate oxide with top metal-gate (MG) electrode, and a bottom gate oxide with bottom MG electrode. In Figure 3.1(b), L_g is the gate length of the DG-MOS capacitor structure. In this chapter, we will use the DG-MOS capacitor structure shown in Figure 3.1(b) to develop the basic theory of multigate MOS capacitors.

In order to discuss the basic operation of an MOS capacitor, let us consider the 2D cross-section of an ideal SG-MOS capacitor structure as shown in Figure 3.2. The structure includes a p-type (or n-type) semiconductor substrate such as silicon, a dielectric layer such as silicon dioxide (SiO_2) gate dielectric, a metal or polysilicon gate, and a gate electrode (G) and a body (back or bulk) electrode (B) for operating the MOS capacitor at the target applied biases V_g and V_b, respectively. Typically, the SiO_2 layer is thermally grown on the silicon substrate with a typical thickness between 1 to 100 nm. The gate metal is formed on the top of the gate dielectric by masking, photolithography, and annealing processes [3–7]. The body electrode is obtained by deposited metal to achieve an ohmic contact. If the substrate conducts sufficiently to support the displacement currents, the structure in Figure 3.2 forms a parallel plate capacitor with G as one electrode, B as the second electrode, and SiO_2 as the dielectric. This structure is referred to as the MOS capacitor system. This

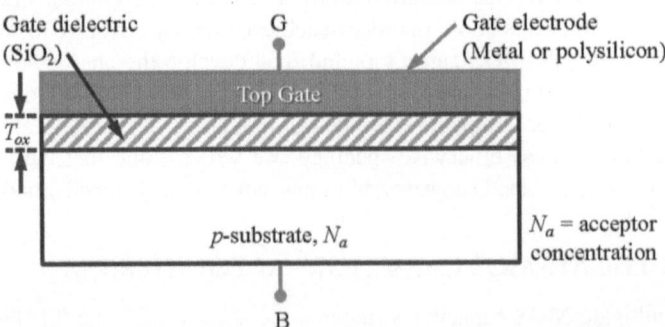

FIGURE 3.2 Two-dimensional cross-section of an ideal MOS capacitor structure using only top gate on a uniformly doped p-type substrate with doping concentration N_a; here, G and B denote the gate and body terminals for applied biases to the gate and body, respectively.

system is in thermal equilibrium with applied static bias and if the change in the voltage is sufficiently slow, it is approximated to be a constant. Thus, using the parallel plate capacitance formulation, we can write the oxide capacitance (C_{ox}) per unit area between the metal and silicon surface as:

$$C_{ox} = \frac{\varepsilon_0 K_{ox}}{T_{ox}} = \frac{\varepsilon_{ox}}{T_{ox}} \tag{3.1}$$

where:
ε_0 is the permittivity of free space or vacuum
K_{ox} is the dielectric constant of oxide
T_{ox} is the gate oxide thickness
$\varepsilon_{ox} = \varepsilon_0 K_{ox}$ is the permittivity of oxide

In this chapter, we will use ε_{ox} or $\varepsilon_0 K_{ox}$ to describe the permittivity of oxide interchangeably for the simplicity of mathematical presentation where applicable.

In order to study the physics of MOS capacitors, first of all, we will discuss the properties of each material individually before forming the MOS systems in Section 3.2.1.

3.2.1 PROPERTIES OF ISOLATED METAL, OXIDE, AND SEMICONDUCTOR MATERIALS

Let us consider the isolated metal, oxide, and semiconductor (p-type silicon) materials before being brought into contact to form an MOS system. The energy band diagram of each material is shown separately in Figure 3.3, where E_0 denotes a

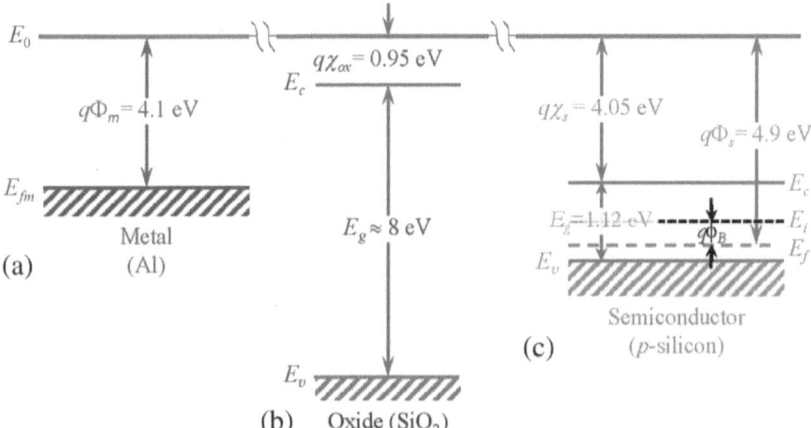

FIGURE 3.3 The energy band diagrams of three separate materials to form an MOS capacitor: (a) metal; (b) ideal SiO_2; and (c) p-type silicon substrate with $N_a = 1 \times 10^{15}$ cm^{-3}; here, E_0 = vacuum energy level (reference energy), E_{fm} = Fermi level in metal; Φ_m = metal workfunction, E_c = bottom-edge of the conduction band, E_v = top-edge of the valence band, E_f = Fermi level in silicon, E_g = forbidden energy (energy gap), E_i = intrinsic energy level in silicon, χ_s = electron affinity in silicon, χ_{ox} = electron affinity in oxide, Φ_s = semiconductor workfunction, ϕ_B = bulk potential in semiconductor, and q = electronic charge.

convenient reference potential energy level, referred to as *the vacuum* or *free electron energy* level. In reality, E_0 is the level at which the Coulombic potential of an isolated positive charge becomes zero. In order to discuss the material properties of SiO_2, it is to be noted that the reported value of bandgap energy for the SiO_2 layer is in the range of 8.0–9.0 eV [2,5–10]. In Figure 3.3, we have used 8.0 eV as the bandgap energy for SiO_2 to discuss the behavior of the MOS capacitor. The relevant material parameters for the metal, oxide, and semiconductor forming an MOS capacitor are shown in Figure 3.3 and the workfunctions of the metal and semiconductor are defined in Section 3.2.1.1.

3.2.1.1   Workfunction

With reference to Figure 3.3(a), Φ_m is defined as the metal workfunction in units of volts (V) (or $q\Phi_m$ in units of energy eV, where q is the electronic charge). Physically, $q\Phi_m$ is the energy required to take an electron across the surface energy barrier of the metal from its Fermi level E_{fm} to E_0. Since the Fermi level E_{fm} of a metal is at its conduction band (CB) level E_c, $q\Phi_m$ is the energy difference between E_0 and E_{fm}, that is,

$$q\Phi_m = E_0 - E_{fm} \qquad (3.2)$$

For a pure metal without impurities and contaminations, the value of Φ_m depends only on the charge distribution of the atomic core or the type of atom involved. In the case of FinFETs, dual workfunction metal-gates (MGs) are engineered with a target value of $\Phi_m \cong 4.05$ V, so that E_{fm} is aligned to the bottom-edge of the silicon CB of a p-type silicon substrate of the MOS capacitor, whereas a value of $\Phi_m \cong 5.17$ V is engineered so that E_{fm} is aligned to the top-edge of the silicon valance band (VB) for an n-type silicon substrate of the MOS capacitor [11]. Since the value of $\Phi_m = 4.10$ V for aluminum, therefore, we will use aluminum MG to discuss the basic characteristics of MOS capacitors on a p-type substrate as shown in Figure 3.3.

In oxides and semiconductors, the height of the surface energy barrier is defined by electron affinity, χ_{ox} and χ_s, respectively, as shown in Figure 3.3(b) and (c), respectively. In Figure 3.3(b) and (c), χ is the energy difference between the vacuum level E_0 and the bottom-edge of CB E_c at the surface. Therefore, $q\chi_s = (E_0 - E_c)$, where χ_s is a fundamental material parameter independent of the presence of impurities or imperfections and only varies from one atomic type to another or is changed by alloy composition. Unlike metals, the Fermi level E_f in a semiconductor is not a constant and depends on the doping concentration of impurities. Since the workfunction is the energy required to take an electron from E_f to E_0, the electron affinity χ_s is used to define the workfunction Φ_s in semiconductors. Thus, for a p-type semiconductor (Figure 3.3(c)) the workfunction (Φ_{sp}) is given by

$$q\Phi_{sp} = q\chi_s + \frac{E_g}{2} + q\phi_{Bp} \quad (p\text{-type semiconductor}) \qquad (3.3)$$

where:
  E_g is the bandgap energy
  ϕ_{Bp} is the bulk or Fermi potential of a p-type semiconductor

Similarly, the workfunction for an *n*-type semiconductor (Fermi level above E_i), Φ_{sn} is given by

$$q\Phi_{sn} = q\chi_s + \frac{E_g}{2} - q\phi_{Bn} \quad (n\text{-type semiconductor}) \tag{3.4}$$

where:

ϕ_{Bn} is the bulk or Fermi potential for an *n*-type semiconductor

If the doping concentrations for both the *n*-type and *p*-type semiconductors are the same, then $|\phi_{Bp}| = |\phi_{Bn}| \equiv \phi_B$ and is given by Equation 2.37 as

$$\phi_B = v_{kT} \ln\left(\frac{N_b}{n_i}\right) \tag{3.5}$$

where:

$v_{kT} (= kT/q)$ is the *thermal voltage* at the ambient temperature T with k and q representing the Boltzmann constant and electronic charge, respectively

n_i is the intrinsic carrier concentration

In order to calculate the value of the Φ_s, the magnitude of ϕ_B is calculated from Equation 3.5. Thus, for a *p*-type silicon with $N_b = N_a = 1 \times 10^{15}$ cm^{-3} at room temperature, 300 °K at which $v_{kT} \cong 0.0259$ V and $n_i = 1.45 \times 10^{10}$ cm^{-3}, we can show that the value of $\phi_B \cong 0.29$ V. Then using $q\chi_s = 4.05$ eV and $E_g = 1.12$ eV, we get from Equation 3.3, $q\Phi_{sp} \equiv q\Phi_s \cong 4.90$ eV for a *p*-type silicon. Since for aluminum $q\Phi_m = 4.10$ eV, therefore, for an aluminum MG-MOS capacitor on a *p*-type substrate, $\Phi_m < \Phi_s$, that is, the energy required to free an electron from the *p*-type silicon is higher than the energy required to free an electron from metal.

In a planar-CMOS technology, degenerately (heavily) doped polysilicon gates are used. In this case, Φ_s is calculated assuming that the Fermi energy lies at the band-edges, that is, E_f is at E_c for an *n*-type polysilicon and E_f is at E_v for a *p*-type polysilicon [6,8]. For a FinFET technology, workfunction engineering is used to achieve the target value of MG workfunction [11]. The workfunctions of commonly used gate material for IC technology are shown in Table 3.1 [5,9].

TABLE 3.1

Workfunction of Different Materials Used as Gate Electrode in Multigate Capacitor Systems

Material	Workfunction (eV)
Al	4.10
Au	5.27
MoSi$_2$	4.73
TiSi$_2$	3.95
Metal with silicon mid-bandgap	4.61
n-type degenerately doped polysilicon	4.05
p-type degenerately doped polysilicon	5.17

3.2.2 METAL, OXIDE, AND SEMICONDUCTOR MATERIALS IN CONTACT FORMING MOS SYSTEMS

Now, let us consider how the energy bands of three materials, shown separately in Figure 3.3, are brought in contact to form an MOS capacitor as shown in Figure 3.4. It can be shown that when different materials are in contact with each other, the workfunction between its two ends depends only on the first and the last materials [10]. Thus, for an MOS capacitor, the workfunction difference between the metal and the semiconductor defines the behavior of the system. The workfunction difference between two materials in contact can be visualized as the contact potential between them. For an MOS capacitor, the workfunction difference between the metal and semiconductor, Φ_{ms} ($= \Phi_m - \Phi_s$), causes distortion in the band structure of the system as shown in Figure 3.4(a). This is because when *three* materials are in contact, E_f is constant at equilibrium and E_0 is continuous; *holes* flow from the p-type semiconductor to metal and *electrons* flow from the metal to p-type semiconductor on contact until a potential is built up to counterbalance the difference in workfunction. However, the current through SiO_2 is very small. Thus, there is a variation in electrostatic potential from one region to another causing band bending in the oxide and silicon. Since metal is an equipotential region, there is no band bending in metal. The Φ_{ms} causes a potential drop (V_{ox}) in the oxide and near the silicon surface due to the band bending. We can compensate for this band bending by applying an external voltage, $V_{fb} = \Phi_{ms}$, that caused the band bending in the first place. Here, V_{fb} is referred to as the *flat band voltage*, and the resulting band structure for an MOS capacitor at flat band condition is shown in Figure 3.4(b). Thus, the condition for flat band voltage at the Si/SiO$_2$ interface is given by

$$V_{fb} = \Phi_m - \Phi_s \equiv \Phi_{ms} \qquad (3.6)$$

where:

   Φ_{ms} is the workfunction difference between the MG electrode and bulk silicon (in units of volts)

3.2.2.1 MOS Systems with MG Workfunction at Silicon Band-Edges

In FinFET technology, workfunction engineering is used to achieve dual workfunction MG-MOS capacitor systems so that E_{fm} is aligned to the bottom-edge of the silicon CB for an MG-MOS capacitor on a p-type substrate, whereas E_{fm} is aligned to the top-edge of silicon VB for an MG-MOS capacitor on an n-type substrate. Thus, for an MG-MOS capacitor system on a p-type silicon body with E_{fm} aligned to the bottom-edge of the silicon CB, the metal workfunction, Φ_m is given by

$$q\Phi_m = q\chi_s \qquad (3.7)$$

On the other hand, the silicon workfunction Φ_s for a p-type silicon body is given by

$$q\Phi_s = \left(q\chi_s + \frac{E_g}{2} + q\phi_B \right) \qquad (3.8)$$

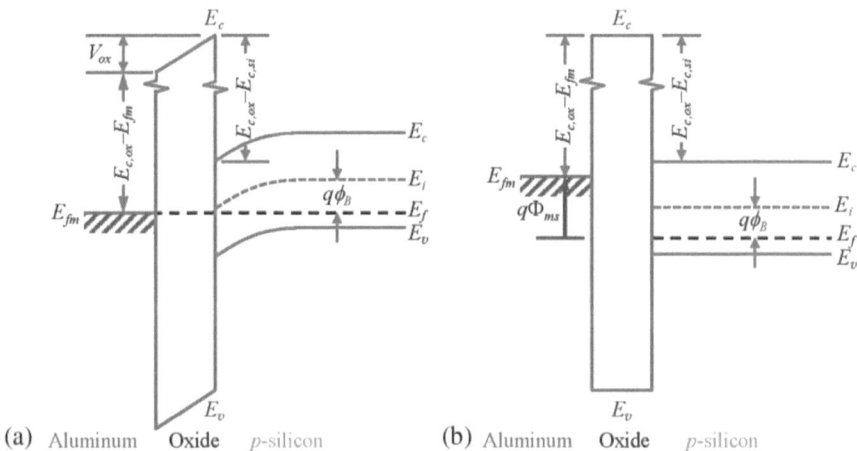

(a) Aluminum Oxide p-silicon   (b) Aluminum Oxide p-silicon

FIGURE 3.4 Aluminum-SiO$_2$-(p-type) silicon MOS capacitor system at equilibrium with applied gate voltage, $V_g = 0$: (a) band bending at the surface due to Φ_{ms} between the metal and p-type silicon; and (b) flat band condition for the structure shown in (a); here, oxide is assumed to be free of any charges. In the figure, $E_{c,ox}$ and $E_{c,si}$ represent the CB-edges of the oxide and silicon, respectively, and V_{ox} is the potential drop in the oxide due to band bending.

In Equations 3.7 and 3.8, all parameters have usual meanings as defined earlier. Therefore, the workfunction difference Φ_{ms} between the metal and p-type silicon is given by

$$q\Phi_{ms} = q\chi_s - \left(q\chi_s + \frac{E_g}{2} + q\phi_B \right) = -\left(\frac{E_g}{2} + q\phi_B \right) \quad (3.9)$$

Now, in Section 3.2.1.1, it is shown that for a lightly doped silicon body with $N_a = 1 \times 10^{15}$ cm^{-3}, the value of $\phi_B \cong 0.29$ V. Then using the values of $q\chi_s = 4.05$ eV, $E_g = 1.12$ eV, and $q\phi_B = 0.29$ eV, we get $q\Phi_m = 4.05$ eV and $q\Phi_s = 4.90$ eV from Equations 3.7 and 3.8, respectively. Thus, for an MG-MOS capacitor on a p-type silicon substrate with $E_{fm} = q\chi_s$ and $\Phi_m < \Phi_s$, the value of $q\Phi_{ms} = -0.85$ eV. Since at equilibrium, E_f is continuous, therefore, the energy bands bend downwards in the oxide and silicon near the surface as shown in Figure 3.5 for a DG-MOS capacitor system.

The difference between the MG E_{fm} and oxide CB ($E_{c,ox}$) at the MG/SiO$_2$ interface is given by $(q\Phi_m - q\chi_{ox})$ eV = 3.10 eV. Similarly, the difference between the silicon CB ($E_{c,si}$) and oxide CB at the silicon/SiO$_2$ interface is 3.10 eV. Thus, due to these differences between the CB levels ($E_{c,ox}$, $E_{c,si}$), there is an abrupt transition in E_c and E_v levels at the material interfaces as shown in Figure 3.5. The workfunction difference, Φ_{ms} between MG and p-type silicon substrate, causes a potential drop in the oxide and near the silicon surface due to the band bending. A typical potential drop in oxide is about 0.4 V. This potential drop depends on the doping level in silicon and can be supported since no current flows through oxide. The values shown in Figure 3.5 for the band bending in the oxide and silicon are obtained by assuming that the oxide is an ideal insulator without any charges. As shown in Figure 3.4(b), we can compensate for this band bending by applying $V_{fb} = \Phi_{ms}$.

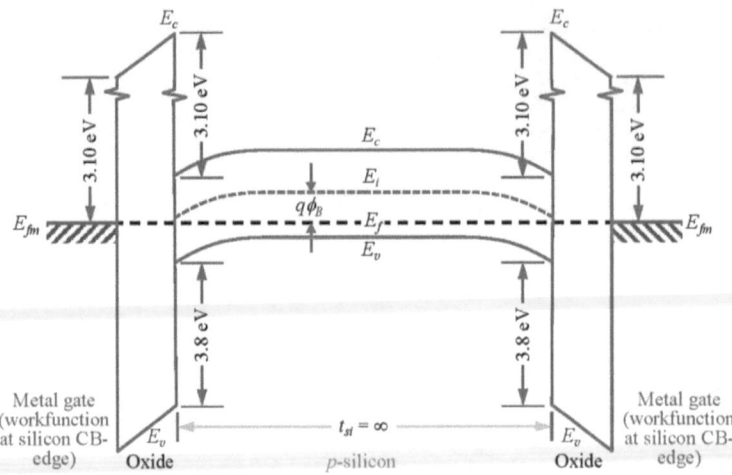

FIGURE 3.5 Energy band diagram of an ideal MG-SiO$_2$-(p-type) silicon DG-MOS capacitor at equilibrium with applied gate voltage, $V_g = 0$ and metal workfunction at the bottom of the silicon conduction band-edge showing band bending at the surface due to Φ_{ms} between the MG and p-type silicon; here, t_{si} is the thickness of the silicon body.

Now, for an MG-SiO$_2$-(p-type) silicon system with MG workfunction $\Phi_m = \chi_s$, Φ_{ms} is given by Equation 3.9. Then considering the values of the material parameters for silicon shown in Figure 3.3(c), we get from Equation 3.9

$$q\Phi_{ms} = -\left(0.56 + q\phi_B\right) \qquad (3.10)$$

Since $\phi_B \cong 0.29$ V for the substrate concentration $N_b = 1 \times 10^{15}$ cm^{-3}, therefore, Φ_{ms} is a negative number.

Similarly, for an MG-SiO$_2$-(n-type) silicon system with metal E_{fm} aligned to the top-edge of silicon VB so that $\Phi_m = \chi_s + E_g$, the Φ_{ms} is given by

$$q\Phi_{ms} = \left(q\chi_s + E_g\right) - \left(q\chi_s + \frac{E_g}{2} - q\phi_B\right) \qquad (3.11)$$

$$q\Phi_{ms} = \left(0.56 + q\phi_B\right) \qquad (3.12)$$

It is to be noted that for advanced FinFET technologies, the channel is undoped (however, not intrinsic) or lightly doped ($\Phi_B \sim 0$), therefore, from Equation 3.10 $\Phi_{ms} \sim -0.56$ V for MOS capacitors with $\Phi_m = \chi_s$ on p-type silicon, whereas, $\Phi_{ms} \sim 0.56$ V for MOS capacitors with $\Phi_m = \chi_s + E_g$ on n-type silicon.

3.2.2.2 MOS Systems with Silicon-Midgap MG Workfunction

For a silicon-midgap MG workfunction, the metal Fermi level E_{fm} is aligned to the silicon intrinsic energy level E_i. Therefore, for such an MG-MOS capacitor on a p-type silicon substrate, the metal workfunction Φ_m is given by

$$q\Phi_m = \left(q\chi_s + \frac{E_g}{2} \right) \tag{3.13}$$

On the other hand, the silicon workfunction Φ_s is given by

$$q\Phi_s = \left(q\chi_s + \frac{E_g}{2} + q\phi_B \right) \tag{3.14}$$

Therefore, Φ_{ms} for a silicon midgap metal-gate on a p-type substrate MOS capacitor is

$$q\Phi_{ms} = \left(q\chi_s + \frac{E_g}{2} \right) - \left(q\chi_s + \frac{E_g}{2} + q\phi_B \right) = -q\phi_B \tag{3.15}$$

Thus, for silicon midgap MG workfunction MOS capacitors on p-type substrate, we can show from Equation 3.13, $q\Phi_m = 4.61$ eV. And, considering a lightly doped p-type substrate with $N_b = 1 \times 10^{15}$ cm^{-3} and $q\phi_B \approx 0.29$ eV, we get $q\Phi_s = 4.90$ eV from Equation 3.14. Then, for an MG-MOS capacitor on p-type silicon substrate with silicon midgap MG workfunction, $\Phi_m < \Phi_s$, the value of $q\Phi_{ms} = -q\phi_B$. Therefore, the energy bands bend downwards in the oxide and silicon near the surface as shown in Figure 3.6.

Now, following the procedure described in Section 3.2.2.1, we can show that the difference between MG E_{fm} and oxide CB at the MG/SiO$_2$ interface is 3.66 eV. And, the difference between the silicon and oxide CB at the silicon/SiO$_2$ interface is 3.10 eV. Thus, due to these differences between the CB levels, there is an abrupt transition in E_c and E_v levels at the material interfaces as shown in Figure 3.6.

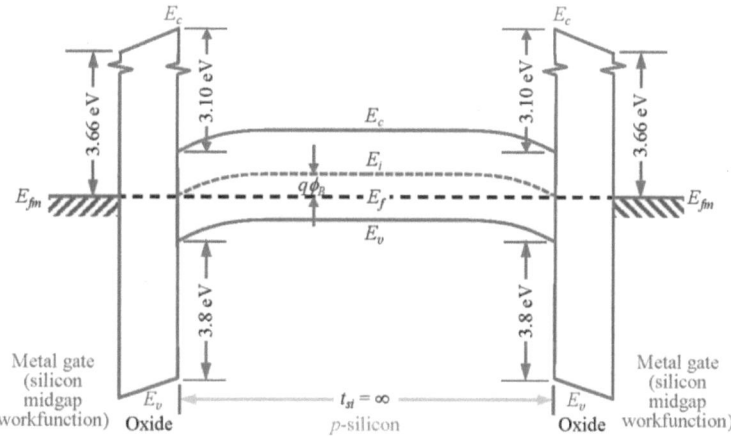

FIGURE 3.6 MG-SiO$_2$-(p-type) silicon DG-MOS capacitor system at equilibrium with applied gate voltage, $V_g = 0$ and silicon midgap MG workfunction showing band bending at the surface due to Φ_{ms} between MG and p-type silicon; t_{si} is the thickness of the silicon body.

Typically, the substrate is undoped or lightly doped ($\phi_B \approx 0$), therefore, for a silicon midgap MG-MOS capacitor, Φ_{ms} is negligibly small and the band in the substrate is nearly flat as shown in the band diagram in Figure 3.6.

Similarly, Φ_{ms} for an MG-MOS capacitor on n-type silicon substrate with silicon midgap workfunction is

$$q\Phi_{ms} = \left(q\chi_s + \frac{E_g}{2} \right) - \left(q\chi_s + \frac{E_g}{2} - q\phi_B \right) = q\phi_B \qquad (3.16)$$

Again, since the substrate is undoped or lightly doped, Equation 3.16 implies that for a silicon-midgap MG-MOS capacitor on an n-type substrate, the value of Φ_{ms} is small and, therefore, the band in the substrate is almost flat. The workfunctions for n^+ and p^+ polysilicon gates on p-type and n-type substrates can be calculated following the above described procedure [12,13].

In our discussions, we assumed an ideal SiO_2 layer without any charges and contamination. In reality, oxide includes several charges that affect the behaviors of MOS capacitors. In Section 3.2.3, we will discuss the effect of non-ideal oxide on MOS capacitors.

3.2.3 Oxide Charges

During the oxide growth process or subsequent IC fabrication processing steps, some impurities or defects are inadvertently incorporated into the oxide. As a result, the oxide is contaminated with various types of charges and traps. Typically, four different types of charge have been identified in thermally grown oxide on a silicon surface as shown in Figure 3.7 [14]. These charges are (1) interface trapped charge Q_{it}, (2) fixed oxide charge Q_f, (3) oxide trapped charge Q_{ot}, and (4) mobile ionic

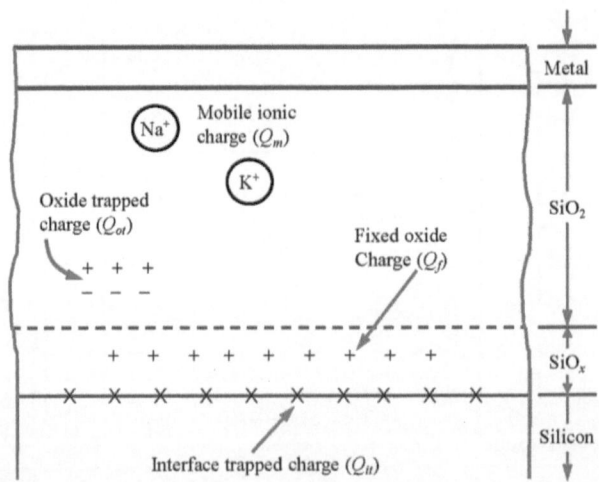

FIGURE 3.7 Types and location of the charges associated with thermally grown silicon dioxide SiO_2 on silicon.

charge Q_m. All of these charges are dependent on IC fabrication processing steps. The detailed description of the origin and techniques of measurements of different oxide charges are available in literature [2,15]. In the following section, the basic properties of these charges are described.

3.2.3.1 Interface Trapped Charge

The interface trapped charge density Q_{it} also referred to as the *surface states, fast states*, or *interface states* exists at the Si/SiO$_2$ interface as shown in Figure 3.7. It is caused by defects at the interface which give rise to *charge traps* or *electronic energy levels* with energy states (E_s) in the silicon bandgap that can capture or emit mobile carriers. These electronic states are due to lattice mismatch at the interface, dangling bonds, the adsorption of external impurity atoms at the silicon surface, and other defects caused by radiation or any bond-breaking process. Q_{it} is the most important type of charge because of its wide-ranging and degrading effect on device characteristics. Under the equilibrium condition, the occupancy of the interface states or traps depends on the position of the Fermi level.

Typically, the interface trap levels with density D_{it} (traps cm^{-2} eV^{-1}) are distributed over energies within the silicon energy gap [2,5–7]. D_{it} varies significantly from process to process and is dependent on crystal orientation. In thermally grown SiO$_2$ on silicon, most of the interface trapped charge is neutralized by low temperature ($\leq$500 °C) hydrogen annealing. D_{it} correlates with the density of available bonds at the surface. Therefore, in $\langle 100 \rangle$ orientation with lower density of silicon atoms (available bonds) at the surface, D_{it} is about an order of magnitude smaller than that in $\langle 111 \rangle$ oriented silicon with higher available bonds at the surface. The value of D_{it} at midgap for the $\langle 100 \rangle$ oriented silicon of the modern MOS VLSI process can be as low as 5×10^9 cm^{-2} eV^{-1}. *Higher values of D_{it} cause instabilities in the MOS transistor behavior.*

3.2.3.2 Fixed Oxide Charge

The fixed charge density Q_f is the immobile charge always present and located within 1 nm transition layer of nonstoichiometric silicon oxide (SiO$_x$) at the boundary between the silicon and SiO$_x$ layer as shown in Figure 3.7. Generally, Q_f is positive and appears to arise from incomplete silicon-to-silicon bonds and depends on the oxidation ambient, temperature and annealing conditions, and silicon orientation. Since the density of atoms at the surface of a silicon crystal depends on the crystal orientation, therefore, Q_f is higher in $\langle 111 \rangle$ silicon than $\langle 100 \rangle$ wafers. However, it is independent of the doping type and concentration in the silicon, oxide thickness, and oxidation time. Q_f can be minimized by annealing in an inert ambient, such as Argon at a temperature in excess of 900 °C. A typical value of Q_f for a carefully treated Si/SiO$_2$ system is about 1×10^{10} cm^{-2} for the $\langle 100 \rangle$ surface. Because of the low values of Q_{it} and Q_f, the $\langle 100 \rangle$ orientation is preferred for silicon MOSFETs.

3.2.3.3 Oxide Trapped Charge

The oxide trapped charge density Q_{ot} is associated with the defects in SiO$_2$. Q_{ot} is located in traps distributed throughout the oxide layer. The oxide traps are usually electrically neutral and are charged by introducing electrons and holes into the oxide through ionizing radiation such as implanted ions, X-rays, electron beams, and so on. The magnitude of Q_{ot} depends on the amount of radiation dose and energy and the field across the oxide during irradiation. Like Q_{it}, the Q_{ot} could be positive (trapped holes) or negative (trapped electrons). Q_{ot} resembles Q_f in that its magnitude is not a function of silicon surface potential and there is no capacitance associated with it.

3.2.3.4   Mobile Ionic Charge

The mobile ionic charge density Q_m is due to sodium (Na$^+$) or other alkali ions that get into the oxide during cleaning, processing, and handling of MOS devices. These ions move very slowly within the oxide; their transport depends strongly on the applied electric field ($\sim$1 MV cm^{-1}) and temperature (30 °C–400 °C). A positive voltage pushes these ions towards the Si/SiO$_2$ interface while a negative voltage draws them towards the gate. A current is observed in the external circuit during ion drift. The drift of ions changes the centroid of charge within the oxide layer resulting *in a shift of the flat band voltage of MOS capacitor system and may cause an unexpected device failure*. Different approaches are used to reduce mobile ion contamination in gate oxide and mitigate the risk of mobile ionic induced device failure [2,5]. The shift in the value of Φ_{ms}-induced V_{fb} due to oxide charges is described in Section 3.2.4.

3.2.4   Effect of Oxide Charges on Energy Band Structure: Flat Band Voltage

In order to determine the total shift in the flat band voltage (ΔV_{fb}) by various oxide charges, let us consider $Q_{ox}(x)$ as the charge per unit area at any point x within the oxide of thickness T_{ox}. Then from Gauss' law [Equation 2.75], we can show

$$E = -\frac{dV}{dx} = -\frac{\Sigma Q_{ox}(x)}{K_{ox}\varepsilon_0}$$

$$\text{or,}\quad dV = -\frac{1}{K_{ox}\varepsilon_0}\int_0^{T_{ox}} Q_{ox}(x)dx = -\frac{1}{C_{ox}}\int_0^{T_{ox}}\frac{1}{T_{ox}}Q_{ox}(x)dx \qquad (3.17)$$

where:
$C_{ox} = K_{ox}\varepsilon_0/T_{ox}$ is the oxide capacitance per unit area as defined in Equation 3.1
dV is the voltage drop in oxide due to the oxide charges

The voltage drop dV defines the shift in the flat band voltage V_{fb} of an MOS capacitor and is given by

$$\Delta V_{fb} = -\frac{1}{K_{ox}\varepsilon_0}\int_0^{T_{ox}} Q_{ox}(x)dx = -\frac{1}{C_{ox}}\int_0^{T_{ox}}\frac{1}{T_{ox}}Q_{ox}(x)dx \qquad (3.18)$$

where:

$$Q_{ox}(x) = Q_{it}(x) + Q_f(x) + Q_{ot}(x) + Q_m(x)$$

Q_f and Q_{it} are located at or near the Si/SiO$_2$ interface ($x = T_{ox}$). On the other hand, $Q_{ot}(x)$ and $Q_m(x)$ are distributed throughout the oxide. Therefore, after integration, we get

$$\Delta V_{fb} = -\frac{Q_{it} + Q_f}{C_{ox}} - \frac{1}{C_{ox}} \int_0^{T_{ox}} \frac{1}{T_{ox}} \left[Q_{ot}(x) + Q_m(x)\right] dx \qquad (3.19)$$

For the simplicity of circuit CAD, Equation 3.19 is expressed as

$$\Delta V_{fb} = -\frac{Q_o}{C_{ox}} \qquad (3.20)$$

where:

Q_o is the *equivalent interface charge* located at the Si/SiO$_2$ interface and causes the same effect as that of the actual charges of unknown distribution

Q_o is always positive for both p- and n-type substrates. ΔV_{fb} is the gate voltage that is needed to cause Q_o to be imaged in the gate electrode so that none is induced in the silicon. However, when gate "floats" or the gate electrode is absent, the oxide charges will seek all their image charges in the silicon.

In Figures 3.4(a), 3.5, and 3.6, we have shown the band bending of an MOS capacitor due to workfunction difference between the metal and semiconductor. The corresponding flat band voltage is given by Equation 3.6. Now, the shift in workfunction due to additional band bending by oxide charges is given by Equation 3.20. Then, combining Equations 3.6 and 3.20, the overall V_{fb} due to Φ_{ms} and Q_o is given by

$$V_{fb} = \Phi_{ms} - \frac{Q_0}{C_{ox}} \qquad (3.21)$$

Typically, Q_0/C_{ox} is much smaller than Φ_{ms} in Equation 3.21. Therefore, for an MOS capacitor on p-type substrate with the value of MG workfunction close to silicon CB-edge, the value of V_{fb} is a negative number since Φ_{ms} is negative from Equation 3.10. On the other hand, for an MOS capacitor with n-substrate and MG workfunction close to silicon VB-edge, V_{fb} is positive since Φ_{ms} is positive from Equation 3.12.

The band bending at the Si/SiO$_2$ interface due to V_{fb} induces a potential on the substrate at the Si/SiO$_2$ interface referred to as the *surface potential*. The surface potential is an important parameter to discuss the performance and mathematical analysis of FET devices. In Section 3.2.5, we define the surface potential of an MOS capacitor.

3.2.5 SURFACE POTENTIAL

Now, let us consider an MG-SiO$_2$-(p-type) silicon MOS capacitor to discuss the effect of band bending at the silicon surface on the surface behavior of MOS capacitors.

We know that the concentration of holes in a p-type substrate (assuming complete ionization of acceptor atoms) is given by [Equations 2.30 and 2.32]

$$p = n_i \exp\left(\frac{E_i - E_f}{kT}\right) \tag{3.22}$$

The band structure of the system can be shown as in Figure 3.8. Since E_f is continuous within the system, it is evident from Figure 3.8 that as the bands bend downwards, the energy difference $(E_i - E_f)$ gradually decreases as we approach the silicon surface at $x = 0$ from the bulk ($x = \infty$). Then from Equation 3.22, the decrease in $(E_i - E_f)$ results in a decrease in the hole concentration p. This implies that the holes are depleted at the surface giving rise to a space charge region for an MOS system with $\Phi_{ms} < \Phi_s$. On the other hand, in the case of an MOS capacitor with $\Phi_m > \Phi_s$, the bands bend upwards and, therefore, the value of $(E_i - E_f)$ increases at the surface resulting in an increase in the hole concentration (accumulation) at the surface.

From Equation 3.22, it is seen that even without an applied external voltage the carrier concentration at the surface of an MOS capacitor differs from that in the bulk due to Φ_{ms} and Q_o. This change in the concentration sets up an electric field at the surface and hence a voltage difference between the silicon surface and bulk. This voltage difference is referred to as the *surface potential* ϕ_s and represents the electrostatic potential at the surface measured from the bulk intrinsic level E_i as shown in Figure 3.8. Thus, ϕ_s is the difference between $E_i(x = 0)$ at the surface and $E_i(x = \infty)$ at a point deep into the substrate. As shown in Figure 3.8, ϕ_s is a measure of the amount

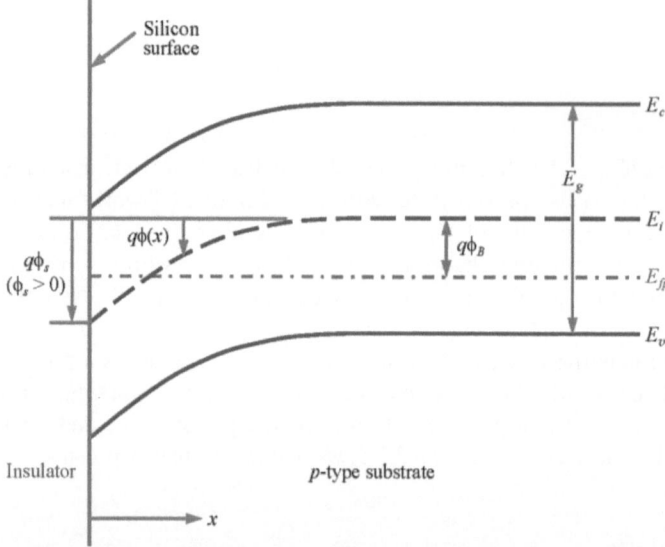

FIGURE 3.8 MOS capacitor system: band bending near the silicon surface showing surface potential ϕ_s at the surface of a p-type silicon; here, x is the distance from the insulator/substrate interface, $x = 0$ into the substrate.

of total band bending at the silicon surface. And, at a depth x into the surface, the potential is given by $\phi(x)$.

The band bending described above can be compensated by applying an external gate voltage V_{fb}, defined in Section 3.2.4 as the *flat band voltage* and is given by Equation 3.21. The condition to achieve the flat bands at the surface is called the *flat band condition*. Thus, V_{fb} *is the applied gate voltage to have zero surface potential with flat energy bands over the entire semiconductor surface.* The flat band condition is often used as a reference state along with V_{fb} as a reference voltage and thus, can be considered as an important figure-of-merit for an MOS capacitor system.

3.3 MOS CAPACITOR UNDER APPLIED BIAS

In Section 3.2, we described the behavior of MOS capacitors without any applied external bias. Now, let us discuss the behavior of MOS capacitors under an applied gate bias V_g as shown in Figure 3.9. The applied V_g is shared among the voltage drop across the oxide V_{ox}, surface potential ϕ_s, and the workfunction difference Φ_{ms} between the metal and semiconductor. Thus,

$$V_g = \begin{cases} V_{ox} + \phi_s + \Phi_{ms}, & \text{for } Q_0 = 0 \\ V_{ox} + \phi_s + V_{fb}, & \text{for } Q_0 \neq 0 \end{cases} \tag{3.23}$$

With reference to charges, an MOS capacitor consists of three different charges under the applied V_g such as: (1) gate charge Q_g due to the applied V_g to the gate, (2) effective interface charge Q_o at the Si/SiO$_2$ interface for a non-ideal insulator as discussed in Section 3.2.4, and (3) the induced charge Q_s in the silicon underneath the gate oxide. Then from the charge neutrality condition we get

$$Q_g + Q_o + Q_s = 0. \tag{3.24}$$

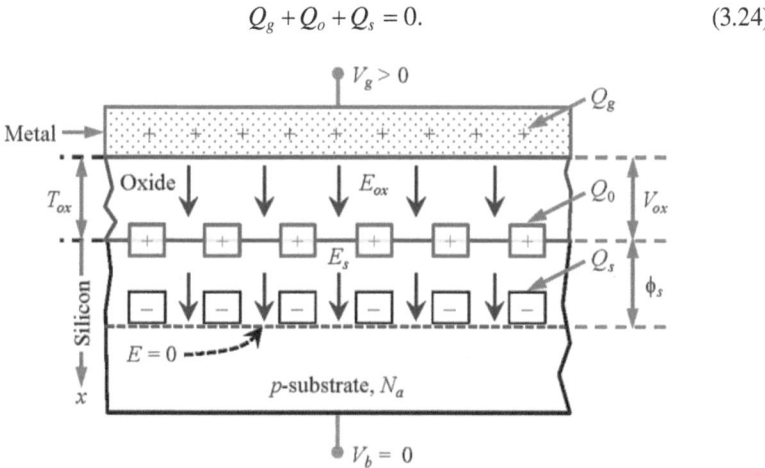

FIGURE 3.9 An MOS capacitor under the applied gate bias V_g showing various charges, electric fields, and potentials; E_{ox} is the electric field in the oxide and E_s is the surface electric field in the substrate.

If the applied voltage $V_g > 0$, then the electric field E_s is directed into the silicon surface at the interface and will induce charge Q_s in the silicon. The density of the induced charge Q_s per unit area can be calculated by applying Gauss' law [Equation 2.75] at the Si/SiO$_2$ interface. Thus, Q_s per unit area is given by

$$Q_s = \varepsilon_0 K_{si} E_s = \varepsilon_{si} E_s = -\varepsilon_{si} \left. \frac{d\phi}{dx} \right|_{x=0} \tag{3.25}$$

where:

K_{si} is the dielectric constant of silicon

$\varepsilon_{si} = \varepsilon_0 K_{si}$ is the permittivity of silicon

E_s is the surface electric field give the potential gradient $\left. \dfrac{d\phi}{dx} \right|_{x=0}$ at the silicon surface, $x = 0$

Again, in this chapter, we will use ε_{si} or $\varepsilon_0 K_{si}$ to describe the permittivity of silicon interchangeably for the simplicity of mathematical presentation where applicable.

Similarly, applying Gauss' law at the MG/SiO$_2$ interface, we get the gate charge Q_g per unit area as

$$Q_g = \varepsilon_{ox} E_{ox} = \varepsilon_0 K_{ox} E_{ox} \tag{3.26}$$

In Equation 3.26, all parameters have their usual meanings as defined earlier.

The fields E_{ox} and E_s are related to terminal charges given by Equation 3.24. Now we know that for an ideal oxide, $Q_o = 0$, therefore, from Equation 3.24,

$$Q_g = -Q_s \tag{3.27}$$

Then substituting the expression for Q_g from Equation 3.26 into Equation 3.27, we get

$$Q_s = -\varepsilon_0 K_{ox} E_{ox} = -\varepsilon_0 K_{ox} \left(\frac{V_{ox}}{T_{ox}} \right) \tag{3.28}$$

or, $\quad Q_s = -C_{ox} V_{ox}$

where:

$$E_{ox} = V_{ox} / T_{ox}$$

V_{ox} is the voltage drop in gate oxide

$C_{ox} = \varepsilon_0 K_{ox}/T_{ox}$ is the gate oxide capacitance per unit area defined in Equation 3.1

Therefore, from Equation 3.28, we get

$$V_{ox} = -\frac{Q_s}{C_{ox}} \tag{3.29}$$

Now, substituting for V_{ox} from Equation 3.29 in Equation 3.23, we get

$$V_g = V_{fb} + \phi_s - \frac{Q_s}{C_{ox}} \tag{3.30}$$

Equation 3.30 relates the applied bias V_g with the surface potential ϕ_s. Since at the flat band condition, $\phi_s = 0$ and $Q_s = 0$, therefore, Equation 3.30 shows that at the flat band condition of MOS capacitors, $V_g = V_{fb}$. And, within the range of applied gate bias, $0 > V_g > 0$ different surface conditions result in an MOS capacitor as discussed in Sections 3.3.1–3.3.3.

3.3.1 ACCUMULATION

To continue our discussions on MG-SiO$_2$-(p-type) silicon MOS capacitors with $\Phi_{ms} < 0$, let us apply a gate voltage $V_g < 0$ with body grounded such that $V_g < V_{fb}$. The negative voltage at the gate creates an upward electric field E_{ox} from the substrate to metal as shown in Figure 3.10(a). Since the applied negative voltage depresses the electrostatic potential of the metal relative to the substrate, electron energies are raised in the metal relative to the substrate. As a result, the Fermi level E_{fm} of the metal *moves up* above its equilibrium position by qV_g as shown in Figure 3.10(b). Since Φ_m and Φ_s do not change with V_g, moving E_{fm} up in energy relative to E_f causes the oxide conduction band to bend upward, consistent with the direction of the field E_{ox} causing gradient in the energy bands [6,8,16].

With reference to charge, the negative voltage at the gate results in a negative charge ($Q_g < 0$) on the gate. This in turn induces an equal amount of positive charge Q_s at the silicon surface. This amount of positive charge in the p-type silicon means excess hole concentration is created at the surface as shown in Figure 3.10(a). These holes are accumulated at the surface and known as the *accumulation* charges. We know from Equation 3.22 that as the hole concentration

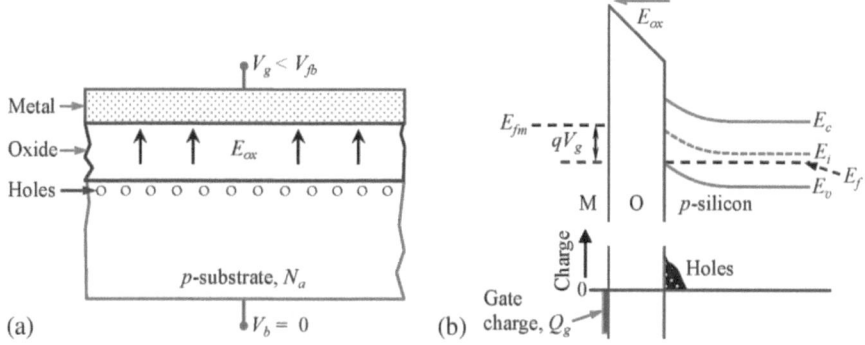

FIGURE 3.10 Effect of an applied voltage, $V_g < V_{fb}$ on a p-type silicon MOS capacitor with $\Phi_{ms} < 0$: (a) the applied negative bias $V_g < V_{fb}$ causes hole accumulation at the silicon surface; (b) corresponding band diagram with upward band bending.

increases at the surface, $(E_i - E_f)$ increases resulting in the bands bending upwards as shown in Figure 3.10(b). Thus, for an MOS capacitor on a *p*-type silicon substrate in accumulation, we have

$$\text{Accumulation} \begin{cases} V_g < V_{fb}, \\ \phi_s < 0, \\ Q_s > 0. \end{cases} \tag{3.31}$$

This biasing condition given by Equation 3.31 is useful in the characterization of MOS capacitor systems in accumulation.

3.3.2 DEPLETION

Now, let us apply a positive gate voltage $V_g > V_{fb}$ with body grounded. This positive V_g will create a downward electric field E_{ox} from the gate into the substrate as shown in Figure 3.11(a). A positive gate voltage raises the potential of the gate, lowering the Fermi level E_{fm} by qV_g. Moving E_{fm} down in energy relative to E_f causes band bending downward in the oxide conduction band in accordance to the direction of E_{ox}.

Again, with reference to charge, a positive voltage at the gate deposits positive charge $(Q_g > 0)$ on the gate. Due to $V_g > 0$, the holes are repulsed away from the silicon surface leaving behind negatively charged acceptor ions. Thus, a positive charge on the gate induces a negative charge Q_s at the surface due to the depletion of holes creating a depletion region of width X_d. This is known as the *depletion* condition. Since the hole concentration decreases at the surface, then from Equation 3.22, $(E_i - E_f)$ must decrease. As a result, E_i slowly approaches E_f, thereby bending the

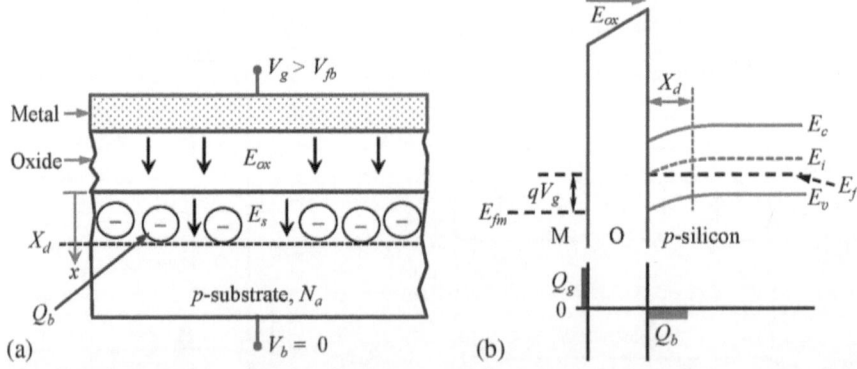

FIGURE 3.11 Effect of an applied voltage, $V_g > V_{fb}$ on a *p*-type silicon MOS capacitor with $\Phi_{ms} < 0$: (a) the applied positive bias $V_{gt} = (V_g - V_{fb})$ depletes the holes from the silicon surface, (b) corresponding band diagram with downward band bending. Here, Q_b is the depletion or bulk charge, Q_g is the gate charge, and X_d is the width of the depletion region, and "⊖" represents the ionized acceptor atoms.

bands downwards near the surface as shown in Figure 3.11(b). Thus, the depletion condition is given by

$$\text{Depletion} \begin{cases} V_g > V_{fb}, \\ \phi_s > 0, \\ Q_s < 0. \end{cases} \tag{3.32}$$

3.3.3 INVERSION

If we further increase the positive gate voltage, the downward band bending will further increase. At a sufficiently high $V_g \gg V_{fb}$ as shown in Figure 3.12(a), the band bending may pull down the midgap energy level E_i below the constant E_f at the silicon surface, that is, $E_f > E_i$. At this condition, the surface behaves like an n-type material with an electron concentration given by [Equation 2.29]

$$n = n_i \exp\left(\frac{E_f - E_i}{kT}\right) \tag{3.33}$$

Thus, the n-surface is formed by inversion of the p-type substrate due to the applied gate voltage. This is known as the *inversion* condition as shown in Figure 3.12(b). In inversion, the total charge Q_s in the semiconductor consists of depletion charge Q_b and the inversion charge Q_i. This inversion condition for an MOS capacitor on a p-type substrate is defined by

$$\text{Inversion} \begin{cases} V_g \gg V_{fb}, \\ \phi_s > 0, \\ Q_s < 0. \end{cases} \tag{3.34}$$

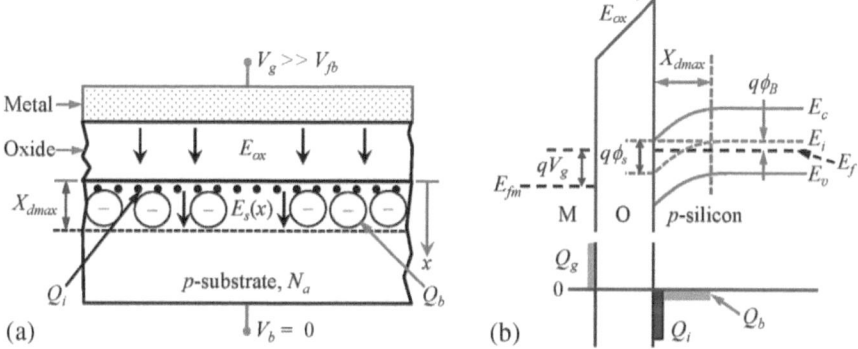

FIGURE 3.12   Effect of an applied voltage, $V_g \gg V_{fb}$ on a p-type silicon MOS capacitor with $\Phi_{ms} < 0$: (a) the large positive bias $V_g \gg V_{fb}$ causes inversion of the p-type surface forming an n-type layer in addition to depletion region; (b) corresponding band diagram with downward band bending. The gate charge, Q_g is compensated by the depletion charge, Q_b and inversion charge, Q_i in the semiconductor; X_{dmax} is the maximum depletion width; "•" represents electrons and "⊖" represents the ionized acceptor atoms.

Under the applied $V_g \gg V_{fb}$, the p-type surface is inverted as soon as E_i is pulled below E_f. However, for small $(E_f - E_i)$, the electron concentration remains very small and the inversion is *weak*. This is referred to as the *weak inversion regime*. If we increase V_g such that $(E_f - E_i)$ at the surface equals $(E_i - E_f)$ at the p-type bulk, the concentration of electrons at the surface will be equal to that of the holes in the bulk. This is called the *strong inversion regime*. On further increase of V_g, the electron concentration will exceed the concentration of the holes in the inversion region. Under the inversion condition, the depth of the inversion region (X_{inv}) into the substrate can be defined from the Si/SiO$_2$ interface to the point of $E_f = E_i$ and is about 3 nm [5,8].

Now, let us discuss how the inversion layer is formed in the substrate. At the onset of inversion, the minority carrier electrons in the p-type substrate of an MOS capacitor system originate from thermally generated electron-hole pairs within the depletion region. The rate of thermal generation depends upon the minority carrier lifetime which is of the order of microseconds. It is found that the time required to form an inversion layer at the surface is about 0.2 sec [7]. Thus, the *formation of the inversion layer is a relatively slow process compared to the time required for the holes (majority carriers) to flow from or to the silicon surface* which is of the order of picoseconds. Once the inversion layer is formed, it shields the underneath depletion layer, thus limiting the maximum width X_{dmax} of the depletion layer.

So far, we have presented a qualitative overview of the basic operation of an MOS capacitor system. In the following section, we will develop MOS capacitor theory that can be extended to develop the operational theory of FinFET devices in the subsequent chapters of this book.

3.4   MULTIGATE MOS CAPACITOR SYSTEMS: MATHEMATICAL ANALYSIS

In this section, we will derive the relation between the surface potential (ϕ_s), electric field (E_s), and charge (Q_s) by solving Poisson's equation for potential (ϕ) near the surface region of the silicon substrate of an MOS capacitor. The one-dimensional (1D) Poisson's equation [Equation 2.72] is given by

$$\frac{d^2\phi}{dx^2} = -\frac{1}{K_{si}\varepsilon_0}\rho(x) \tag{3.35}$$

where:

$\rho(x)$ is the charge density at any point x near the surface along the depth of the substrate

Again, the charge density $\rho(x)$ is given by

$$\rho(x) = q\left[p(x) - n(x) + N_d^+(x) - N_a^-(x) \right] \tag{3.36}$$

where:

$p(x)$ is the hole concentration

$n(x)$ is the electron concentration

$N_d^+(x)$ is the ionized donor concentration in the semiconductor substrate

$N_a^-(x)$ is the ionized acceptor concentration in the semiconductor substrate

Then substituting for $\rho(x)$ from Equation 3.36 into Equation 3.35, we get

$$\frac{d^2\phi}{dx^2} = -\frac{q}{K_{si}\varepsilon_0}\left[p(x)-n(x)+N_d^+(x)-N_a^-(x)\right] \qquad (3.37)$$

In order to solve Equation 3.37 for $\phi(x)$ at any point x near the surface region of the substrate, we need to express $p(x)$ and $n(x)$ in terms of $\phi(x)$ as described in the following section.

3.4.1 Poisson's Equation

In order to solve Equation 3.37 for electrostatic potential, $\phi(x)$ at any point x near the surface of the substrate, we transform Equation 3.37 in terms of the intrinsic potential ϕ_i, Fermi potential ϕ_f, and $\phi(x)$. We know that in an n-type semiconductor with doping concentration N_d, the majority carrier *electron* concentration n is given by [Equations 2.29 and 2.31]

$$n \cong N_d^+ = n_i \exp\left(\frac{E_f-E_i}{kT}\right) = n_i \exp\left(\frac{q(\phi_i-\phi_f)}{kT}\right) \qquad (3.38)$$

where:

$\phi_f = -E_f/q$ is the Fermi potential

$\phi_i = -E_i/q$ is the intrinsic potential

And the minority carrier concentration p_n in an n-type semiconductor is given by [Equation 2.33]

$$p_n \cong \frac{n_i^2}{N_d^+} \qquad (3.39)$$

Similarly, the majority carrier concentration in a p-type semiconductor with doping concentration N_a is given by [Equations 2.30 and 2.32]

$$p \cong N_a^- = n_i \exp\left(\frac{E_i-E_f}{kT}\right) = n_i \exp\left(\frac{q(\phi_f-\phi_i)}{kT}\right) \qquad (3.40)$$

And the minority carrier concentration n_p in a p-type semiconductor is given by [Equation 2.34]

$$n_p \cong \frac{n_i^2}{N_a^-} \qquad (3.41)$$

In order to develop a generalized theory of MOS capacitors on both the n-type and p-type substrates, we define N_b as the substrate concentration such that

$$N_b = \begin{cases} N_a, & \text{for } p\text{-type substrate} \\ N_d, & \text{for } n\text{-type substrate} \end{cases} \tag{3.42}$$

Then from Equations 3.38 and 3.40, we can show that the bulk potential is

$$\phi_B = \left| \phi_f - \phi_i \right| = v_{kT} \ln\left(\frac{N_b}{n_i} \right)$$

$$\text{or,} \quad N_b = n_i \exp\left(\frac{\phi_B}{v_{kT}} \right) \tag{3.43}$$

Note that the bulk potential ϕ_B is also referred to as the Fermi potential with reference to $\phi_i = 0$.

Now, in order to express $\rho(x)$ in terms of band bending $\phi(x)$ at any point x near the surface of a semiconductor, we consider the band structure of an MG-SiO$_2$-(p-type) silicon MOS capacitor as shown in Figure 3.13.

From Figure 3.13, the amount of band bending at any point x near the semiconductor surface with reference to vacuum level E_0 is given by

$$\phi(x) = \phi_i(x) - \phi_i(x \to \infty) \tag{3.44}$$

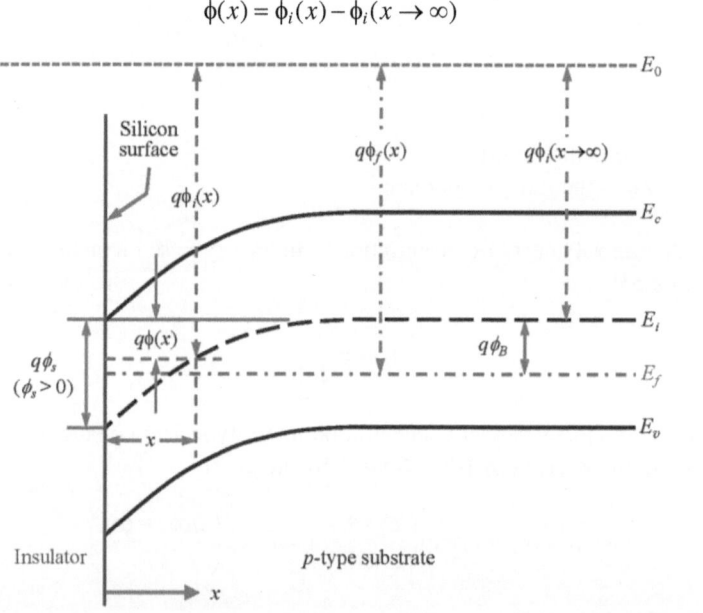

FIGURE 3.13 Equilibrium band structure of a p-type MOS capacitor showing band bending in the substrate; here, $x = 0$ at the Si/SiO$_2$ interface and increases with the depth into the substrate; ϕ_s is the surface potential representing the total band bending at the surface with reference to flat bands; and $\phi_B = (\phi_f - \phi_i)$ is the bulk potential.

where:

$\phi(x = 0) = \phi_s$ is the surface potential at $x = 0$

$\phi_i(x \longrightarrow \infty) = \phi_i$ is the intrinsic potential at $x \longrightarrow \infty$ (deep into the substrate)

It is observed from Figure 3.13 that $\phi_i(x) > \phi_i$ when bands bend downward near the oxide/substrate interface resulting in $\phi(x) > 0$. Thus, it is clear that $\phi(x) > 0$ when bands bend downward in the *depletion* and *inversion* condition, whereas $\phi(x) < 0$ when bands bend upward in the *accumulation* condition.

Now, let us express $p(x)$, $n(x)$, $N_d^+(x)$, and $N_a^-(x)$ in Poisson's Equation 3.37 in terms of potential $\phi(x)$ to solve for ϕ_s and E_s at the surface of the substrate of a multigate MOS capacitor. For a p-type substrate, the majority carrier concentration $p(x)$ at any point x is obtained by substituting for $\phi_i(x)$ from Equation 3.44 into Equation 3.40 so that

$$p(x) = n_i \exp\left(\frac{(\phi_f - \phi_i(x))}{v_{kT}}\right) = n_i \exp\left[\frac{\phi_f - \left(\phi_i(x \to \infty) + \phi(x)\right)}{v_{kT}}\right] \qquad (3.45)$$

Since $[\phi_f - \phi_i(x \longrightarrow \infty)] = (\phi_f - \phi_i) = \phi_B$, then we can express Equation 3.45 as

$$p(x) = n_i \exp\left[\frac{\left(\phi_f - \phi_i\right) - \phi(x)}{v_{kT}}\right] = n_i \exp\left(\frac{\phi_B - \phi(x)}{v_{kT}}\right) \qquad (3.46)$$

Now, from Equations 3.43 and 3.46, we can express the majority carrier concentration at any point x in a p-type substrate in terms of the intrinsic carrier concentration n_i as well as substrate doping concentration N_a as

$$p(x) = \begin{cases} n_i \exp\left(\dfrac{\phi_B - \phi(x)}{v_{kT}}\right); & n_i\text{-dependence} \\[4mm] N_a \exp\left(-\dfrac{\phi(x)}{v_{kT}}\right); & N_a\text{-dependence} \end{cases} \qquad (3.47)$$

Then substituting Equation 3.47 into 3.41, we get the minority carrier electron concentration at any point x near the surface of a p-type substrate as

$$n(x) \cong \frac{n_i^2}{p(x)} = \frac{n_i^2}{N_a} \exp\left(\frac{\phi(x)}{v_{kT}}\right) \qquad (3.48)$$

Again, from Equation 3.43, we can show that for a p-type substrate

$$\frac{n_i^2}{N_a} = \begin{cases} n_i \exp\left(-\dfrac{\phi_B}{v_{kT}}\right); & n_i\text{-dependence} \\[4mm] N_a \exp\left(-\dfrac{2\phi_B}{v_{kT}}\right); & N_a\text{-dependence} \end{cases} \qquad (3.49)$$

Therefore, the minority carrier concentration in a p-type substrate given by Equation 3.48 can also be written as

$$
n(x) = \begin{cases} n_i \exp\left(\dfrac{\phi(x) - \phi_B}{v_{kT}}\right); & n_i\text{-dependence} \\[3ex] N_a \exp\left(\dfrac{\phi(x) - 2\phi_B}{v_{kT}}\right); & N_a\text{-dependence} \end{cases} \tag{3.50}
$$

Now, substituting the expressions for $p(x)$ and $n(x)$ from Equations 3.47 and 3.50, respectively, into Equation 3.36, we get

$$
\rho(x) = q\left[n_i e^{-\frac{(\phi(x) - \phi_B)}{v_{kT}}} - n_i e^{\frac{\phi(x) - \phi_B}{v_{kT}}} + N_d^+(x) - N_a^-(x) \right] \tag{3.51}
$$

Again, assuming complete ionization of acceptor atoms, for a p-type substrate, we get from Equation 3.42, $p \cong N_a^- \equiv N_a$ and $n \cong N_d^+ \equiv N_d$, respectively. Therefore, for a uniformly doped substrate, the total charge density is given by

$$
\rho(x) = q\left[\left(n_i e^{-\frac{\phi(x) - \phi_B}{v_{kT}}} - N_a(x) \right) - \left(n_i e^{\frac{\phi(x) - \phi_B}{v_{kT}}} - N_d(x) \right) \right] \tag{3.52}
$$

It is to be noted that the first term within the square brackets of $\rho(x)$ in Equation 3.52 represents the charge density in a p-type substrate due to the majority carrier concentration, and the second term represents that due to minority carrier concentration. Now, substituting Equation 3.52 into Equation 3.37, we get Poisson's equation in terms of band bending potential $\phi(x)$ at a depth x near the surface of the p-type substrate as

$$
\frac{d^2\phi(x)}{dx^2} = -\frac{q}{\varepsilon_0 K_{si}}\left[\left(n_i e^{-\frac{\phi(x) - \phi_B}{v_{kT}}} - N_a(x) \right) - \left(n_i e^{\frac{\phi(x) - \phi_B}{v_{kT}}} - N_d(x) \right) \right] \tag{3.53}
$$

We will solve Equation 3.53 for $\phi(x)$ to characterize the multigate MOS capacitor behavior under different operating conditions. Again, we notice from Equation 3.53 that the first term inside the square brackets is due to the majority carrier charge density, whereas the second term is that due to minority carriers in the semiconductor substrate.

3.4.2 ELECTROSTATIC POTENTIALS AND CHARGE DISTRIBUTION

3.4.2.1 Induced Charge in Semiconductor

In order to solve Equation 3.53 for $\phi(x)$, we assume that the substrate of the multigate MOS capacitors is undoped or lightly doped so that the majority carrier concentration is negligibly smaller than the minority carrier concentration. Also, we assume that the ionized dopant concentration of the substrate is significantly higher than the opposite type of dopant; that is, for a p-type substrate $N_a \gg N_d$. Therefore, for a uniformly doped p-type substrate with, $N_b = N_a(x)$, we can safely consider only the minority carrier term at any point x near the surface given by Equation 3.53 and acceptor doping concentration N_a. Then for a DG-MOS capacitor shown in Figure 3.1(b), we can express Poisson's Equation 3.53 as

$$\frac{d^2\phi(x)}{dx^2} = \frac{q}{K_{si}\varepsilon_0}\left[n_i e^{\frac{\phi(x)-\phi_B}{v_{kT}}} + N_b \right]$$

$$\text{or,} \quad \frac{d^2\phi(x)}{dx^2} = \frac{qn_i}{K_{si}\varepsilon_0}\left[e^{\frac{\phi(x)-\phi_B}{v_{kT}}} + e^{\frac{\phi_B}{v_{kT}}} \right]$$

(3.54)

In Equation 3.54, we have used Equation 3.43 for $N_b = n_i \exp\left(\phi_B / v_{kT}\right)$. Now, in order to solve Equation 3.54 for potential distribution in silicon, we use the mathematical identity

$$\frac{d}{dx}\left(\frac{d\phi}{dx}\right)^2 = 2\frac{d\phi}{dx}\frac{d^2\phi}{dx^2}$$

(3.55)

Then multiplying both sides of Equation 3.54 by $2\dfrac{d\phi(x)}{dx}$, we get

$$2\frac{d\phi(x)}{dx}\frac{d^2\phi(x)}{dx^2} = \frac{qn_i}{K_{si}\varepsilon_0}\left[e^{\frac{\phi(x)-\phi_B}{v_{kT}}} + e^{\frac{\phi_B}{v_{kT}}} \right]\left(2\frac{d\phi(x)}{dx}\right)$$

(3.56)

Now, using Equation 3.55 on the left-hand side of Equation 3.56, we get

$$\frac{d}{dx}\left(\frac{d\phi(x)}{dx}\right)^2 = \frac{2qn_i}{K_{si}\varepsilon_0}\left[e^{\frac{\phi(x)-\phi_B}{v_{kT}}} + e^{\frac{\phi_B}{v_{kT}}} \right]\frac{d\phi(x)}{dx}$$

(3.57)

In order to integrate Equation 3.57, let us consider an ideal symmetric DG-MOS capacitor structure with substrate thickness t_{si}, gate oxide thickness T_{ox}, and metal gate with common gate voltage V_g as shown in Figure 3.14. In Figure 3.14, x is along the thickness of the structure with $x = 0$ at the center point with potential $\phi_0(x = 0)$ and $x = \pm t_{si}/2$ at the two Si/SiO$_2$ interfaces with potential $\phi_s(x = \pm t_{si}/2)$.

We integrate Equation 3.57 from the center point, $x = 0$ to a point x along the body of thickness t_{si} to find the potential distribution along the thickness of the structure. Then from Figure 3.14, the boundary conditions for integration are

$$\begin{cases} \phi(x) = \phi_0, \dfrac{d\phi(x)}{dx} = 0; & \text{at } x = 0 \\[2mm] \phi(x) = \phi(x), \dfrac{d\phi(x)}{dx} = \dfrac{d\phi(x)}{dx}; & \text{at } x = x \end{cases}$$

(3.58)

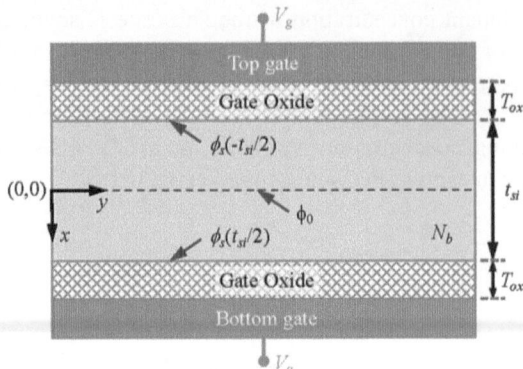

FIGURE 3.14 An ideal symmetric DG-MOS capacitor structure with a p-type silicon body: t_{si}, T_{ox}, and N_b are the body thickness, gate oxide thickness, and the body doping concentration, respectively; x is along the depth and $x = 0$ at the middle of the body so that at the SiO_2/Si interface of the top gate $x = -t_{si}/2$ and at the Si/SiO_2 interface of the bottom gate $x = t_{si}/2$; ϕ_s is the surface potential; and ϕ_0 is the center potential at $x = 0$.

Now, we integrate Equation 3.57 from the center point, $x = 0$ to at any point x toward the surface using the boundary conditions in Equation 3.58 to get

$$\int_0^{\frac{d\phi(x)}{dx}} d\left(\frac{d\phi(x)}{dx}\right)^2 = \frac{2qn_i}{K_{si}\varepsilon_0} \int_{\phi_0}^{\phi(x)} \left[e^{\frac{\phi(x)-\phi_B}{v_{kT}}} + e^{\frac{\phi_B}{v_{kT}}} \right] d\phi \qquad (3.59)$$

After integration and simplification, we can show

$$\left(\frac{d\phi(x)}{dx}\right)^2 = \frac{2qn_i v_{kT}}{K_{si}\varepsilon_0} \left[\left(e^{\frac{\phi(x)}{v_{kT}}} - e^{\frac{\phi_0}{v_{kT}}} \right) \cdot e^{-\frac{\phi_B}{v_{kT}}} + \left(\frac{\phi(x)-\phi_0}{v_{kT}} \right) e^{\frac{\phi_B}{v_{kT}}} \right] \qquad (3.60)$$

or, $$\frac{d\phi(x)}{dx} = \pm \sqrt{\frac{2qn_i v_{kT}}{K_{si}\varepsilon_0}} \left[\left(e^{\frac{\phi(x)}{v_{kT}}} - e^{\frac{\phi_0}{v_{kT}}} \right) \cdot e^{-\frac{\phi_B}{v_{kT}}} + \left(\frac{\phi(x)-\phi_0}{v_{kT}} \right) e^{\frac{\phi_B}{v_{kT}}} \right]^{1/2} \qquad (3.61)$$

where:
the "+" sign applies for $0 \leq x \leq t_{si}/2$
the "−" sign applies for $-t_{si}/2 \leq x \leq 0$

We know that at $x = \pm t_{si}/2$, $\phi(x) = \phi_s$ and $-d\phi/dx = E_s$ (surface electric field); therefore, in terms of the surface potential ϕ_s, we can write Equation 3.61 as

$$\left. \frac{d\phi(x)}{dx} \right|_{x=\pm t_{si}/2} = \pm \sqrt{\frac{2qn_i v_{kT}}{\varepsilon_0 K_{si}}} \left[\left(e^{\frac{\phi_s}{v_{kT}}} - e^{\frac{\phi_0}{v_{kT}}} \right) \cdot e^{-\frac{\phi_B}{v_{kT}}} + \left(\frac{\phi_s-\phi_0}{v_{kT}} \right) e^{\frac{\phi_B}{v_{kT}}} \right]^{1/2} \qquad (3.62)$$

$$= E_s(x) \equiv E_s$$

Then from Equation 3.62, the total charge per unit area induced in silicon (equal and opposite to the charge on the metal gate) can be obtained by Gauss' law: $Q_s = -\varepsilon_0 K_{si} E_s$. Thus, the charge per unit area at the surface of the MOS capacitor substrate is given by

$$Q_s = \pm\sqrt{2qK_{si}\varepsilon_0 n_i v_{kT}}\left[\left(e^{\frac{\phi_s}{v_{kT}}} - e^{\frac{\phi_0}{v_{kT}}}\right)\cdot e^{-\frac{\phi_B}{v_{kT}}} + \left(\frac{\phi_s - \phi_0}{v_{kT}}\right)e^{\frac{\phi_B}{v_{kT}}}\right]^{1/2} \quad (3.63)$$

Equation 3.63 is valid for all regions of multigate MOS capacitor operations: accumulation, depletion, and inversion. The + sign indicates the induced charge is positive for accumulation and the − sign is for depletion and inversion charge in a p-type substrate. Equation 3.63 can also be expressed as

$$Q_s = \pm\frac{2v_{kT}K_{si}\varepsilon_0}{L_{di}}\left[\left(e^{\frac{\phi_s}{v_{kT}}} - e^{\frac{\phi_0}{v_{kT}}}\right)\cdot e^{-\frac{\phi_B}{v_{kT}}} + \left(\frac{\phi_s - \phi_0}{v_{kT}}\right)e^{\frac{\phi_B}{v_{kT}}}\right]^{1/2} \quad (3.64)$$

where:
L_{di} is the intrinsic Debye length defined by

$$L_{di} = \sqrt{\frac{2K_{si}\varepsilon_0 v_{kT}}{qn_i}} \quad (3.65)$$

Again, on the right-hand side of Equations 3.61–3.64, the first term within the square brackets is due to the minority carrier charge in the p-type substrate, whereas the second term is due to the majority carrier bulk doping concentration, $N_b = N_a$ only.

We can use Equation 3.63 to qualitatively describe the dependence of Q_s on ϕ_s in different regions of MOS capacitor operation.

1. When $\phi_s < 0$, the MOS capacitor is in the accumulation mode and the inversion carrier term is negligible. Then the dominant term in Equation 3.63 is $\sqrt{(\phi_s - \phi_0)/v_{kT}}$. Therefore, Q_s varies with ϕ_s as

$$Q_s \approx \sqrt{2qK_{si}\varepsilon_0 n_i(\phi_s - \phi_0)}\exp\left(\frac{\phi_B}{2v_{kT}}\right), \quad \text{(accumulation)} \quad (3.66)$$

Since for a lightly doped body ($\phi_B \approx 0$), we get

$$Q_s \approx \sqrt{2qK_{si}\varepsilon_0 n_i(\phi_s - \phi_0)}, \quad \text{(accumulation)} \quad (3.67)$$

To further simplify Q_s (accumulation), we define E_{avg} as the average electric field in the region between $x = t_{si}/2$ to the center point at $x = 0$, then using Gauss' law, we can write

$$E_{avg} = -\frac{d\phi(x)}{dx} = \frac{Q_s}{K_{si}\varepsilon_0} \quad (3.68)$$

Then assuming that ϕ_s varies linearly from ϕ_0 (at $x = 0$) to ϕ_s (at $x = t_{si}/2$) for an ultrathin-body fully depleted substrate, we can write Equation 3.68 as

$$\frac{d\phi(x)}{dx} = \frac{\phi_s - \phi_0}{t_{si}/2} = \frac{Q_s}{K_{si}\varepsilon_0}$$

$$\text{or,} \quad \phi_s - \phi_0 = \frac{Q_s t_{si}}{2K_{si}\varepsilon_0}$$

(3.69)

Now, substituting for $(\phi_s - \phi_0)$ from Equation 3.69 into Equation 3.66 and using the expression for $N_b = n_i \exp(\phi_B/v_{kT})$ from Equation 3.43, we can show after simplification

$$Q_s \approx qN_b t_{si}; \quad \text{(accumulation)} \tag{3.70}$$

Thus, the accumulation charge is proportional to the thickness of the ultra-thin-body DG-MOS capacitors.

2. When $V_{fb} < \phi_s < 2\phi_B$ such that the DG-MOS capacitor is in the depletion and weak inversion regime, the term that dominates Equation 3.63 is $\sqrt{\phi_s}$ and therefore, Q_s varies with ϕ_s as

$$Q_s \sim \sqrt{(\phi_s - \phi_0)e^{\frac{\phi_B}{v_{kT}}}}; \quad \text{(depletion and weak inversion)} \tag{3.71}$$

For an undoped or lightly doped substrate ($\phi_B \approx 0$) and for an ultrathin-body symmetric DG-MOS capacitor, assuming $\phi_s \gg \phi_0$, Q_s varies with ϕ_s as

$$Q_s \sim \sqrt{\phi_s}; \quad \text{(depletion and weak inversion)} \tag{3.72}$$

3. When $\phi_s > 2\phi_B$, the DG-MOS capacitor structure is in the strong inversion regime and to a first approximation, the majority carrier term due to N_a in Equation 3.63 can be neglected and for a lightly doped substrate, $\phi_B \approx 0$. Therefore, Q_s varies with ϕ_s as

$$Q_s \sim \sqrt{e^{\phi_s/v_{kT}} - e^{\phi_0/v_{kT}}}; \quad \text{(strong inversion)} \tag{3.73}$$

The accumulation, depletion, and inversion conditions described by Equations 3.67, 3.72, and 3.73, respectively, are for the p-type substrates. For the n-type substrates, these conditions will be reversed.

3.4.2.2 Formulation of Surface Potential

In order to determine the potential distribution within the body of a DG-MOS capacitor, let us consider a symmetrical structure shown in Figure 3.14. Then we solve Poisson's equation given by Equation 3.54 as repeated below

$$\frac{d^2\phi(x)}{dx^2} = \frac{q}{K_{si}\varepsilon_0}\left[n_i\exp\left(\frac{\phi(x)-\phi_B}{v_{kT}}\right) + N_b\right] \tag{3.74}$$

If the silicon body thickness t_{si} is less than the width of the depletion region, then for a certain gate bias V_g, the entire t_{si} is fully depleted and consequently, the inversion carriers are spread through the entire body. Thus, $Q_i \gg Q_b$ and therefore, we can safely neglect the term containing N_b in Equation 3.74. Then the potential $\phi(x)$ at any point x along the thickness of the substrate can be obtained by solving simplified Poisson's equation given by

$$\frac{d^2\phi(x)}{dx^2} = \frac{qn_i}{K_{si}\varepsilon_0}\exp\left(\frac{\phi(x)-\phi_B}{v_{kT}}\right) \tag{3.75}$$

Again, in order to solve Equation 3.75 for $\phi(x)$, we use the mathematical identity given in Equation 3.55 to get

$$\frac{d}{dx}\left(\frac{d\phi(x)}{dx}\right)^2 = \frac{2qn_i}{K_{si}\varepsilon_0}\left[e^{\frac{\phi(x)-\phi_B}{v_{kT}}}\right]\frac{d\phi(x)}{dx} \tag{3.76}$$

Now, we integrate Equation 3.76 from the center point, $x = 0$ to a point x to find the potential distribution along the thickness of the structure shown in Figure 3.14. Since for a symmetric DG-MOS capacitor the vertical component of the electric field, $E(x) = 0 = d\phi/dx$ at the center, $x = 0$ where $\phi(x = 0) = \phi_0$, we use the following boundary conditions [Equation 3.58] for integration

$$\begin{cases} \phi(x) = \phi_0, \dfrac{d\phi(x)}{dx} = 0; & \text{at } x = 0 \\[3mm] \phi(x) = \phi(x), \dfrac{d\phi(x)}{dx} = \dfrac{d\phi(x)}{dx}; & \text{at } x = x \end{cases} \tag{3.77}$$

Then, using the boundary conditions given by Equation 3.77, we can express Equation 3.76 as

$$\int_0^{\frac{d\phi}{dx}} d\left(\frac{d\phi(x)}{dx}\right)^2 = \frac{2qn_i}{K_{si}\varepsilon_0}\int_{\phi_0}^{\phi(x)} e^{\frac{\phi(x)-\phi_B}{v_{kT}}} d\phi(x) \tag{3.78}$$

After integrating Equation 3.78 and using $n_i\exp(-\phi_B/v_{kT}) = n_i^2/N_b$ from Equation 3.49, we get

$$\left(\frac{d\phi(x)}{dx}\right)^2 = \frac{2q}{K_{si}\varepsilon_0}v_{kT}\frac{n_i^2}{N_b}\left(e^{\frac{\phi(x)}{v_{kT}}} - e^{\frac{\phi_0}{v_{kT}}}\right) \tag{3.79}$$

or,

$$\frac{d\phi(x)}{dx} = \pm\sqrt{\frac{2v_{kT}q}{K_{si}\varepsilon_0}\frac{n_i^2}{N_b}\left(e^{\frac{\phi(x)}{v_{kT}}} - e^{\frac{\phi_0}{v_{kT}}}\right)} \qquad (3.80)$$

where:
the "+" sign is for $x > 0$
the "−" sign is for $x < 0$

We can obtain $\phi(x)$ after the *second* integration of Equation 3.80. In order to perform the integration of Equation 3.80, let us consider the general mathematical solution of simplified Poisson's Equation 3.75. The differential Equation 3.75 has two mathematical solutions: one is a trigonometric function (proportional to $1/\cos^2 \xi$) and another is a hyperbolic function [17]. Now, let us consider the trigonometric solution [18] of simplified Poisson's Equation 3.75 as

$$e^{\frac{\phi(x)}{v_{kT}}} = e^{\frac{\phi_0}{v_{kT}}} \sec^2 \xi \qquad (3.81)$$

where:
ξ is a spatial parameter $f(x)$

Then substituting Equation 3.81 into Equation 3.80, we can show after simplification

$$\frac{d\phi(x)}{dx} = \pm\sqrt{\frac{2v_{kT}q}{K_{si}\varepsilon_0}\frac{n_i^2}{N_b}e^{\frac{\phi_0}{v_{kT}}}\left(\sec^2 \xi - 1\right)}$$

$$\qquad (3.82)$$

or, $\quad \frac{d\phi(x)}{dx} = \pm\sqrt{\frac{2v_{kT}q}{K_{si}\varepsilon_0}\frac{n_i^2}{N_b}e^{\frac{\phi_0}{v_{kT}}}\tan^2 \xi}$

where:

$$\left(\sec^2 \xi - 1\right) = \tan^2 \xi$$

We can also obtain an expression for $[d\phi(x)/dx]$ by differentiating (trigonometric solution) Equation 3.81 with respect to x, to get

$$\frac{d}{dx}\left(e^{\frac{\phi(x)}{v_{kT}}}\right) = \frac{d}{dx}\left(e^{\frac{\phi_0}{v_{kT}}} \sec^2 \xi\right) \qquad (3.83)$$

After differentiation and simplification, we can show

$$\frac{1}{v_{kT}}e^{\frac{\phi(x)}{v_{kT}}}\frac{d\phi(x)}{dx} = e^{\frac{\phi_0}{v_{kT}}}\sec^2 \xi \tan \xi \frac{d\xi}{dx} \times 2 \qquad (3.84)$$

Now, using $e^{\frac{\phi(x)}{v_{kT}}} = e^{\frac{\phi_0}{v_{kT}}} \sec^2 \xi$ from Equation 3.81 into Equation 3.84, we get after simplification

$$\frac{d\phi(x)}{dx} = 2v_{kT} \tan \xi \frac{d\xi}{dx} \tag{3.85}$$

The expressions for $[d\phi(x)/dx]$ in Equations 3.82 and 3.85 are obtained from the solution of the same Equation 3.80 and therefore, are equal. Then equating Equations 3.82 and 3.85, we get after simplification

$$2v_{kT} \tan \xi \frac{d\xi}{dx} = \pm \sqrt{\frac{2v_{kT}q}{K_{si}\varepsilon_0} \frac{n_i^2}{N_b} e^{\frac{\phi_0}{v_{kT}}} \tan^2 \xi} \tag{3.86}$$

Therefore, after simplification, we can express Equation 3.86 as

$$\frac{d\xi}{dx} = \pm \sqrt{\frac{q}{2v_{kT}K_{si}\varepsilon_0} \frac{n_i^2}{N_b} e^{\frac{\phi_0}{v_{kT}}}} \tag{3.87}$$

Now, we integrate Equation 3.87 from $x = 0$ to any point x; since $\xi = f(x)$, therefore, at $x = 0$; $\xi = 0$; and at $x = x$; $\xi = \xi$. Thus, we get

$$\int_0^\xi d\xi = \sqrt{\frac{q}{2v_{kT}K_{si}\varepsilon_0} \frac{n_i^2}{N_b} e^{\frac{\phi_0}{v_{kT}}}} \int_0^x dx \tag{3.88}$$

$$\xi = \sqrt{\frac{q}{2v_{kT}K_{si}\varepsilon_0} \frac{n_i^2}{N_b} e^{\frac{\phi_0}{v_{kT}}}} \cdot x \tag{3.89}$$

Again, from Equation 3.81, we get

$$e^{\frac{\phi(x)}{v_{kT}}} = \frac{e^{\frac{\phi_0}{v_{kT}}}}{\cos^2 \xi} \tag{3.90}$$

Therefore,

$$\cos \xi = \frac{1}{\sqrt{e^{\frac{\phi(x)-\phi_0}{v_{kT}}}}} \tag{3.91}$$

or,

$$\xi = \cos^{-1}\left(\frac{1}{\sqrt{e^{\frac{\phi(x)-\phi_0}{v_{kT}}}}}\right) \tag{3.92}$$

Then we can eliminate ξ using Equation 3.89 and Equation 3.92 to get

$$\cos^{-1}\left(\frac{1}{\sqrt{e^{\frac{\phi(x)-\phi_0}{v_{kT}}}}}\right) = \sqrt{\frac{q}{2v_{kT}K_{si}\varepsilon_0}\frac{n_i^2}{N_b}e^{\frac{\phi_0}{v_{kT}}}} \cdot x$$

or, $$\frac{1}{\sqrt{e^{\frac{\phi(x)-\phi_0}{v_{kT}}}}} = \cos\left[\sqrt{\frac{q}{2v_{kT}K_{si}\varepsilon_0}\frac{n_i^2}{N_b}e^{\frac{\phi_0}{v_{kT}}}} \cdot x\right] \tag{3.93}$$

or, $$\frac{1}{e^{\frac{\phi(x)-\phi_0}{v_{kT}}}} = \left\{\cos\left[\sqrt{\frac{q}{2v_{kT}K_{si}\varepsilon_0}\frac{n_i^2}{N_b}e^{\frac{\phi_0}{v_{kT}}}} \cdot x\right]\right\}^2$$

or, $$e^{\frac{\phi(x)-\phi_0}{v_{kT}}} = \left\{\cos\left[\sqrt{\frac{q}{2v_{kT}K_{si}\varepsilon_0}\frac{n_i^2}{N_b}e^{\frac{\phi_0}{v_{kT}}}}.x\right]\right\}^{-2} \tag{3.94}$$

Taking the natural log of both sides of Equation 3.94, we get the expression for $\phi(x)$ after simplification as

$$\phi(x) = \phi_0 - 2v_{kT}\ln\left[\cos\left(\sqrt{\frac{q}{2K_{si}\varepsilon_0 v_{kT}}\frac{n_i^2}{N_b}\exp\left(\frac{\phi_0}{v_{kT}}\right)} \cdot x\right)\right] \tag{3.95}$$

where:
ϕ_0 is the potential at the center of the body as shown in Figure 3.14

Equation 3.95 is valid for the entire range $-t_{si}/2 \le x \le t_{si}/2$. Note that the right-hand side of Equation 3.95 is always positive. From Equation 3.95, the surface potential $\phi_s \equiv \phi(x = \pm t_{si}/2)$ is given by

$$\phi_s\left(x = \frac{t_{si}}{2}\right) = \phi_0 - 2v_{kT}\ln\left[\cos\left(\sqrt{\frac{q}{2K_{si}\varepsilon_0 v_{kT}}\frac{n_i^2}{N_b}\exp\left(\frac{\phi_0}{v_{kT}}\right)} \cdot \frac{t_{si}}{2}\right)\right] \tag{3.96}$$

In order to express ϕ_s, in terms of the applied gate bias V_g and gate oxide thickness T_{ox} of DG-MOS capacitors, we use the boundary condition given by Gauss' law in Equation 3.27 at the Si/SiO$_2$ interface of the structure shown in Figure 3.14. Thus, at $x = \pm t_{si}/2$, we get

$$K_{ox}\varepsilon_0 E_{ox} = K_{si}\varepsilon_0 E_s \tag{3.97}$$

In Equation 3.97, all parameters have usual meanings as defined earlier. Equation 3.97 can be written as

$$K_{ox}\varepsilon_0 \frac{V_{ox}}{T_{ox}} = \pm K_{si}\varepsilon_0 \frac{d\phi}{dx}\Big|_{x=\pm t_{si}/2} \tag{3.98}$$

Again, from Equation 3.23 we get

$$V_{ox} = V_g - V_{fb} - \phi_s \tag{3.99}$$

Therefore, substituting for V_{ox} from Equation 3.99 into Equation 3.98, we can write the boundary condition as

$$K_{ox}\varepsilon_0 \frac{V_g - V_{fb} - \phi_s}{T_{ox}} = \pm K_{si}\varepsilon_0 \frac{d\phi}{dx}\Big|_{x=\pm t_{si}/2} \quad ; \quad \text{(interms of dielectric constant)}$$

$$\text{or,} \quad \varepsilon_{ox} \frac{V_g - V_{fb} - \phi_s}{T_{ox}} = \pm \varepsilon_{si} \frac{d\phi}{dx}\Big|_{x=\pm t_{si}/2} \quad ; \quad \text{(interms of permittivity)} \tag{3.100}$$

Now, substituting for $d\phi/dx$ at $x = \pm t_{si}/2$ from Equation 3.80 in Equation 3.100, we get

$$K_{ox}\varepsilon_0 \frac{V_g - V_{fb} - \phi_s}{T_{ox}} = K_{si}\varepsilon_0 \sqrt{\frac{2v_{kT}q}{K_{si}\varepsilon_0} \frac{n_i^2}{N_b} \left(e^{\frac{\phi_s}{v_{kT}}} - e^{\frac{\phi_0}{v_{kT}}} \right)} \tag{3.101}$$

It is to be noted that $V_{fb} = \Phi_{ms} - Q_0/C_{ox}$ [Equation 3.21] with Φ_{ms} as the workfunction difference between the gate electrode and intrinsic silicon. Since for an undoped or lightly doped substrate, $\phi_B \approx 0$, and for an ideal gate oxide, the interface charge, $Q_0 = 0$; then $V_{fb} \approx \Phi_{ms}$ [Equation 3.21]. Therefore,

$$V_{fb} \cong \Phi_{ms} = \begin{cases} -E_g/2q & \text{for MG CB aligned to the intrinsic silicon CB [Equation 3.10]} \\ +E_g/2q; & \text{for for MG CB aligned to the intrinsic silicon VB [Equation 3.12]} \\ 0; & \text{for midgap workfunction [Equation 3.15]} \end{cases}$$

For any given V_g, Equations 3.96 and 3.101 are coupled equations that can be solved for ϕ_s and ϕ_0.

In deriving Equation 3.48 for minority carrier concentration $n(x)$ in a p-type substrate and Equations 3.80 for $(d\phi/dx)$, 3.95 for $\phi(x)$, 3.96 for ϕ_s, and 3.101 boundary condition for solution of ϕ_s as a function of V_g, we have used bulk potential ϕ_B. However, for DG-MOS capacitors, typically, the substrate is undoped or lightly doped and $\phi_B \approx 0$ [Equation 3.43] resulting in $n_i^2/N_b = n_i$ [Equation 3.49]. Therefore, for DG-MOS capacitors on undoped or lightly doped p-type substrate, the relevant mathematical expressions can be written as

$$n(x) \cong n_i \exp\left(\frac{\phi(x)}{v_{kT}}\right) \tag{3.102}$$

$$\frac{d\phi(x)}{dx} = \pm \sqrt{\frac{2v_{kT}qn_i}{K_{si}\varepsilon_0} \left(e^{\frac{\phi(x)}{v_{kT}}} - e^{\frac{\phi_0}{v_{kT}}} \right)} \tag{3.103}$$

$$\phi(x) = \phi_0 - 2v_{kT} \ln\left[\cos\left(\sqrt{\frac{qn_i}{2K_{si}\varepsilon_0 v_{kT}}} \exp\left(\frac{\phi_0}{v_{kT}}\right) \cdot x\right)\right] \tag{3.104}$$

$$\phi_s\left(x = \frac{t_{si}}{2}\right) = \phi_0 - 2v_{kT} \ln\left[\cos\left(\sqrt{\frac{qn_i}{2K_{si}\varepsilon_0 v_{kT}}} \exp\left(\frac{\phi_0}{v_{kT}}\right) \cdot \frac{t_{si}}{2}\right)\right] \tag{3.105}$$

$$K_{ox}\varepsilon_0 \frac{V_g - V_{fb} - \phi_s}{T_{ox}} = K_{si}\varepsilon_0 \sqrt{\frac{2v_{kT}qn_i}{K_{si}\varepsilon_0}\left(e^{\frac{\phi_s}{v_{kT}}} - e^{\frac{\phi_0}{v_{kT}}}\right)} \tag{3.106}$$

Again, for any given V_g, Equations 3.105 and 3.106 are coupled equations that can be solved for Φ_s and ϕ_0.

Numerical solutions to the above equations have been reported [19]. The plots of electrostatic potential $\phi(x)$ and electron volume density $n(x)$ as a function of position x in the undoped silicon body for three different values of ϕ_0 are shown in Figures 3.15 for DG-MOS capacitors with $t_{si} = 20$ nm, $T_{ox} = 2$ nm, and $V_{fb} = 0$.

Figure 3.16 shows the plots from the solutions for ϕ_s and ϕ_0 as a function of V_g.

3.4.2.3 Threshold Voltage

In order to derive an expression for threshold voltage (V_{th}), let us compute inversion charge (Q_i) from the expression for the total charge in semiconductor given by Equation 3.63 and repeated below

$$Q_s = \pm\sqrt{2qK_{si}\varepsilon_0 n_i v_{kT}}\left[\left(e^{\frac{\phi_s}{v_{kT}}} - e^{\frac{\phi_0}{v_{kT}}}\right) \cdot e^{-\frac{\phi_B}{v_{kT}}} + \left(\frac{\phi_s - \phi_0}{v_{kT}}\right)e^{\frac{\phi_B}{v_{kT}}}\right]^{1/2} \tag{3.107}$$

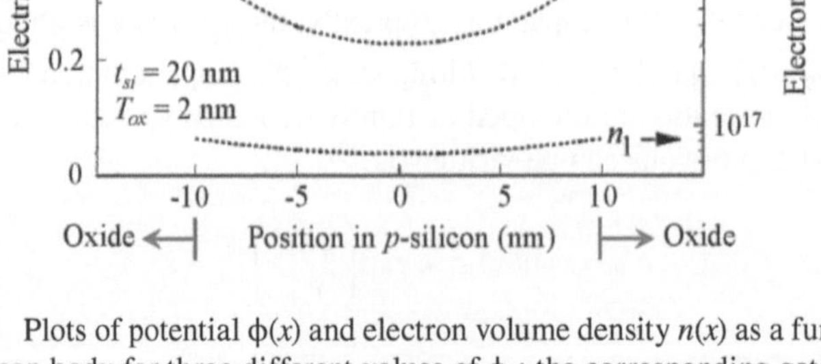

FIGURE 3.15 Plots of potential $\phi(x)$ and electron volume density $n(x)$ as a function of position in the silicon body for three different values of ϕ_0; the corresponding gate voltages are: $V_{g1} - V_{fb} = 0.412$ V for ϕ_1 and n_1, $V_{g2} - V_{fb} = 0.573$ V for ϕ_2 and n_2, and $V_{g3} - V_{fb} = 0.845$ V for ϕ_3 and n_3 [19].

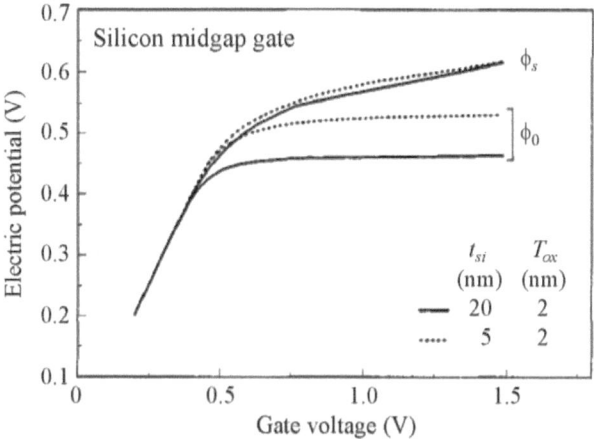

FIGURE 3.16 Plots showing the solutions of ϕ_s and ϕ_0 from the coupled Equations 3.105 and 3.106 versus V_g for two sets of values of t_{si} and T_{ox}; here $V_{fb} = 0$ assuming undoped or lightly doped body [19].

As discussed in Section 3.4.2.1, the first term within the square brackets of Equation 3.107 is due to minority carriers and represents the inversion charge, whereas the second term is due to the majority carrier dopant atom N_b and represents the bulk charge Q_b. Therefore, from Equation 3.107, the total bulk charge Q_b is given by

$$Q_b = \pm\sqrt{2qK_{si}\varepsilon_0 n_i \left(\phi_s - \phi_0\right)e^{\frac{\phi_B}{v_{kT}}}} \qquad (3.108)$$

We know that for an undoped or lightly doped substrate, $\phi_B \approx 0$ [Equation 3.43] and in the weak inversion regime, the bulk charge, $Q_b \ll Q_i$ in the semiconductor surface; therefore, from Equation 3.107, we can write the expression for Q_i as

$$Q_i \cong -2\sqrt{2qK_{si}\varepsilon_0 n_i v_{kT}}\left[\left(e^{\frac{\phi_s}{v_{kT}}} - e^{\frac{\phi_0}{v_{kT}}}\right)\right]^{1/2}$$

$$\text{or,} \quad Q_i = -\sqrt{8K_{si}\varepsilon_0 q n_i v_{kT}\left(e^{\frac{\phi_s}{v_{kT}}} - e^{\frac{\phi_0}{v_{kT}}}\right)} \qquad (3.109)$$

where:

    factor "2" arises from the two (top and bottom) Si/SiO$_2$ interfaces

Again, from Equation 3.28,

$$-C_{ox}V_{ox} = Q_s \cong Q_i$$

$$\therefore \quad V_{ox} = -\frac{Q_i}{C_{ox}} \qquad (3.110)$$

Then substituting for V_{ox} from Equation 3.110 into Equation 3.99 we can show

$$\phi_s \cong V_g - V_{fb} + \frac{Q_i}{C_{ox}} \tag{3.111}$$

Since for any value of V_g below the threshold voltage V_{th}, the mobile charge density Q_i is low, therefore, from Equation 3.109, we get $\phi_s \approx \phi_0$. Then, from Equation 3.111, we get

$$\phi_s \approx \phi_0 \approx V_g - V_{fb} \tag{3.112}$$

Thus, both ϕ_s and ϕ_0 closely follow V_g in the weak inversion regime and the entire substrate under the gate is inverted. This is called the *volume inversion* [20]. However, Equation 3.111 shows that as V_g increases, ϕ_s and the total Q_i increases exponentially with V_g at an inverse slope of 60 mV/decade as shown in Figure 3.17 [19].

When the CB of the undoped body is bent close to the silicon midgap MG CB or by $\approx E_g/2$, the terms due to minority carriers on the right-hand side of Equations 3.105 and 3.106 are no longer negligible. As the mobile charge near the silicon surfaces screens the gate field from the center of the silicon film, ϕ_s and ϕ_0 become de-coupled, i.e., there is no further volume inversion. This is illustrated by the case of ϕ_3 and n_3 in Figure 3.15, and also by the behavior of ϕ_s and ϕ_0 in Figure 3.16 for $V_g > 0.5$ V. Since the angle of the cosine function in Equation 3.105 cannot exceed $\pi/2$, ϕ_0 is pinned to an upper bound to $\phi_{0,sat}$, given by

$$\sqrt{\frac{qn_i}{2K_{si}\varepsilon_0 v_{kT}}\exp\left(\frac{\phi_{0,sat}}{v_{kT}}\right)} \cdot \frac{t_{si}}{2} = \frac{\pi}{2}$$

$$\text{or,} \quad \frac{qn_i}{2K_{si}\varepsilon_0 v_{kT}}\exp\left(\frac{\phi_{0,sat}}{v_{kT}}\right)\frac{t_{si}^2}{4} = \frac{\pi^2}{4} \tag{3.113}$$

$$\text{or,} \quad \exp\left(\frac{\phi_{0,sat}}{v_{kT}}\right) = \frac{2\pi^2 K_{si}\varepsilon_0 v_{kT}}{qn_i t_{si}^2}$$

After simplification of Equation 3.113, we can show

$$\phi_{0,sat} = v_{kT}\ln\left(\frac{2\pi^2 K_{si}\varepsilon_0 v_{kT}}{qn_i t_{si}^2}\right) \tag{3.114}$$

The expression for $\phi_{0,sat}$ derived in Equation 3.114 represents the saturation condition of ϕ_0 shown in Figure 3.16. On the other hand, ϕ_s continues to increase slowly as governed by Equation 3.106 and $\phi_s \gg \phi_0$, then neglecting the $\exp(\phi_0/v_{kT})$ term in the square root of Equation 3.106, we get

$$K_{ox}\varepsilon_0 \frac{V_g - V_{fb} - \phi_s}{T_{ox}} = \sqrt{2K_{si}\varepsilon_0 v_{kT}qn_i\left(e^{\frac{\phi_s}{v_{kT}}}\right)} \tag{3.115}$$

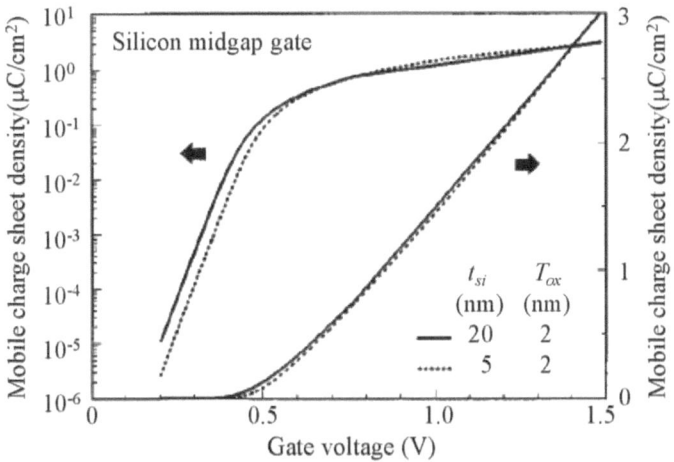

FIGURE 3.17 Sheet density of mobile charge Q_i given by Equation 3.109 in both logarithmic (left) and linear (right) scales versus gate voltage V_g for two sets of values of t_{si}; here, we assumed $V_{fb} = 0$ [19].

Let us define the threshold voltage V_{th} of a DG-MOS capacitor as the gate voltage V_g at which $\phi_0 = \phi_{0,sat}$. With this condition for V_{th}, it is evident from Equation 3.115 that the above-threshold behavior of MOS capacitors with $\phi_s > \phi_{0,sat}$ is independent of silicon thickness t_{si}. This is also observed from the similarity between the two curves of different t_{si} in Figure 3.17.

In order to derive an expression for V_{th}, we determine the expression for ϕ_s from Equation 3.115 at the threshold condition: $\phi_0 = \phi_{0,sat} \ll \phi_s$. Then from Equation 3.115, we get

$$C_{ox}\left(V_g - V_{fb} - \phi_s\right) = \sqrt{2K_{si}\varepsilon_0 v_{kT} q n_i}\, e^{\frac{\phi_s}{2v_{kT}}} \tag{3.116}$$

After simplification of Equation 3.116, we can write the expression for ϕ_s at threshold condition as

$$\phi_s = 2v_{kT} \ln\left[\frac{C_{ox}\left(V_g - V_{fb} - \phi_s\right)}{\sqrt{2K_{si}\varepsilon_0 v_{kT} q n_i}}\right] \tag{3.117}$$

Again, for $V_g < V_{th}$, we get from Equation 3.111

$$C_{ox}\left(V_g - V_{fb} - \phi_s\right) \approx -\frac{Q_i}{2}$$

$$\text{or,} \quad V_g = -\frac{Q_i}{2C_{ox}} + \left(V_{fb} + \phi_s\right) \tag{3.118}$$

where:
factor "2" arises from the two surfaces (Si/SiO$_2$ interfaces)

In Equation 3.118, the value of $(Q_i/2C_{ox}) < (V_{fb} + \phi_s)$ in the subthreshold region is defined by $V_g \leq V_{th}$; then in the subthreshold regime we get from Equation 3.118

$$V_g \approx \left(V_{fb} + \phi_s\right) \equiv V_{th} \tag{3.119}$$

Now, substituting the expression for ϕ_s from Equation 3.117 into Equation 3.119, we get

$$V_{th} = V_{fb} + 2v_{kT} \ln\left[\frac{C_{ox}V_{gt}}{\sqrt{2K_{si}\varepsilon_0 v_{kT} q n_i}}\right] \tag{3.120}$$

where:
$V_{gt} = V_g - (V_{fb} + \phi_s)$ is the gate overdrive voltage for the region of operation

Again, from Equation 2.14, we get

$$n_i^2 = N_c N_v \exp\left(-E_g / kT\right) \tag{3.121}$$

where:
N_c is the effective density of states of conduction band
N_v is the effective density of states of valance band

Then using Equation 3.121, V_{th} of a DG-capacitor given by Equation 3.120 can also be calculated from the following expression

$$V_{th} = V_{fb} + \frac{E_g}{2q} + 2v_{kT} \ln\left[\frac{C_{ox}V_{gt}}{\sqrt{2K_{si}\varepsilon_0 v_{kT} q \sqrt{N_c N_v}}}\right] \tag{3.122}$$

In V_{th} Equation 3.120 and Equation 3.122, $V_{fb} \cong \Phi_{ms}$, for an ideal defect-free gate oxide.

The value of V_{th} is insensitive to the exact choice of V_{gt}. The reported data show that V_{th} is insensitive to V_{gt}, T_{ox}, and ambient temperature [19].

3.4.2.4   Surface Potential Function

In Section 3.4.2.2, we solved coupled Equation 3.105 and Equation 3.106 for surface potential ϕ_s and center potential ϕ_0. In this section, we will derive a surface potential function using a transform variable defined as $\beta = f(\phi_s)$ using the generalized solution of Poisson's equation derived in Section 3.4.2.2.

In order to derive an expression for β, we use the generalized solution of Poisson's equation given by Equations 3.95 and 3.96 to obtain the relation between the gate voltage V_g and surface potential ϕ_s

$$\phi(x) = \phi_0 - 2v_{kT} \ln\left[\cos\left(\sqrt{\frac{q}{2K_{si}\varepsilon_0 v_{kT}} \frac{n_i^2}{N_b} \exp\left(\frac{\phi_0}{v_{kT}}\right)} \cdot x\right)\right] \tag{3.123}$$

$$\phi\left(x = \frac{t_{si}}{2}\right) = \phi_0 - 2v_{kT} \ln\left[\cos\left(\sqrt{\frac{q}{2K_{si}\varepsilon_0 v_{kT}} \frac{n_i^2}{N_b} \exp\left(\frac{\phi_0}{v_{kT}}\right)} \cdot \frac{t_{si}}{2}\right)\right] \quad (3.124)$$

Thus, to calculate the surface potential through a single continuous equation, we define the transformation variable β as the argument of the cosine function in $\phi(x = t_{si}/2)$ in Equation 3.124 so that [21]

$$\beta = \sqrt{\frac{q}{2K_{si}\varepsilon_0 v_{kT}} \frac{n_i^2}{N_b} \exp\left(\frac{\phi_0}{v_{kT}}\right)} \cdot \frac{t_{si}}{2}$$

$$\text{or,} \quad \frac{2\beta}{t_{si}} = \sqrt{\frac{q}{2K_{si}\varepsilon_0 v_{kT}} \frac{n_i^2}{N_b} \exp\left(\frac{\phi_0}{v_{kT}}\right)} \quad (3.125)$$

Now, we define [18]

$$a \equiv \sqrt{\frac{q}{2K_{si}\varepsilon_0 v_{kT}} \frac{n_i^2}{N_b}}$$

$$b \equiv \sqrt{\frac{q}{2K_{si}\varepsilon_0 v_{kT}} \frac{n_i^2}{N_b} \exp\left(\frac{\phi_0}{v_{kT}}\right)} = \frac{2\beta}{t_{si}} \quad (3.126)$$

Then, from Equation 3.126, we can show

$$a \exp\left(\frac{\phi_0}{2v_{kT}}\right) = b = \frac{2\beta}{t_{si}}$$

$$\text{or,} \quad \exp\left(\frac{\phi_0}{2v_{kT}}\right) = \frac{b}{a} = \frac{1}{a}\frac{2\beta}{t_{si}} \quad (3.127)$$

Then, we get from Equation 3.127 as

$$\phi_0 = 2v_{kT} \ln\left(\frac{b}{a}\right) = 2v_{kT} \ln\left(\frac{1}{a}\frac{2\beta}{t_{si}}\right) \quad (3.128)$$

Now, substituting for a and b from Equation 3.127 and ϕ_0 from Equation 3.128 into Equation 3.123 we can write for $\phi(x)$ as

$$\phi(x) = 2v_{kT} \ln\left(\frac{1}{a}\frac{2\beta}{t_{si}}\right) - 2v_{kT} \ln\left[\cos\left(\frac{2\beta}{t_{si}} \cdot x\right)\right] \quad (3.129)$$

Then, we can further simplify Equation 3.129 as follows

$$\phi(x) = -2v_{kT}\left\{-\ln\left(\frac{1}{a}\frac{2\beta}{t_{si}}\right)+\ln\left[\cos\left(\frac{2\beta}{t_{si}}\cdot x\right)\right]\right\}$$

$$= -2v_{kT}\ln\left\{a\frac{t_{si}}{2\beta}\cos\left(\frac{2\beta}{t_{si}}\cdot x\right)\right\}$$

(3.130)

Now, substituting the expressions for a from Equation 3.126 into Equation 3.130, we get

$$\phi(x) = -2v_{kT}\ln\left\{\frac{t_{si}}{2\beta}\sqrt{\frac{q}{2K_{si}\varepsilon_0 v_{kT}}\frac{n_i^2}{N_b}}\cos\left(\frac{2\beta}{t_{si}}x\right)\right\}$$

(3.131)

Then rearranging Equation 3.131, we can show

$$\phi(x) = -2v_{kT}\left[-\ln\beta-\ln\left(\frac{2}{t_{si}}\sqrt{\frac{2K_{si}\varepsilon_0 v_{kT}}{q}\frac{N_b}{n_i^2}}\right)+\ln\left\{\cos\left(\frac{2\beta}{t_{si}}x\right)\right\}\right]$$

(3.132)

From Equation 3.132, we can show that the surface potential, at $x = \pm t_{si}/2$ (Si/SiO$_2$ interface)

$$\phi\left(x=\pm\frac{t_{si}}{2}\right) = -2v_{kT}\left[-\ln\beta-\ln\left(\frac{2}{t_{si}}\sqrt{\frac{2K_{si}\varepsilon_0 v_{kT}}{q}\frac{N_b}{n_i^2}}\right)+\ln\left(\cos\beta\right)\right]$$

(3.133)

And, (at $x=t_{si}/2$)

$$\frac{d\phi(x)}{dx} = 2v_{kT}\frac{2}{t_{si}}\beta\tan\beta$$

(3.134)

In order to derive an expression for β relating to the gate voltage V_g and surface potential ϕ_s, we use the boundary condition given by Equation 3.100 as

$$\varepsilon_{ox}\frac{V_g - V_{fb} - \phi(x=\pm t_{si}/2)}{T_{ox}} = \pm\varepsilon_{si}\left.\frac{d\phi}{dx}\right|_{x=\pm t_{si}/2}$$

(3.135)

Then substituting for $\phi(x = \pm t_{si}/2)$ from Equation 3.133 and $d\phi(x = \pm t_{si}/2)/dx$ from Equation 3.134 into Equation 3.135, we get

$$\frac{\varepsilon_{ox}}{T_{ox}}\left[V_g - V_{fb} + 2v_{kT}\left\{-\ln\beta-\ln\left(\frac{2}{t_{si}}\sqrt{\frac{2K_{si}\varepsilon_0 v_{kT}}{q}\frac{N_b}{n_i^2}}\right)+\ln\left(\cos\beta\right)\right\}\right] = K_{si}\varepsilon_0\left(2v_{kT}\frac{2}{t_{si}}\beta\tan\beta\right)$$

$$\text{or, } \frac{V_g - V_{fb}}{2v_{kT}} - \ln\beta - \ln\left(\frac{2}{t_{si}}\sqrt{\frac{2K_{si}\varepsilon_0 v_{kT}}{q}\frac{N_b}{n_i^2}}\right) + \ln\left(\cos\beta\right) = \frac{2K_{si}\varepsilon_0 T_{ox}}{K_{ox}\varepsilon_0 t_{si}}\beta\tan\beta$$

(3.136)

After simplification of Equation 3.136, we can show

$$\frac{V_g - V_{fb}}{2v_{kT}} - \ln\left(\frac{2}{t_{si}}\sqrt{\frac{2K_{si}\varepsilon_0 v_{kT}}{q}\frac{N_b}{n_i^2}}\right) = \ln\beta - \ln\left(\cos\beta\right) + \frac{2K_{si}\varepsilon_0 T_{ox}}{K_{ox}\varepsilon_0 t_{si}}\beta\tan\beta \quad (3.137)$$

Thus, through a change of variable, the expression for the surface potential ϕ_s can be represented by the function

$$f(\beta) = \ln\beta - \ln\left(\cos\beta\right) - \frac{V_g - V_{fb}}{2v_{kT}} + \ln\left(\frac{2}{t_{si}}\sqrt{\frac{2K_{si}\varepsilon_0 v_{kT}}{q}\frac{N_b}{n_i^2}}\right) + \frac{2K_{si}\varepsilon_0 T_{ox}}{K_{ox}\varepsilon_0 t_{si}}\beta\tan\beta = 0$$

$$(3.138)$$

Equation 3.138 can be solved for β at a given value of V_g to determine the surface potential and determine the characteristics of DG-MOS capacitors. We have shown that for an undoped or lightly doped silicon substrate $n_i^2 / N_b \approx n_i$, therefore, for an undoped or lightly doped substrate, Equation 3.138 can be expressed as

$$f(\beta) = \ln\beta - \ln\left(\cos\beta\right) - \frac{V_g - V_{fb}}{2v_{kT}} + \ln\left(\frac{2}{t_{si}}\sqrt{\frac{2K_{si}\varepsilon_0 v_{kT}}{qn_i}}\right) + \frac{2K_{si}\varepsilon_0 T_{ox}}{K_{ox}\varepsilon_0 t_{si}}\beta\tan\beta = 0$$

$$(3.139)$$

Equation 3.138 and Equation 3.139 are used to determine the FinFET device performance with appropriate modifications for the lateral electric field due to applied drain voltage as discussed in Chapter 10 [8].

3.4.2.5 Unified Expression for Inversion Charge Density

In Section 3.4.2.1, we derived the expression [Equation 3.63] for the total induced charge in a p-type semiconductor substrate of multigate MOS capacitors given by

$$Q_s = \pm\sqrt{2qK_{si}\varepsilon_0 n_i v_{kT}}\left[\left(e^{\frac{\phi_s}{v_{kT}}} - e^{\frac{\phi_0}{v_{kT}}}\right)\cdot e^{-\frac{\phi_B}{v_{kT}}} + \left(\frac{\phi_s - \phi_0}{v_{kT}}\right)e^{\frac{\phi_B}{v_{kT}}}\right]^{1/2} \quad (3.140)$$

As discussed in Section 3.4.2.1, on the right-hand side of Equation 3.140, the first term within the square brackets is due to the minority carrier electrons in the p-type substrate, whereas the second term is due to the majority carrier bulk doping concentration N_a. For a *lightly doped body*, the bulk charge $Q_b \ll Q_i$; therefore, neglecting the bulk charge term in Equation 3.140, we can express the inversion charge density in a p-type substrate of multigate MOS capacitor as

$$Q_i = \sqrt{2qK_{si}\varepsilon_0 n_i v_{kT}}\left[\left(e^{\frac{\phi_s}{v_{kT}}} - e^{\frac{\phi_0}{v_{kT}}}\right)\cdot e^{-\frac{\phi_B}{v_{kT}}}\right]^{1/2} \quad (3.141)$$

Equation 3.141 can be further simplified as

$$Q_i = \sqrt{2qK_{si}\varepsilon_0 n_i v_{kT}}\, e^{\frac{\phi_s-\phi_B}{2v_{kT}}} \sqrt{1-e^{\frac{\phi_0-\phi_s}{v_{kT}}}} \qquad (3.142)$$

In *strong inversion* $\phi_s \gg \phi_0$, therefore, $\sqrt{1-e^{\frac{\phi_0-\phi_s}{v_{kT}}}}$ approaches 1.

In *weak inversion*, we can simplify the term $\sqrt{1-e^{\frac{\phi_0-\phi_s}{v_{kT}}}}$, assuming linear potential profile from $x = 0$ (center point) to $x = \pm t_{si}/2$ (surface). Then if E_{avg} is the average electric field in the region between $x = t_{si}/2$ to the center point at $x = 0$, then using Gauss' law [Equation 3.68], we can write

$$E_{avg} = -\frac{d\phi}{dx} = -\frac{Q_i}{K_{si}\varepsilon_0} \qquad (3.143)$$

If we assume that potential varies linearly from center potential ϕ_0 to the surface potential ϕ_s, then Equation 3.143 can be expressed as

$$\frac{d\phi}{dx} = \frac{\phi_s-\phi_0}{t_{si}/2} = \frac{Q_i}{K_{si}\varepsilon_0} \qquad (3.144)$$

Thus, the inversion charge is given by

$$\phi_s - \phi_0 = \frac{Q_i}{2\left(K_{si}\varepsilon_0/t_{si}\right)} = \frac{Q_i}{2C_{si}} \qquad (3.145)$$

where:
$C_{si} = K_{si}\varepsilon_0/t_{si}$; is the depletion capacitance of silicon body

Now, substituting Equation 3.145 into Equation 3.142, we get

$$Q_i = \sqrt{2qK_{si}\varepsilon_0 n_i v_{kT}}\, e^{\frac{\phi_s-\phi_B}{2v_{kT}}} \sqrt{1-e^{\frac{-Q_i}{2C_{si}v_{kT}}}} \qquad (3.146)$$

We can further simplify Equation 3.146 by Taylor series expansion of the term $\exp\left[-Q_i/(2C_{si}v_{kT})\right]$ to keep the first term and neglect the higher order terms since $Q_i/(2C_{si}v_{kT}) < 1$ in the weak inversion regime. Then, we get

$$\sqrt{1-e^{\frac{-Q_i}{2C_{si}v_{kT}}}} = \sqrt{1-\frac{1}{\exp\left[Q_i/(2C_{si}v_{kT})\right]}} = \sqrt{1-\frac{1}{1+Q_i/(2C_{si}v_{kT})+\cdots}} = \sqrt{\frac{Q_i}{Q_i+2C_{si}v_{kT}}} \qquad (3.147)$$

Therefore, using Equation 3.147 we can write the expression for the inversion charge density $Q_i(LD)$ for a lightly doped substrate as

$$Q_i(LD) = \sqrt{2qK_{si}\varepsilon_0 n_i v_{kT}} \, e^{\frac{\phi_s - \phi_B}{2v_{kT}}} \sqrt{\frac{Q_i}{Q_i + 2C_{si}v_{kT}}} \tag{3.148}$$

Equation 3.148 is an implicit equation in Q_i and is solved iteratively with coupled Equations 3.96 and 3.101 to compute V_g dependence of inversion charge. It is reported that the numerical device simulation data show that a modified form of Equation 3.148 fits the real data on MOS capacitors, as given below [23]

$$Q_i(LD) = \sqrt{2qK_{si}\varepsilon_0 n_i v_{kT}} \, e^{\frac{\phi_s - \phi_B}{2v_{kT}}} \sqrt{\frac{Q_i}{Q_i + 5C_{si}v_{kT}}} \tag{3.149}$$

Now, let us derive an expression for Q_i in *heavily doped* DG-MOS capacitor systems. We know that the second term on the right-hand side of Equation 3.140 represents the bulk charge Q_b and is given by Equation 3.108 as

$$Q_b = \pm\sqrt{2qK_{si}\varepsilon_0 n_i \left(\phi_s - \phi_0\right) e^{\frac{\phi_B}{v_{kT}}}} \tag{3.150}$$

In the *strong inversion* regime of heavily doped DG-MOS capacitors, ϕ_s term is large and $Q_i \gg Q_b$. However, in the *weak inversion* regime $Q_i \ll Q_b$ and therefore, $(\phi_s - \phi_0)$ is a small perturbation in the total surface potential ϕ_s and $\phi_s \gg \phi_0$. Using this assumption, we can simplify Equation 3.140 for a heavily doped body as

$$Q_s = \pm\sqrt{2qK_{si}\varepsilon_0 n_i v_{kT} \left(e^{\frac{\phi_s}{v_{kT}}}\right) \cdot e^{-\frac{\phi_B}{v_{kT}}} + 2qK_{si}\varepsilon_0 \phi_s n_i e^{\frac{\phi_B}{v_{kT}}}} \tag{3.151}$$

And

$$Q_b = \begin{cases} -\sqrt{2qK_{si}\varepsilon_0 v_{kT}\phi_s \left(n_i e^{\frac{\phi_B}{v_{kT}}}\right)}; & \text{in terms of } \phi_B \text{ and } n_i \\ -\sqrt{2qK_{si}\varepsilon_0 v_{kT}N_a\phi_s}; & \text{in terms of } N_a \end{cases} \tag{3.152}$$

In the second expression of Equation 3.152, we have used Equation 3.43 for $N_a = n_i \exp\left(\dfrac{\phi_B}{v_{kT}}\right)$. Then we can express Equation 3.151 as

$$Q_s = \pm\sqrt{2qK_{si}\varepsilon_0 n_i v_{kT} \left(e^{\frac{\phi_s}{v_{kT}}}\right) \cdot e^{-\frac{\phi_B}{v_{kT}}} + Q_b^2} \tag{3.153}$$

We know that the total charge in the semiconductor is given by $Q_s = Q_i + Q_b$; therefore, Equation 3.153 can be expressed as

$$Q_i + Q_b = \pm \sqrt{2qK_{si}\varepsilon_0 n_i v_{kT} \left(e^{\frac{\phi_s}{v_{kT}}} \right) \cdot e^{-\frac{\phi_B}{v_{kT}}} + Q_b^2} \qquad (3.154)$$

After taking the square to both sides of Equation 3.154, we get

$$\left(Q_i + Q_b\right)^2 = 2qK_{si}\varepsilon_0 n_i v_{kT} \left(e^{\frac{\phi_s}{v_{kT}}} \right) \cdot e^{-\frac{\phi_B}{v_{kT}}} + Q_b^2$$

$$\text{or,} \quad Q_i^2 + 2Q_iQ_b + Q_b^2 = 2qK_{si}\varepsilon_0 n_i v_{kT} \left(e^{\frac{\phi_s}{v_{kT}}} \right) \cdot e^{-\frac{\phi_B}{v_{kT}}} + Q_b^2$$

$$(3.155)$$

$$\text{or,} \quad Q_i\left(Q_i + 2Q_b\right) = 2qK_{si}\varepsilon_0 n_i v_{kT} \left(e^{\frac{\phi_s}{v_{kT}}} \right) \cdot e^{-\frac{\phi_B}{v_{kT}}}$$

$$\text{or,} \quad Q_i = 2qK_{si}\varepsilon_0 n_i v_{kT} \left(e^{\frac{\phi_s}{v_{kT}}} \right) \cdot e^{-\frac{\phi_B}{v_{kT}}} \cdot \frac{1}{\left(Q_i + 2Q_b\right)}$$

Now, multiplying both sides of Equation 3.155 by Q_i, we get the inversion charge density after simplification as

$$Q_i(HD) \approx \sqrt{2qn_iK_{si}\varepsilon_0 v_{kT}} \cdot \exp\left(\frac{\left(\phi_s - \phi_B\right)}{2v_{kT}} \right) \cdot \sqrt{\frac{Q_i}{Q_i + 2Q_b}}, \qquad (3.156)$$

From the similarities of charge expressions in Equation 3.149 for $Q_i(LD)$ and 3.156 for $Q_i(HD)$, a unified expression can be used to calculate the inversion charge density for a wide range of devices as a function of Q_b and is given by

$$Q_i = \sqrt{2qn_iK_{si}\varepsilon_0 v_{kT}} \cdot \exp\left(\frac{\left(\phi_s - \phi_B\right)}{2v_{kT}} \right) \sqrt{\frac{Q_i}{Q_i + Q_0}}, \qquad (3.157)$$

where: $Q_0 = 2Q_b + 5C_{si}v_{kT}$, with $C_{si} = K_{si}\varepsilon_0/t_{si}$; Q_b is the fixed depletion charge and for an ultrathin-body is given by qN_bt_{si}. It is reported that the unified charge density model agrees very well with inversion charge density calculated using exact equation for a wide range of body doping concentration [22,23].

3.5 QUANTUM MECHANICAL EFFECT

Generally, inversion-carriers must be treated quantum-mechanically as a 2D gas [1,8]. According to Quantum Mechanical (QM) model, the inversion layer carriers occupy discrete energy bands as shown in Figure 3.18(a) for a single gate structure and the peak distribution is about 1–3 nm away from the surface as shown in Figure 3.18(b). Thus, near the silicon surface, the inversion layer charges are confined to a potential well bounded by (a) oxide barrier height at the Si/SiO₂ interface of about 3.1 eV; and (b) barrier height of about $E_g/2$ due to bend silicon conduction band at the surface as shown in Figure 3.18(a).

Due to QM confinement of inversion layer electrons in the *p-type* silicon surface, the *electron* energy levels are grouped in discrete sub-bands of energy E_j; where $j = 1, 2, 3, ...$ quantized states as shown in Figure 3.18(a). Each E_j corresponds to a quantized level for electron motion in the normal direction. The net result of QM effect is that the inversion layer density peaks below the SiO₂/Si interface with about zero value at the surface contrary to the classical inversion carrier distribution as shown in Figure 3.18(b) for a single-gate MOS capacitor structure. Figure 3.18(c) shows the overall inversion carrier distribution in silicon body of a DG-MOS capacitor (solid line) along with the inversion carrier profiles for each gate (broken lines). Therefore, for accurate computation of inversion carrier distribution at the silicon surface, we have to solve both Schrödinger and Poisson equations self-consistently with boundary conditions: $\phi(x) = 0$ for $x < 0$ in the oxide; and $\phi(x) = 0$ at $x = \infty$ deep into the silicon substrate.

As observed from Figure 3.18(b) and (c), the silicon surface is depleted of mobile carriers due to inversion layer quantization. This depletion region in silicon can be considered as an insulating layer of silicon increasing the effective gate oxide thickness. This increase in the effective gate thickness (EOT) is given by

$$\Delta T_{ox} = \frac{\varepsilon_{ox}}{\varepsilon_{si}} \Delta z \qquad (3.158)$$

where:

Δz is the centroid of inversion charge, that is, the peak of the charge away from the surface

Since the peak of the inversion charge is away from the surface due to QM effect, a higher V_g over-drive is required to produce the same level of inversion charge

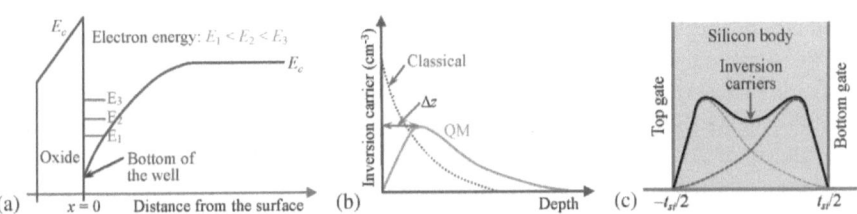

FIGURE 3.18 Inversion layer quantization in MOS systems: (a) minority carrier electron in a potential well of a single-gate MOS capacitor system on a *p*-type silicon substrate; the potential well is bounded by potential barrier at the Si/SiO₂ interface and conduction band bending; (b) typical minority carrier electron concentration in silicon surface as a function of silicon depth for classical and QM model for a single-gate structure; here, Δz is the centroid of inversion charge; t_{si} is the thickness of silicon body; (c) qualitative distribution of inversion carriers for a symmetric double-gate structure due to QM effect, solid lines inversion carrier profile of DG-MOS capacitors, and broken lines show the inversion carrier profiles due to each gate.

density predicted by the classical theory. In other words, QM effect can be considered to reduce the net inversion charge density. Thus, the inversion layer quantization can be modeled as *bandgap widening* effect due to an increase in the effective bandgap energy E_g by an amount ΔE_g [24]. Then from Equation 2.14 relating E_g and the intrinsic carrier concentration, we can show that the intrinsic carrier concentration n_i^{QM} due to QM effect is given by

$$n_i^{QM} = n_i^{CL} \exp\left(-\frac{\Delta E_g}{2kT}\right) \tag{3.159}$$

where:

$\Delta E_g = \left(E_g^{QM} - E_g^{CL}\right)$ is the increase in the apparent value of E_g due to QM effect; here, E_g^{QM} is the energy gap due to QM effect; and E_g^{CL} and n_i^{CL} are the energy gap and intrinsic carrier concentration, respectively, without the QM effect denoting the classical expression.

Equation 3.159 shows that the inversion layer quantization decreases the intrinsic concentration compared to the classical value. We know from Equations 3.102 and 3.157 that the inversion carrier concentration as well as charge Q_i is proportional to n_i. Thus, Q_i decreases due to QM effect. This decrease in Q_i due to QM effect has severe consequences on MOS transistor device performance as we will discuss in Chapters 6 and 10.

3.6 SUMMARY

This chapter presented the basic structure and operation of multigate MOS capacitor systems to build the foundation for developing FinFET device theory and performance. We have discussed the basic MOS structure by considering the energy band model of metal, oxide, and semiconductors. The basic operation of an MOS capacitor system is discussed at equilibrium and under biasing conditions. The important parameter of the MOS structure is the flat-band voltage V_{fb}. The significance of V_{fb} and the workfunction difference between the metal and semiconductor for MOS operation is clearly discussed using the energy band diagram. Analytical expressions for multigate MOS capacitor systems are derived to discuss the accumulation, depletion, and inversion mode operations of MOS capacitor structures. A unified surface potential function is developed to analyze the characteristics of multigate MOS capacitors. Finally, a unified inversion charge expression is presented to account for the substrate doping effect in multigate MOS capacitors.

REFERENCES

1. J.-P. Colinge (ed.), *FinFETs and Other Multi-Gate Transistors*, Springer, New York, 2008.
2. E.H. Nicollian and J.R. Brews, *MOS (Metal Oxide Semiconductor) Physics and Technology*, John Wiley & Sons, New York, 1982.
3. N. Collaert (ed.), *CMOS Nanoelectronics*, Pan Stanford Publishing, Singapore, 2013.

4. J.D. Plummer, M.D. Deal, and P.B. Griffin, *Silicon VLSI Technology: Fundamentals, Practice and Modeling*, Prentice Hall, Upper Saddle River, NJ, 2000.

5. S.M. Sze, *Semiconductor Physics and Technology*, John Wiley & Sons, New York, 1982.

6. N.D. Arora, *MOSFET Models for VLSI Circuit Simulation: Theory and Practice*, Springer – Verlag, Vienna, 1993.

7. R.S. Muller and T.I. Kamins with M. Chan, *Device Electronics for Integrated Circuits*, 3rd edition, John Wiley & Sons, New York, 2003.

8. S.K. Saha, *Compact Models for Integrated Circuit Design: Conventional Transistors and Beyond*, CRC Press, Taylor & Francis Group, Boca Raton, FL, 2015.

9. T.P. Chow and A.J. Steckl, "Refractory metal silicides: Thin film properties and processing technology," *IEEE Transactions on Electron Devices*, 30(11), pp. 1480–1497, November 1983.

10. Y. Tsividis, *Operation and Modeling of the MOS Transistors*, 2nd edition, Oxford University Press, New York, 1999.

11. P. Ranade, H. Takeuchi, T.-J. King, and C. Hu, "Work function engineering of molybdenum gate electrodes by nitrogen implantation," *Electrochemical and Solid-State Letters*, 4(11), pp. G85–G87, 2001.

12. N. Lifshitz, "Dependence of the work function difference between the polysilicon gate and silicon substrate on the doping level in polysilicon," *IEEE Transactions on Electron Devices*, 32(3), pp. 617–621, 1985.

13. T. Kamins, *Polycrystalline Silicon for IC Application*, 2nd edition, Kluwer Academic Publisher, Boston, MA, 1998.

14. B.E. Deal, "Standardized terminology for oxide charge associated with thermally oxidized silicon," *IEEE Transactions on Electron Devices*, 27(3), pp. 606–608, 1980.

15. D.K. Schroder, *Semiconductor Materials and Device Characterization*, John Wiley & Sons, New York, 1990.

16. B.G. Streetman and S. Banerjee, *Solid State Electronic Devices*, 7th edition, Prentice Hall, Inc., Englewood Cliffs, NJ, 2014.

17. J. He, X. Xi, C.-H. Lin, M. Chan, A. Niknejad, and C. Hu, "A non-charge-sheet analytical theory for undoped symmetric double-gate MOSFETs from the exact solution of Poisson's equation using SPP approach," *Proceedings of the NSTI Nanotech*, 2, pp. 124–127, 2004.

18. N. Pandey and Y.S. Chauhan, Private communications, December, 2018.

19. Y.Taur, "An analytical solution to a double-gate MOSFET with undoped body," *IEEE Electron Device Letters*, 21(5), pp. 245–247, 2000.

20. F. Balestra, S. Cristoloveanu, M. Benachir, J. Brini, and T. Elewa, "Double-gate silicon-on-insulator transistor with volume inversion: A new device with greatly enhanced performance," *IEEE Electron Devices Letters*, 8(9), pp. 410–412, 1987.

21. Y.Taur, X. Liang, W. Wang, and H. Lu, "A continuous, analytical drain-current model for DG MOSFETs," *IEEE Electron Device Letters*, 25(2), pp. 107–109, February 2004.

22. M.V. Dunga, C.-H. Lin, X. Xi, D.D. Lu, A.M. Niknejad, and C. Hu, "Modeling advanced FET technology in a compact model," *IEEE Transactions on Electron Devices*, 53(9), pp. 1971–1978, 2006.

23. M.V. Dunga, "Nanoscale CMOS modeling," Ph.D. dissertation, Electrical Engineering and Computer, Science, University of California Berkeley, Berkeley, CA, 2008.

24. M.J. van Dort, P.H. Woerlee, and A.J. Walker, "A simple model for quantisation effects in heavily-doped silicon MOSFETs at inversion conditions," *Solid-State Electronics*, 37(3), pp. 411–414, 1994.

4 Overview of FinFET Device Technology

4.1 INTRODUCTION

In Chapter 3 we discussed that in a multiple-gate or multigate metal-oxide-semiconductor (MOS) capacitor system, an inversion condition can be reached by a certain applied bias to the gates forming minority carrier concentration (e.g., electron) in the majority carrier (e.g., *p*-type) thin silicon body. Under this inversion condition, the thermally generated minority carriers form the inversion charge within the volume of the ultrathin silicon body. However, it is difficult to sustain this minority carrier charge in an undoped or a lightly doped majority carrier body from the thermal generation of carriers without a steady source of carrier supply. Therefore, a heavily doped minority carrier region (e.g., *n*+ region in a *p*-type body), called the *source*, is added to one end of the silicon body of the multigate MOS structure as a terminal for a steady supply of minority carriers at the inversion condition. And, another heavily doped region with the same doping-type as the source region, called the *drain*, is added as a terminal at the other end of the multigate MOS capacitor body to form a multiple-gate MOS field-effect transistor (FET) or MOSFET. These source and drain terminals contact the two opposite ends of the inverted silicon body so that a potential difference can be applied across the body and cause a current flow in the multiple-gate MOSFET structure. Such a multiple-gate MOSFET device with ultrathin vertical silicon body, called the "fin," on silicon pedestal, sidewall gate stack, and a source and a drain at the two ends of the gate length is called the "fin field-effect transistor" or "FinFET." Thus, a FinFET includes an ultrathin vertical silicon fin on a silicon substrate with a thin insulating layer such as SiO_2 grown on the sidewalls, a conducting metal layer, called the *gate electrode* deposited on the top of the gate oxide, and heavily doped source and drain regions formed from one end of the fin to the nearest gate edge and from the far edge of the gate to the far end of the fin, respectively. In reality, the gate can be placed on two, three, or four sides of the channel or wrapped around the channel as discussed in Chapter 1 [1,2].

Typically, a FinFET can be designed on a bulk-silicon substrate or silicon-on-insulator (SOI) substrate as described in Section 1.4 [1–3]. A bulk-FinFET is a four-terminal device with *gate*, *source*, *drain*, and *substrate or body*, whereas an SOI-FinFET is a three-terminal device with *gate*, *source*, and *drain* with floating *body*. Thus, bulk-FinFETs are more familiar to integrated circuit (IC) design engineers and the fabrication steps of the devices are compatible with those of the conventional planar complementary metal-oxide-semiconductor (CMOS) devices fabricated on bulk-silicon wafers [4]. The body terminal in bulk-FinFETs offer more flexibility of device operation in very large scale integrated (VLSI) circuits and systems. Bulk-FinFETs eliminate the problems associated with SOI-FinFETs such as expensive wafer cost, high defect density, floating body effect, and poor heat dissipation [5]. The heat generated in the channel can be transferred to the substrate through the fin body that is connected to the substrate of the bulk-FinFETs. In addition, both the bulk and SOI FinFETs offer the same scalability while bulk-FinFETs have better heat dissipation characteristics [5].

A FinFET device is symmetrical and cannot be distinguished without the applied bias. In bulk-FinFET device architecture, shallow trench isolations (STI) are used to isolate various devices fabricated on the same substrate. In advanced VLSI circuits, n-channel FinFETs (p-type fin with n^+ source-drain) and p-channel FinFETs (n-type fin with p^+ source-drain) are fabricated together and is referred to as non-planar CMOS technology. In this chapter, an overview of the fabrication processes of FinFET devices in non-planar CMOS technology is presented to appreciate the operation of FinFET devices discussed in Chapters 5–10 of this book.

4.2 FinFET MANUFACTURING TECHNOLOGY

Figure 4.1 shows three-dimensional (3D) cross-sections of ideal FinFET structures with gate oxide on three sides (sidewalls and top) of the ultrathin-body fin on bulk-silicon substrate (Figure 4.1(a)) and SOI substrate (Figure 4.1(b)). As shown in Figure 4.1, the basic technology parameters for both the bulk and SOI FinFET devices include gate oxide thickness (T_{ox}), gate length (L_g), fin body thickness (t_{fin}), and fin height (H_{fin}). Additional technology parameters for bulk-FinFETs include field oxide or STI of thickness (T_{fox}) for isolation between neighboring devices. For FinFET manufacturing, the existing CMOS manufacturing technology is adopted [6–11]. However, some specific process steps or modules require additional restrictions and optimization, e.g., controlling the fin critical dimension (CD) and H_{fin}. In addition, the manufacturing flows for the bulk-FinFETs (Figure 4.1(a)) and SOI-FinFETs (Figure 4.1(b)) have subtle differences as shown in Figure 4.2. Furthermore, each

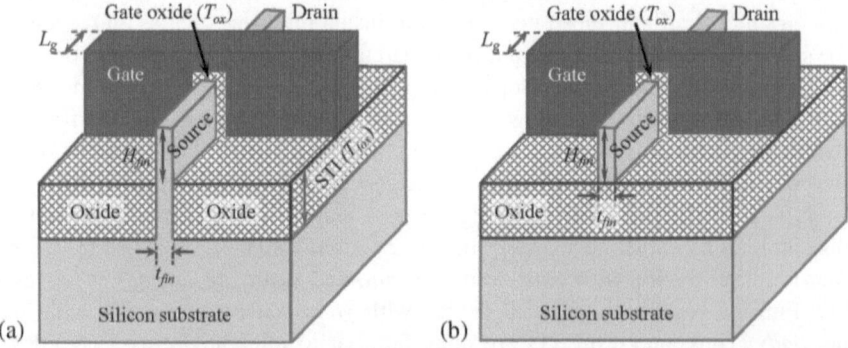

FIGURE 4.1 Ideal three-dimensional FinFET device structures with gate oxide on three sides (sidewalls and top) of the ultrathin fin body: (a) bulk-FinFET on silicon substrate; and (b) SOI-FinFET on silicon-on-insulator substrate. Here t_{fin}, H_{fin}, and L_g, are the fin thickness, fin height, and gate length of the devices, respectively; T_{ox} is the gate oxide thickness and T_{fox} is the field-oxide thickness of STI for device-to-device isolation.

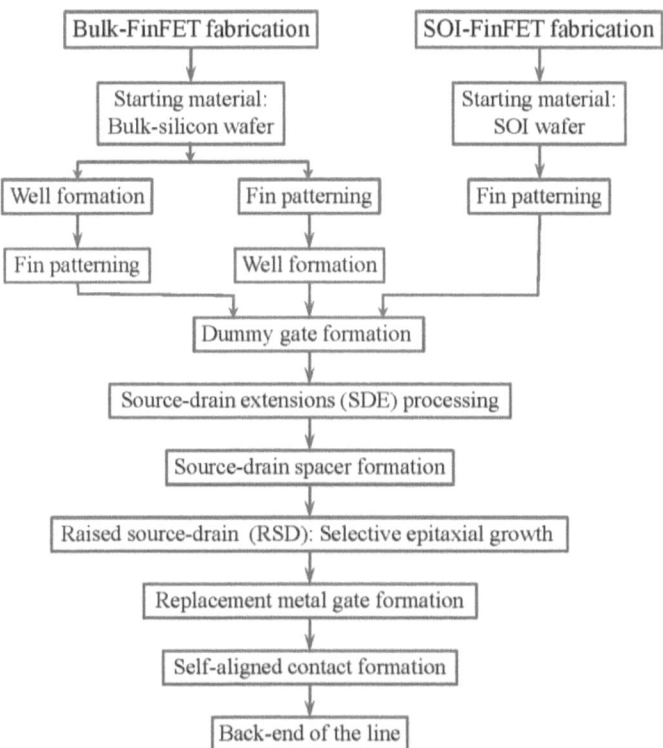

FIGURE 4.2 A typical flowchart for the fabrication of the bulk and SOI FinFET devices showing the major process modules of a representative VLSI technology.

IC manufacturer has its own proprietary fabrication processes different from that discussed in this chapter [6–11].

Figure 4.2 shows a typical flowchart of the major process modules for the manufacturing of bulk and SOI FinFETs at the nanometer node VLSI technology. Apart from the fin patterning and well formation, the major process flows for both the bulk and SOI FinFETs are identical. In the following section, we will present a brief overview of a typical FinFET fabrication technology. First of all, the bulk-FinFET fabrication process is overviewed. Then the major differences between the bulk and SOI FinFETs fabrication processes are highlighted.

4.3   BULK-FinFET FABRICATION

The fabrication process discussed in the following section is only to illustrate a representative FinFET manufacturing technology [7–12] and highlight the basic features of FinFET devices. In reality, a complementary FinFET or non-planar CMOS manufacturing technology is more complex than described in this section. Since the channel width of a FinFET is determined by the fin height H_{fin} (and also by fin thickness, t_{fin}) of the ultrathin-body fin, multiple-fin device architecture is used in FinFET layout to achieve the target drain current [13].

4.3.1 STARTING MATERIAL

As shown in Figure 4.2, the starting material for the fabrication of FinFET ICs is $\langle 100 \rangle$ heavily doped p^+ silicon wafers. The front-side of the wafers is a lightly doped or undoped layer of epitaxial silicon with thickness determined by the target H_{fin} of FinFET device technology. The starting wafers are cleaned and a thin layer of pad or screen oxide is grown on the surface for the well implantation process as shown in the flowchart in Figure 4.2.

4.3.2 WELL FORMATION

As outlined in Figure 4.2, there are two alternative methods to form wells in FinFET manufacturing for VLSI circuits: (1) at the beginning of the fabrication process; or (2) after the fin patterning.

4.3.2.1 p-Well Formation

In the well first FinFET fabrication process, a zero-level mask and subsequent etching process are used to define an alignment notch in the wafer with pad oxide. The wafers are then primed in hexamethyldisilazane (HMDS) to improve photoresist wetting and adhesion to oxidized silicon surface followed by photoresist coating and *top anti-reflective coating* (TARC). After the development of photoresist, p-well mask is used to expose the p-well region for ion implantation, hereafter referred to as the "implant," and cover the n-well region by photoresist pattern. Then the exposed p-well region of the wafers is implanted with p^+ (boron) anti-punchthrough (APT) dopants at the channel bottom region to suppress the leakage current [14] followed by p-type (boron) dopant implant to define the p-well region for n-channel FinFET (nFinFET) fabrication.

4.3.2.2 n-Well Formation

After the p-well formation, the photoresist is stripped and the wafers are cleaned. The wafers are then primed in HMDS followed by photoresist coating and TARC. Then the lithography and masking steps are performed so that the photoresist pattern covers the p-well region and exposes the n-well region. Next, the exposed region of the wafers is implanted with n^+ (phosphorus) APT at the channel bottom region to suppress the leakage current [14] followed by n-type dopant (phosphorus) implant to define the n-well region for p-channel FinFET (pFinFET) fabrication.

After the n-well ion implant processing step, the photoresist is stripped off and the wafers are cleaned followed by the removal of the initial pad oxide grown on the silicon front surface. After post-implant cleaning, the wafers are annealed using the rapid thermal annealing (RTA) process to activate the implanted dopants. Finally, the wafers are cleaned and a thin pad oxide is grown on the surface for silicon fin patterning in the next processing step.

4.3.3 FIN PATTERNING: SPACER ETCH TECHNIQUE

After the well formation, an array of ultrathin silicon fins is formed using the *spacer etch technique*, referred to as the *self-aligned double patterning* (SADP) [7,13,15–17].

It is to be noted that the dimensions of such ultrathin ($\approx$7 nm) vertical structures are beyond the resolution of available optical (193 nm ArF Immersion) lithography. Although a high resolution extreme ultraviolet (EUV) lithography stepper shows great potential to print such fine lines, it is challenging to pattern such small fins in high volume production [17,18]. The major processing steps in patterning multiple-fins using SADP technique include: (1) carbon hard mask or mandrel patterning; (2) offset spacer formation; and (3) silicon fin formation as described below.

4.3.3.1 Mandrel Patterning

In this process step, a stack of layers is deposited over the previously grown pad oxide on the wafer: first of all, a thick layer of silicon nitride (Si_3N_4) using the *chemical vapor deposition* (CVD) process; then a thicker amorphous carbon sacrificial layer using the CVD process to support the fabrication of oxide spacers for fin patterning and is referred to as the "mandrel," which is removed after the spacer formation; a thin layer of dielectric anti-reflective coating that forms the *bottom anti-reflective coating* (BARC); and finally, photoresist coating and TARC.

After the deposition of the above layers on the wafers, the photoresist and masking steps are performed to expose multiple photoresist patterns on the BARC of the wafers. These photoresist strips are used as the mask to pattern the amorphous carbon mandrel layer using a highly anisotropic etch as shown in Figure 4.3(a). The Si_3N_4 layer deposited on the wafer is used as the etch stop layer (ESL) to ensure that the etch process stops on the front surface of the Si_3N_4 layer. Then the photoresist patterns and BARC are stripped in ionized oxygen plasma and the wafers with carbon mandrel patterns are cleaned to prepare the top surface for oxide spacer formation.

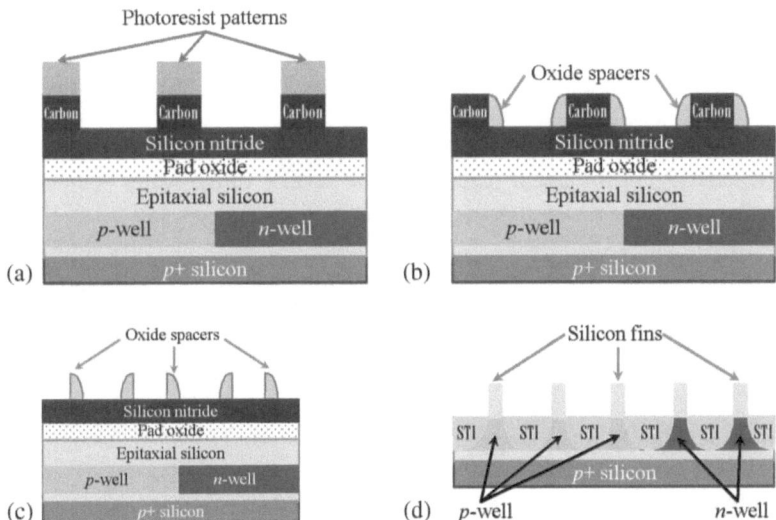

FIGURE 4.3 Two-dimensional cross-section of wafer for fin patterning: (a) mandrel patterning; (b) after selective oxide etch defining thin oxide spacers separated by mandrel strips; (c) after mandrel removal exposing oxide spacers; and (d) after TEOS etch to expose silicon fins along with the formation of STI between fins.

4.3.3.2   Oxide Spacer Formation

After mandrel patterning, a thin blanket layer of SiO_2 is deposited on the wafer using the CVD process. The deposited oxide layer surrounds the amorphous carbon mandrel at both ends of the carbon layer. In order to achieve straight lines at both ends of the oxide spacers, a photolithography process using a *cut mask* is used to trim the oxide spacers at both ends of the mandrel. Then a highly anisotropic etch process that is selective to oxide is used to form thin oxide spacers on the sidewalls of the amorphous carbon layer as shown in Figure 4.3(b).

After the oxide spacers formation, the mandrel strips between the oxide spacers are removed using a selective etch process as shown in Figure 4.3(c). The oxide spacers and the SiCN (silicon carbo-nitride) hard mask are unaffected by this etch process. After mandrel removal, the wafers are cleaned in a Megasonic bath [19]. Then using the oxide spacers as the hard mask, the underlying Si_3N_4 layer is removed by a highly anisotropic etching process to form the oxide spacer on the substrate.

4.3.3.3   Silicon Fin Formation

After the mandrel removal, the oxide spacers and nitride mask are used as patterns in an anisotropic etch that cuts down through the Si_3N_4, pad oxide, and epitaxial silicon into the well regions forming epitaxial silicon fins. The etching pressure, rate, and energy are extremely critical to create a sloped fin to achieve a smooth curved interface at the bottom of the fin. Furthermore, during the fin patterning process, the oxide spacers are etched and the shape of the Si_3N_4 hard mask strips underneath the oxide spacers are affected at the top. This etch process must terminate inside of the wells without an ESL to facilitate the process and is very challenging [17]. After cleaning the wafers, an ultrathin layer of thermal silicon dioxide (SiO_2) known as the "trench liner" is grown in the trenches to relieve the stress in the silicon around the upper and lower corners of the trenches [20]. Note that the Si_3N_4 is a diffusion barrier to oxygen preventing SiO_2 growth underneath. The etch process is followed by formation of exposed silicon fins and STI.

Silicon fin patterns and STI formation: The oxide trench liner is removed in an isotropic etch. After the linear oxide etch, the surface of the wafer is cleaned and a thick layer of tetraethyl orthosilicate (TEOS) oxide is deposited using the CVD process. Then the TEOS is densified by annealing the wafers to make it more resistant to wet etch. After TEOS densification, the wafers are polished using *chemical-mechanical planarization* (CMP) process using Si_3N_4 layer as the CMP stopper.

After the planarization of the surface, the residual Si_3N_4 layer is removed exposing the pad oxide grown on the wafer after the well formation (Section 4.3.2). The TEOS oxide is then etched back using a highly selective oxide etch to remove all of the oxide surrounding the fins and completely expose the fins as shown in Figure 4.3(d). This is a very challenging etch process to define the fin/well boundary [17,18]. The remaining oxide between the fins forms the STI (Figure 4.3(d)).

Fin removal: The SADP lithography process creates extra fins in bulk-FinFET wells that are removed to achieve the target drive current of FinFET devices at any technology node. In order to remove any odd fin, an ultrathin layer of thermal oxide is grown over the exposed silicon for ESL in the etching process. Next, a thin layer of SiCN hard mask is deposited over the wafer.

In the first step of the odd fin removal process, the wafers with deposited thermal oxide and SiCN are patterned with photoresist and developed. After photolithography and masking step, the photoresist pattern exposes target fin for removal. In the lithography process, a thin BARC layer is formed at the SiCN/photoresist interface. Then the exposed portion of the BARC layer is removed using an anisotropic etching process. In the next processing step, the exposed silicon fin is etched away using wet etch (typically, tetra-methyl-ammonium hydroxide or TMAH). Finally, the photoresist and BARC layers are etched away in an isotropic etch and a layer of oxide is grown on the exposed silicon base of the removed fin. After the fin removal process, the SiCN and oxide layers over the fins are etched away.

4.3.4 ALTERNATIVE WELL FORMATION PROCESS

The alternative well formation approach after the fin patterning in a front-end FinFET fabrication technology enables the formation of uniform fin height H_{fin} after the STI etch-back processing step and mitigates the well/STI interface matching issue. The measured H_{fin} is then used for an accurate estimation of well implant energy to make the top of the wells coplanar with that of the STI. *This is the most common method for the formation of wells in a bulk-FinFET process flow* [15–17]. In this alternative well formation process flow, the SADP technique described in Section 4.3.3 is used for silicon fin patterning directly after the screen oxide growth on starting wafers (Section 4.3.1). The typical steps for the alternative well implant process including APT implants are similar to those described in Section 4.3.2 to form p-well and n-well regions for nFinFETs and pFinFETs, respectively. After the well formation, the wafers are cleaned and subjected to the RTA process to activate the implanted dopants in the wells. Next, the wafers are prepared for gate definition.

4.3.5 GATE DEFINITION: POLYSILICON DUMMY GATE FORMATION

After odd fin removal and cleaning processes (Section 4.3.3.3) or alternative well formation process (Section 4.3.4), the fins are coated with an ultrathin oxide ESL and cleaned to define polysilicon FinFET gates. The major process steps include deposition, polish, and lithography processes to pattern amorphous silicon gate electrodes as described below.

In the first step of gate definition, a thick blanket layer of undoped amorphous silicon is deposited on the substrate with patterned fins using the CVD process. Then the amorphous silicon layer is polished back using CMP to form a smooth planar surface. After the planarization of the amorphous silicon surface, a thick layer of amorphous carbon hard mask is deposited using CVD. Then a thin layer of BARC is spun across the wafer. The wafers are then patterned with photoresist using gate mask to define the gate electrodes. Following the photoresist gate patterning, the carbon hard mask is etched using a highly anisotropic etching process to pattern the photoresist with hard mask underneath.

In the second step, the photoresist is etched away and the wafers are cleaned. Then, a highly anisotropic etch process is used to etch amorphous silicon and transfer the gate pattern from the hard mask into the amorphous silicon. After the amorphous silicon etch step, the hard mask is removed resulting in the target number of continuous gate electrodes over the patterned fins on both the p-well and n-well. In order to form CMOS pair with nFinFET and pFinFET, the continuous gate electrodes are terminated to include a target set of parallel fins in the p-well and n-well, respectively by repeating the lithography processes.

In this step of forming CMOS pairs, a thick layer of amorphous carbon hard mask layer is deposited by the CVD process. Then a thin layer of BARC is spun across the wafer followed by resist deposition and the lithography process. Initially the lithography and photo mask processes are used to etch the hard mask using a highly anisotropic etch process. Then the photoresist is stripped and the wafers are cleaned. Next, a highly anisotropic etch recipe is used to transfer the pattern from the hard mask into the amorphous silicon. This etch terminates the gate electrodes to include the target set of fins in n-well and p-well to form a CMOS pair. *Note that the amorphous silicon gate electrodes have been subjected to the litho-etch-litho-etch* (LELE) process.

The minimum dimension of the amorphous silicon gate electrode defines the length L_g of a FinFET device at a technology node. And, the width W of a FinFET device is defined by ($t_{fin} + 2H_{fin}$). In reality, W of a FinFET can be increased by using gate electrode over multiple fins, thus, increasing the transistor drive current. Also, W can be increased by fabricating taller fins to increase H_{fin}. However, the ratio of H_{fin} and t_{fin} is a critical parameter affecting FinFET device performance.

4.3.6   SOURCE-DRAIN EXTENSIONS PROCESSING

After the formation of CMOS pairs, the photoresist, BARC, and the carbon hard mask layers are removed from the wafers exposing the fully formed polysilicon gate electrodes. Then wafers are cleaned and an ultrathin thermal oxide is grown on gate electrodes followed by the deposition of an ultrathin ($\approx$1.5 nm) of CVD oxide to define the source-drain extension (SDE) *offset spacer* for source-drain extension processing [21].

4.3.6.1   *n*FinFET Source-Drain Extension Formation

After the oxide spacer formation, first a photoresist and then a TARC layer are deposited on the wafers. Next, the wafers are patterned with photoresist to cover the n-wells where the pFinFETs are fabricated. A dual Arsenic implant (typically, 2E15@1 Kev, ±10 degrees) is used to dope the fins in the p-well regions of the wafer to form SDE region of the nFinFET devices. This dual implant process is performed at +10 degrees and −10 degrees to ensure adequate coverage of the tall fins which are located alongside very tall photoresist coated structures. After the implant, the TARC and photoresist layers are removed and the wafers are cleaned to prepare for pFinFET SDE formation.

4.3.6.2   *p*FinFET Source-Drain Extension Formation

Again, photoresist and TARC layers are deposited on the wafers. Then the wafers are patterned with photoresist to cover the p-wells where the nFinFETs are fabricated. A dual boron SDE implant specification (typically, 2E15@1 Kev, ±10 degrees) is used to dope the fins in the *n*-well regions of the wafer to form SDE region for the *p*Fin-FETs. Similar to the *n*FinFET SDE process, the dual implant step for *p*FinFET SDE is performed at +10 degrees and −10 degrees to ensure adequate coverage of the tall fins which are located alongside very tall photoresist coated structures. After *p*Fin-FET SDE implant, the photoresist is removed and the wafers are cleaned to prepare for dopant activation.

In order to activate the dopants for both the *n*FinFET and *p*FinFET SDE implants, the wafers are subjected to spike anneal followed by flash or laser anneal. During the annealing process, the SDE implants also diffuse under the polysilicon gate oxide in the fin-channel for a certain distance, thus, creating overlaps of the SDE implant in the ultrathin fin-channel under both ends of the polysilicon gate. This overlap is controlled by the thickness of the offset spacer which defines the effective channel length L_{eff} of the FinFET devices.

After SDE dopant activation, the wafers are prepared for raised source-drain (RSD) formation as outlined below.

4.3.7 RAISED SOURCE-DRAIN PROCESSING

For RSD formation, a thick layer of Si_3N_4 is deposited on the wafers. Then a highly anisotropic etching process is used to form the Si_3N_4 spacers on the sidewalls of the polysilicon gates. However, even a highly anisotropic etching process leaves behind nitride residue on the sidewalls of the gate electrodes and fins. This is due to the two tall orthogonal vertical structures (gate electrodes and the fins) in the FinFET fabrication technology. The nitride spacer located on the fins is undesirable and must be removed for device reliability [18]. Finally, a highly anisotropic etch is used to remove the Si_3N_4 residue from the entire horizontal surfaces forming nitride side-wall spacers along the sides and ends of the gate electrodes and the fins. However, it is challenging to minimize the residual nitride spacer along the edge of the fins to maintain the target thickness of Si_3N_4 spacers along the gate electrode. After the Si_3N_4 spacer formation, RSD regions are formed for both types of devices.

4.3.7.1 SiGe *p*FinFET Raised Source-Drain Formation

After the Si_3N_4 spacer-etch process, a thin (≈30 nm) layer of SiCN hard mask is deposited over the surface of the wafers to cover all the underlying structures. Then the wafers are primed in HMDS followed by TARC and photoresist coat. After photoresist development and the masking process, the *n*FinFET region and the gate electrodes of the *p*FinFET region are covered in photoresist. After the lithography process, an anisotropic SiCN-specific etch is used to strip away the SiCN over the exposed *p*FinFET fins and the photoresist is stripped, and the wafers are cleaned. Then the *p*FinFET fins outside the Si_3N_4 spacer are etched away using a highly anisotropic etch process. The SiCN hard mask protects the *n*FinFET fins and the *p*FinFET gate electrodes.

After the *p*FinFET fin etch, the SiGe non-rectangular RSD regions are formed on the exposed silicon surface of the *p*FinFET source-drain fins by a selective epitaxial growth (SEG) of SiGe since every other part of the wafer is covered with nitride, or oxide, or SiCN hard mask. After SiGe growth, the remaining SiCN hard mask is

etched away using an etchant that is selective to SiCN and prepares the wafers for
nFinFET RSD formation.

4.3.7.2 SiC nFinFET Raised Source-Drain Formation

After the pFinFET RSD formation, a thin layer of SiCN hard mask is deposited over
the surface of the wafer to cover all of the underlying structures. Then the wafers are
primed in HMDS followed by TARC and photoresist coat. After the lithography and
masking process steps, the pFinFET region and the gate electrodes of the nFinFET
region are covered in photoresist. The nFinFET RSD can be formed in two different
ways: (1) grow epitaxial tip on the fins without nFinFET fin removal; or (2) by fin
removal similar to pFinFET RSD formation process.

Epitaxial tip on fin-top: In this nFinFET RSD formation, an anisotropic SiCN-
specific etchant is used to strip away the SiCN over the exposed silicon base of the
nFinFET fins. Then the photoresist is stripped away and the wafers are cleaned.
Next, the SiO_2 layer over the nFinFET fins is etched away to be followed by a deposition of silicon on the wafers using the SEG process. (Note that every other region
of the wafers is masked by nitride, or oxide, or SiCN hard mask.) After RSD tip
formation, the remaining SiCN hard mask is etched away. The epitaxial silicon caps
on the top of the nFinFET fins are targeted to maximize the contact surface area of
the nFinFET fins and provide a larger surface for the tungsten (W) contacts, thus,
lowering the contact resistance and improving transistor speed.

SiC nFinFET RSD formation: For SiC RSD formation, the SiCN layer is deposited directly and is patterned with BARC and photoresist. Then the exposed SiCN is
etched away in a highly anisotropic etch to expose the underlying fins of the nFin-
FET devices. Next, an isotropic SiCN-specific etch is used to strip away the SiCN
over the exposed nFinFET transistor fins. The photoresist is then removed and the
wafers are cleaned. After cleaning the wafer, the nFinFET fins outside the Si_3N_4
spacer are etched away in an anisotropic etch. The SiCN hard mask protects the
pFinFET fins and the nFinFET gate electrodes. Then the SiC layer is formed on the
exposed silicon surface of the n-well fins to form non-rectangular nFinFET RSD.
After the SiC growth process, the remaining SiCN hard mask is etched away to prepare the wafers for silicidation.

4.3.7.3 Raised Source-Drain Silicidation

For RSD silicidation, the wafers are implanted with a silicon pre-amorphization
implant. This step amorphizes the surface of the silicon and facilitates the formation
of a more uniform, low resistance silicide in the next processing step [22]. After the
silicon pre-amorphization implant, the wafers are prepared for RSD metallization
and local contact formation.

Aluminum implant in SiGe RSD: First of all, the oxide layer on the gate electrodes as well as any native oxide on the surface of the RSD regions of the wafers
are stripped away. Then the wafers are patterned with a photoresist layer to expose
the pFinFET devices while covering the nFinFET devices. The patterned wafers are
implanted with p-type dopants, aluminum (Al). The implant energy is chosen so that
Al is located at the interface between the SiGe and the titanium (Ti) layer deposited
in the next process step. This implant reduces the RSD contact resistance for the

*p*FinFET devices since the Al segregates to the top of the SiGe and being a *p*-type dopant, it lowers the Schottky barrier height for holes (typically, from about 0.4 eV to about 0.12 eV) and increases drive current by up to 19% [22, 23]. Next the photoresist layer is stripped away and the wafers are cleaned.

Titanium silicide formation: After the Al implant on the *p*FinFET RSD regions, the wafers are cleaned and a blanket thin layer of cold titanium (Ti) is deposited on the wafer using physical vapor deposition (PVD). Since Ti is a good getter, the oxygen and other contaminants are removed by the Ti layer [24]. Next, the wafers are rapidly heated to form titanium silicide ($TiSi_2$). After $TiSi_2$ formation, an additional thin layer of Ti is deposited while the wafer is at high temperature to form $TiSi_2$. This process produces in-situ $TiSi_2$ without voids and reduces production cycle time.

Unreacted titanium strip: After the annealing process, there are two types of Ti present on the transistor structure: unreacted Ti and reacted $TiSi_2$. The unreacted Ti located on the spacer sidewalls and on the top of the STI is etched away with highly selective etch. This leaves behind the reacted salicide on the top of the gate electrodes and over the epitaxial coated source-drain regions un-affected.

Oxide/nitride etch-stop layer deposition: After the silicidation process, the wafers are cleaned and a thin layer of SiO_2 followed by a thin layer of Si_3N_4 are deposited as an ESL for the next processing step of replacement metal gate and high-*k* gate dielectric.

4.3.8   REPLACEMENT METAL GATE FORMATION

In the previous sections, we have defined the gate geometry and processed SDE and RSD regions using polysilicon gate. This section outlines the process steps to replace the polysilicon gate with metal gate and high-*k* dielectric gate oxide. The major fabrication process steps include: (1) polysilicon gate removal; (2) high-*k* gate dielectric deposition; and (3) metal gate deposition and workfunction engineering.

4.3.8.1   Polysilicon Dummy Gate Removal

In the first step of replacement metal gate formation, a thick layer of phosphorus doped glass known as phospho-silicate glass (PSG) is deposited over the wafers with ESL using the plasma enhanced CVD (PECVD) process. This layer forms the first half of the pre-metal-dielectric (PMD). It is to be noted that the sharp corners and close proximity of the SiGe crystals "pinches-off" the PECVD disposition resulting in voids between the crystals. The void-free PECVD deposition is challenging, however, the small voids do not pose a significant problem on device performance. The PSG is then polished back using CMP to a lower thickness ($\approx$120 nm).

After CMP, the top of the gate electrode is exposed and the amorphous silicon gate electrode is etched away using a very high selectivity silicon etch TMAH. This creates a cavity in the region of removed amorphous silicon. The entire cavity is lined with spacer oxide and the exposed fins are coated with an oxide ESL. *Note that this region of the fins does not have any SDE implants since they were covered by the amorphous silicon gate electrodes.*

After the removal of the amorphous silicon gate, the ESL inside the gate cavity is etched away exposing the inside walls of the cavity and the silicon fin.

4.3.8.2 High-k Gate Dielectric Deposition

The gate dielectric consists of an ultrathin SiO_2 interfacial layer and a high-k dielectric layer over the interfacial layer. The interfacial layer is thermally grown on silicon fins using a low temperature oxidation process [18]. This forms the bottom interface layer below the high-k dielectric and ensures smooth interface between the high-k material and silicon and prevents degradation of electron mobility.

After bottom interface SiO_2 gate layer growth, an ultrathin layer of hafnium oxide (HfO_2) high-k dielectric is deposited using atomic layer deposition (ALD) [18]. The high-k material is a blanket layer that covers the entire wafer; however, it is only required in the gate cavity over the fins.

4.3.8.3 Metal Gate Formation

pFinFET workfunction metal (TiN) deposition: After gate oxide deposition, the thin *p*FinFET workfunction metal gate is deposited using ALD. It consists of an ultrathin highly conformal layer of TiN (titanium nitride) that fills both the *p*FinFET and *n*FinFET cavities as well as coats the surface of the wafer. After TiN deposition, an ultrathin TaN ESL is deposited across the wafer using ALD. Next, a highly conformal thick layer of TiN is deposited using the ALD process to fill both the *p*FinFET and *n*FinFET cavities as well as coat the surface of the wafers.

A layer of photoresist is patterned on the wafer to cover only the *p*FinFET region and expose the TiN layer over the *n*FinFET region. After photoresist patterning to protect *p*FinFET gate metallization, the exposed TiN layer over the *n*FinFET gate fins is etched away using TaN as an ESL to this etch. After TiN removal from the *n*FinFET gate cavity, the photoresist is stripped and the wafers are cleaned to prepare for *n*FinFET workfunction metal gate.

nFinFET workfunction metal (TiAl) deposition: After the resist strip, a thin TiAl metal gate is deposited using advanced self-ionizing physical vapor deposition (SIPVD) technique. Then another highly conformal thin TiAl layer is deposited to cover the horizontal surfaces of both the *p*FinFET and *n*FinFET cavities and coat the surface of the wafer.

After TiAl deposition across the wafer, an anneal is performed to cause the Al in the TiAl to diffuse through the TaN barrier and creates the TiAlN (titanium-aluminum nitride) *n*FinFET workfunction metal on the top of the high-k dielectric in the *n*FinFET region. During the annealing process, Al diffuses rapidly into the TiN located on the *n*FinFET device region of the *gate cavity* forming TiAlN workfunction metal for *n*FinFETs devices. However, the thick TiN layer over the *p*FinFET region only blocks the diffusion of Al into the TiN *p*FinFET workfunction metal. Thus, the *p*FinFET workfunction metal does not become TiAlN.

Tungsten back fill: After anneal and metal gate workfunction engineering, a thick layer of highly conductive tungsten (W) is deposited using the CVD process to fill the gate cavities. Finally, the tungsten is polished back so that it is coplanar with the top of the gate electrodes.

4.3.9 Self-Aligned Contact Formation

After W deposition and planarization, self-aligned local contacts are formed for interconnection of the transistors. In the first step of the local contact formation process, a recession is created in the *gate electrode cavity* by etch-back of W and metals in the surrounding gate cavity. Next, a thin layer of silicon oxynitride (SiON) is deposited on the wafers using the CVD process to fill the cavity of the recessed gate electrode with SiON. Then the SiON is polished back using CMP to make SiON coplanar with the PSG surface outside the gate electrode regions. Next, a thick layer of PSG is deposited on the wafers to complete the PMD deposition process and metallization for the formation of self-aligned local contacts for the fabricated devices.

4.3.9.1 Metallization

In the metallization process for local contact formation, a photoresist pattern is used to define the position of the contact trenches. Then the photolithography, masking, and anisotropic etch processes are used to cut down the contact holes to the gate electrode and the source-drain regions of FinFETs. Next the photoresist is removed and the wafers are cleaned to remove contaminants from the trenches as well as remaining polymeric residue and carbon contaminants. Then a thin Ti-liner followed by an ultrathin titanium nitride TiN barrier layer are deposited using the ion-metal plasma (IMP) physical vapor deposition (PVD) process [25]. Next, the wafers are annealed using the RTA process to react to the Ti/TiN layer and set the resistance of the Ti-liner.

Tungsten deposition and polish-back: After the deposition of the Ti/TiN barrier layer, a thin W seed is deposited in-situ on the wafers to line in the interior of the trenches and ensure conformal void-free bulk tungsten deposition in the trenches. Next, a thick layer of tungsten is deposited on the wafers by the CVD process. Then the wafers are polished using CMP so that the tungsten plugs are polished back to the top surface of the PSG, and the surface of the TEOS oxide is smoothed out to make the surface of the PSG coplanar with the top of the W-plugs.

Ta/TaN barrier metal patterning: After W deposition and CMP, a photoresist pattern is defined to determine the location of the contact trenches. Then lithography, masking, and etching operations are performed to cut down the contact holes to the gate electrode and the transistor source-drain regions. After opening the contact holes, the photoresist is stripped and the wafers are cleaned and appropriate processing steps are used to remove contaminants in the trenches and clean out any polymeric residue and carbon contaminants. Next, a thin TaN layer followed by a thicker Ta layer is deposited on the wafers using the IMP PVD process as the barrier metal to copper (Cu) trench contacts.

Copper fill: After the Ta/TaN barrier metal patterning, an ultrathin Cu seed is deposited in-situ on the wafers to line the interior of the trenches and ensure conformal void-free bulk-Cu deposition. Then a thick layer of bulk copper is deposited using electrochemical deposition process. Next, the Cu is subjected to low temperature annealing to anneal out the defects and reduce resistance. After the annealing step, the Cu is polished back using CMP which also removes the Ta that was on the upper surfaces of the TEOS. Then appropriate processing steps are used to smooth out the surface of the TEOS oxide and ensure that the surface of the TEOS is coplanar with the top of the copper lines. Finally, the wafers with Cu local contact metal are cleaned for back-end of the line (BEOL) fabrication processes.

4.4   SOI-FinFET PROCESS FLOW

As shown in the flowchart (Figure 4.2) for FinFET fabrication, the SOI-FinFET fabrication process eliminates the requirements for the formation of wells and STI to isolate neighboring FinFET devices. Thus, the major differences between the SOI-FinFET and bulk-FinFET fabrication are the starting material and fin patterning.

4.4.1   STARTING MATERIAL

The starting wafers for SOI-FinFET fabrication consist of a layer of epitaxial silicon on a buried oxide (BOX) over a p-type base silicon substrate (Figure 4.1(b)). The thickness of the epitaxial silicon layer defines the height H_{fin} of the silicon body of a FinFET device. Similar to the bulk-FinFET process (Section 4.3.1), the wafers are cleaned and a thin layer of screen oxide is grown on the surface of the epitaxial silicon layer for fin patterning.

4.4.2   FIN PATTERNING: SPACER ETCH TECHNIQUE

As discussed in Section 4.3.3, the major processing steps in patterning multiple-fins using SADP technique include: (1) carbon hard mask or mandrel patterning; (2) offset spacer formation; and (3) fin formation as described below.

4.4.2.1   Mandrel Patterning

The mandrel patterning includes the deposition of a stack of layers over the screen oxide of the wafer: first of all, a thick layer of Si_3N_4 using CVD; then a thicker CVD amorphous carbon mandrel sacrificial layer to support the fabrication of oxide spacers for fin patterning; next a thin layer of dielectric BARC; and finally, photoresist coating and TARC on the top.

After the deposition of the stack of the above layers on the wafers, the photoresist and masking steps are used to create multiple photoresist patterns on the BARC of the wafers. These photoresist strips are used as the mask to pattern amorphous carbon mandrel layer using a highly anisotropic etch. The Si_3N_4 layer deposited on the wafer is used as the ESL to ensure that the etch process stops on the front surface of the Si_3N_4 layer. Then the photoresist patterns and BARC are stripped in ionized oxygen plasma and the wafers with carbon mandrel patterns are cleaned to prepare the top surface for oxide spacer formation.

4.4.2.2   Oxide Spacer Formation

After the mandrel patterning, a thin blanket layer of SiO_2 is deposited on the wafer using the CVD process. This oxide layer covers the wafers as well as surrounds the amorphous carbon mandrel at both ends of the carbon layer. In order to achieve straight lines at both ends of the oxide spacers, a photolithography process using a *cut mask* is used to trim the oxide spacers at both ends of the mandrel. Then a highly anisotropic etch process selective to oxide is used to form thin oxide spacers on the sidewalls of the amorphous carbon layer.

After the oxide spacers formation, the mandrel strips between the oxide spacers are removed using a selective etch process applicable to amorphous carbon. The oxide spacers and the SiCN hard mask are unaffected by this etch process. After the mandrel removal, the wafers are cleaned in a Megasonic bath [19]. Then, using the oxide spacers as the hard mask, the underlying Si_3N_4 layer is removed by a highly anisotropic etching process to form the oxide spacer on the substrate. The wafers are then subjected to a Megasonic clean.

4.4.2.3 Silicon Fin Formation

After mandrel removal, the oxide spacers and Si_3N_4 hard mask patterns are used in a highly anisotropic etch to cut down through Si_3N_4, pad oxide, and epitaxial silicon to stop on the top of the BOX layer. The etching pressure, rate, and energy are extremely critical to create a sloped fin to achieve a smooth curved interface at the bottom of the fin. Furthermore, during the fin patterning process, the oxide spacers are partially etched and the shape of the Si_3N_4 hard mask strips underneath the oxide spacers are damaged at the top. Then the wafers are subjected to a Megasonic clean. In this etch process BOX is used as an ESL. Thus, the etch process to pattern silicon fins is significantly simpler than that of the bulk-FinFET fin formation (Section 4.3.3.3). As shown in Figure 4.2, there are no wells in an SOI flow and the BOX forms the STI. In addition, all of the fins are at a uniform height which greatly reduces the variability of device performance due to process variability [26,27]. Then, Si_3N_4 hard mask and pad oxide are removed exposing the epitaxial silicon fins standing on the top of the BOX on the base silicon substrate. Finally, any odd fin is removed following the process step described in Section 4.3.3.3.

After fin patterning, the remainder of the process flow is the same as that of the bulk-FinFET without the well and STI process modules.

4.4.3 COMPARISON OF BULK-SILICON FinFET AND SOI-FinFET FABRICATION TECHNOLOGY

From the discussion on bulk-FinFET and SOI-FinFET fabrication process flow, it is obvious that the major differences between bulk-FinFET and SOI-FinFET front-end CMOS fabrication processes are process complexities and hence cost. First of all, the bulk-FinFET fabrication requires formation of wells and STI to isolate each fin and therefore, each device on the substrate. The formation of STI requires deep etch to cut through the epitaxial silicon substrate to form trenches without the presence of an ESL. It is challenging to maintain the precise control of this etch process. Furthermore, it is challenging to maintain a uniform fin height as the trench filled TEOS oxide is etched back to the boundary between n-well and p-well to define H_{fin}, thus, causing variation in H_{fin} [18].

On the other hand, in the SOI-FinFET fabrication process each fin is patterned on the epitaxial silicon layer on the top of a BOX and they are already electrically isolated from each other. Therefore, wells and STI processes are eliminated. Again, H_{fin} is defined by the thickness of the starting epitaxial silicon layer on the BOX. Thus, a simple etch process with the top of the BOX as the ESL is used to pattern fins of uniform height. As a result, the variation in H_{fin} is less in SOI-FinFET fabrication process flow compared to the bulk-FinFET technology. Therefore, the SOI fabrication sequence reduces the process complexity and cost by eliminating a series of challenging fabrication steps. However, SOI-FinFETs suffer from high defect density, floating body effect, and poor heat dissipation [5]. Furthermore, the raised source-drain formation by SEG is comparatively challenging in the SOI-FinFET fabrication flow.

Although the SOI-FinFET process flow presents a simpler front-end FinFET fabrication technology, the manufacturing cost is substantially higher than the bulk-FinFET technology due to the higher price of the SOI wafers compared to the bulk-silicon wafers. Therefore, cost-simplicity trade-off must be considered for volume production of FinFET fabrication technology.

4.5 SUMMARY

In this chapter an overview of the basic FinFET fabrication process flow is presented to introduce a representative FinFET device manufacturing technology. For the completeness of discussions, the bulk-FinFET as well as SOI-FinFET technology is highlighted. However, the bulk-FinFET process flow is described in some detail and only the differentiating features of the SOI-FinFET process flow are outlined. The major motivation of this chapter is to introduce the FinFET technology to appreciate the complexities of device fabrication and understand the device performance described in Chapters 5–10 of this book. Thus, some of the concepts may be different from the real silicon processing and presented often without explanation, or sketches, or justification, or references. Many companies and research laboratories build chips with final devices similar to the devices briefed in this chapter, however, with quite different process details. As, for example, the differences between the commercial process flows and that outlined here are the number and specifications of masks used. In commercial process with proprietary flows, the number of masks and process details varies from company to company. Some of the reasons for these differences between commercial process flows are equipment specific to the companies and application specific to the target technology node. Trade-offs in technology complexities and device performance may lead an individual company to process flow quite different than one we have overviewed. In spite of these differences between process flows from company to company, the final device structure and performances are comparable from all commercial manufacturers.

REFERENCES

1. J.-P. Colinge (ed.), *FinFETs and Other Multi-Gate Transistors*, Springer, New York, 2008.
2. J.P. Colinge, M.H. Gao, A. Romano-Rodriguez, H. Maes, and C. Clays, "Silicon-on-insulator 'Gate-all-around device.'" In: *IEEE Electron Devices Meeting Technical Digest*, pp. 595–598, 1990.
3. S.K. Saha, *Compact Models for Integrated Circuit Design: Conventional Transistors and Beyond*, CRC Press, Taylor & Francis Group, Boca Raton, MA, 2015.
4. J.D. Plummer, M.D. Deal, and P.B. Griffin, *Silicon VLSI Technology: Fundamentals, Practice and Modeling*, Prentice Hall, Upper Saddle River, NJ, 2000.
5. J.-H. Lee, T.-S. Park, E. Yoon, and J.J. Park, "Simulation study of a new body-tied FinFETs (Omega MSOFETs) using bulk Si wafers." In: *Proceedings of the Si Nanoelectronics Technical Digest*, pp. 102–110, 2003.

6. A. Gupta, M. Shrivastava, M.S. Baghini, *et al.*, "Part I: High-voltage MOS device design for improved static and RF performance," *IEEE Transactions on Electron Devices*, 62(10), pp. 3168–3175, 2015.

7. C. Auth, A. Aliyarukunju, M. Asoro, *et al.*, "A 10 nm high performance and low-power CMOS Technology Featuring 3rd generation FinFET transistors, self-aligned quad patterning, contact over active gate and cobalt local interconnects." In: *IEEE International Electron Devices Meeting Technical Digest*, pp. 673–676, 2017.

8. J. Markoff, *Intel Increases Transistor Speed by Building Upward*, New York Times, May 4, 2011. https://www.nytimes.com/2011/05/05/science/05chip.html.

9. C. Auth, C. Allen, A. Blattner, *et al.*, "A 22-nm-high performance and low-power CMOS technology featuring fully-depleted Tri-gate transistors, self-aligned contacts and high density MIM capacitors." In: *Symposium on VLS Technology*, pp. 131–132, 2012.

10. R. Merritt, *TSMC Taps ARM's V8 on Road to 16-nm FinFET*, October 16, 2012. www.eetimes.com/document.asp?doc_id=1262655.

11. D. McGrath, *Globalfoundries Looks to Leapfrog Fab Rival with New Process*, September 20, 2012. www.eetimes.com/document.asp?doc_id=1262552.

12. ThresholdSystems, Inc., Home page, https://secure.thresholdsystems.com/Home.aspx.

13. Y.-K. Choi, T.-J. King, and C. Hu, "A spacer patterning technology for nanoscale CMOS," *IEEE Transactions on Electron Devices*, 49(3), pp. 436–441, 2002.

14. K. Okano, T. Izumida, H. Kawasaki, *et al.*, "Process integration technology and device characteristics of CMOS FinFET on bulk silicon substrate with sub-10 nm fin width and 20 nm gate length." In: *IEEE Electron Devices Meeting Technical Digest*, pp. 721–724, 2005.

15. X. Huang, W.-C. Lee, C. Kuo, *et al.*, "Sub 50-nm FinFET: PMOS." In: *IEEE International Electron Devices Meeting Technical Digest*, pp. 67–70, 1999.

16. D. Hisamoto, W.-C. Lee, J. Kedzierski, *et al.*, "FinFET—A self-aligned double-gate MOSFET scalable to 20 nm," *IEEE Transactions on Electron Devices*, 47(12), pp. 2320–2326, 2000.

17. F.G. Pikus and A. Torres, "Advanced multi-patterning and hybrid lithography techniques." In: *Proceedings of the Asia and South Pacific Design Automation Conference*, pp. 611–616, 2016.

18. H.H. Radamson, Y. Zhang, X. He, *et al.*, "The challenges of advanced CMOS process from 2D to 3D," *Applied Sciences*, 7(10), p. 1047, 2017.

19. Z. Han, M. Keswani, and S. Raghavan, "Megasonic cleaning of blanket and patterned samples in carbonated ammonia solutions for enhanced particle removal and reduced feature damage," *IEEE Transactions on Semiconductor Manufacturing*, 26(3), pp. 400–405, 2013.

20. F. Nouri, O. Laparra, H. Sur, *et al.*, "Optimized shallow trench isolation for Sub-0.18-um ASIC technologies." In: *Proceedings of the SPIE Conference on Microelectronic Device Technology II*, vol. 3506, pp. 156–166, 1998.

21. S. Saha, "Design considerations for 25 nm MOSFET devices," *Solid-State Electronics*, 45(10), pp. 1851–1857, 2001.

22. H. Yu, M. Schaekers, S.A. Chew, *et al.*, "Titanium (Germano-)silicides featuring 10−9 $\Omega \cdot cm2$ contact resistivity and improved compatibility to advanced CMOS technology." In: *Proceedings of 18th International Workshop on Junction Technology*, pp. 1–5, 2018.

23. M. Sinha, R.T.P. Lee, K.-M. Tan, *et al.*, "Novel aluminum segregation at NiSi/p+-Si source/drain contact for drive current enhancement in P-channel FinFETs," *IEEE Electron Device Letters*, 30(1), pp. 85–87, 2009.

24. V.L. Stout and M.D. Gibbons, "Gettering of gas by titanium," *Journal of Applied Physics*, 26(12), pp. 1488–1492, 1955.

25. G.A. Dixit, W.Y. Hsu, A.J. Konecni, *et al.*, "Ion Metal Plasma (IMP) deposited titanium liners for 0.25/0.18 μm multilevel interconnections." In: *IEEE Electron Devices Meeting Technical Digest*, pp. 357–360, 1996.

26. K.J. Kuhn, M.D. Giles, D. Becher, *et al.*, "Process technology variation," *IEEE Transactions on Electron Devices*, 58(8), pp. 2197–2208, 2011.

27. S.K. Saha, "Modeling process variability in scaled CMOS technology," *IEEE Design & Test of Computers*, 27(2), pp. 8–16, 2010.

5 Large Geometry FinFET Device Operation

5.1 INTRODUCTION

In Chapter 4, we presented the basic fabrication processes of fin field-effect transistor (FinFET) devices. It is shown that a FinFET structure consists of an ultrathin vertical silicon fin on a substrate as the *channel*, a thin dielectric layer of thickness T_{ox} on the sidewalls as well as on top of the fin as the *gate*, a conducting metal layer of length L_g on the top of the dielectric as the *gate electrode*, and the heavily doped regions uncovered by gate-stack at the two ends of the fin body as the source and drain, respectively. The source and drain regions encroach the channel under the gate at both ends of the gate length. The length of the fin between the source and drain under the gate is defined as the *channel* of the FinFET. The channel and source-drain are doped with opposite types of impurities (e.g., *p*-type channel with n^+-type source-drain or *n*-type channel with p^+-type source-drain). The source and drain terminals contact the two opposite ends of the channel so that a potential difference can be applied across the channel to cause a current flow in the FinFET structure with appropriate gate bias. In this chapter, the basic theory of this current flow from the source to drain in FinFET devices is presented. In Chapter 1, we also discussed that if a thick dielectric layer is deposited on the top of the ultrathin channel of a FinFET structure described in Chapter 4, the FinFET device can be operated as a double-gate (DG) metal-oxide-semiconductor field-effect transistor (MOSFET) device. Therefore, in this chapter, the electrostatic performance of FinFET devices for any combination of static biasing conditions is discussed by considering FinFETs as the DG-MOSFETs.

The mathematical formulation of the electrostatic behavior of multiple-gate or multigate MOSFETs with all possible types of gate configurations and substrate doping concentrations has been the subject of numerous investigations [1–10]. The major objective of this chapter is to build the basic foundation for the development of FinFET device theory and drain current expression that will be used in Chapter 10 to discuss the FinFET compact models for computer-aided circuit design (CAD) [11,12]. Throughout this chapter, we will assume ideal FinFET devices with sufficiently *long* and *wide* channel so that the real device effects are negligibly small and the channel length is longer than the fin thickness. Unless otherwise stated, we will also assume that the fin body is *undoped* (however, not intrinsic), or uniformly lightly doped *p*-type silicon. We will introduce the basic drain current equation in a systematic way. However, before describing the mathematical formulations, let us review the basic features of FinFET devices described in Chapters 1 and 4 for better appreciation of the rigorous mathematical steps of the complex three-dimensional (3D) FinFET device structure.

5.2 BASIC FEATURES OF FinFET DEVICES

The three-dimensional structure of an ideal DG-FinFET device on a bulk-silicon substrate is shown in Figure 5.1(a) and the two-dimensional (2D) cross-section of the structure along the cutline XX' is shown in Figure 5.1(b). As shown in Figure 5.1, a thick masking oxide of thickness T_{mask} is used to electrically operate the devices as a DG-FinFET with sidewall gates only. Thus, a DG-FinFET can be characterized by a gate or channel length L_g; fin height H_{fin}; fin thickness or width t_{fin}; gate oxide thickness T_{ox}; body or substrate doping N_b; and source-drain regions of junction X_j (not shown in Figure 5.1) underneath the gate oxide. In bulk-FinFET fabrication, shallow trench isolation (STI) with field oxide thickness T_{fox} as shown in Figure 5.1(a) is used to isolate between devices fabricated on the same substrate.

From Figure 5.1, we can easily visualize a DG-FinFET structure as the DG-MOSFET device structure. For illustration, let us consider the active transistor region from gate-to-gate along the x-axis and up to H_{fin} along the z-axis of the 2D-FinFET cross-section shown on Figure 5.1(b) to get the structure shown in Figure 5.2(a); where: y-axis (source-drain and L_g) is perpendicular to the surface. First of all, we rotate this 2D-active transistor structure, Figure 5.2(a), by 90 degrees to the right (clockwise) around the y-axis to bring the left-hand sidewall gate as the top gate and right-hand sidewall gate as the bottom gate representing the zx-plane (x-axis is vertical, z-axis is parallel to the surface, and y-axis, source-drain, and L_g is perpendicular to the surface) as shown in Figure 5.2(b). Finally, rotate the structure Figure 5.2(b) by 90 degrees to the left (counterclockwise) around the x-axis to get the DG-MOSFET structure (yx-plane) as shown in Figure 5.2(c). Thus, a DG-FinFET can safely be represented by a DG-MOSFET.

Again, under an appropriate biasing condition, the current in a FinFET device flows from the source to drain through the entire surface of each fin height H_{fin} (under the gate-stack) from the corresponding sidewall gate, therefore, the width (W) of a DG-FinFET device is given by $2H_{fin}$. On the other hand, for a triple-gate FinFET device, the current also flows from the top gate of the fin body so that $W = 2H_{fin} + t_{fin}$. However, the mathematical formulation of DG-MOSFET is applicable to the

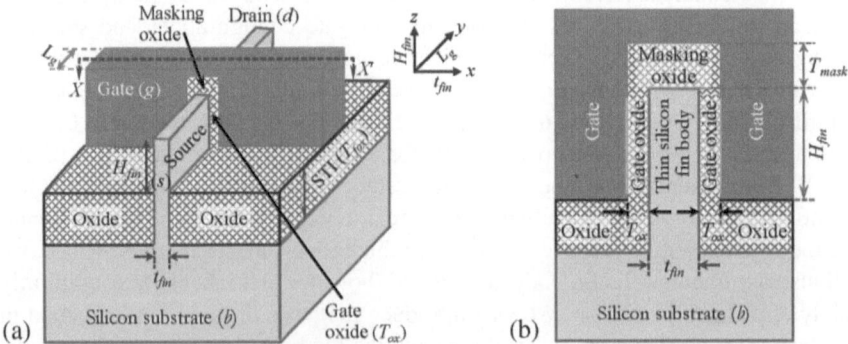

FIGURE 5.1 Ideal DG-FinFET on bulk substrate structure: (a) 3D structure; (b) 2D cross-sectional view along the cutline XX' of the 3D structure representing xz-plane; here, L_g is the gate length and t_{fin} is the fin width of the device.

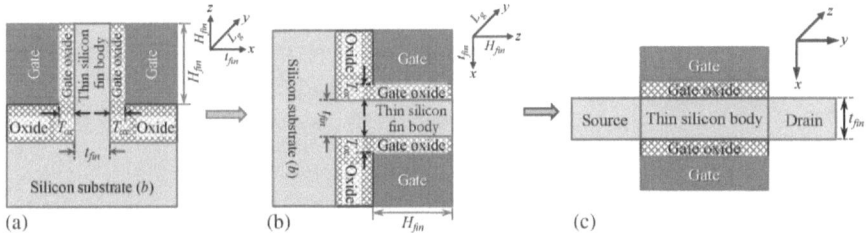

FIGURE 5.2 2D cross-section of a DG-FinFET structure: (a) active region of the device from gate-to-gate and fin height; (b) active region of the device after 90-degree rotation clockwise (left to right) showing fin height (along z-axis) and fin thickness (along x-axis), zx-plane with source-drain and gate perpendicular to the surface (along y-axis); (c) yx-plane of the DG-FinFET structure in (b) after 90-degree counterclockwise (from right to left) rotation around x-axis resulting in a DG-MOSFET device showing source-drain regions with body terminal perpendicular into or behind the surface.

triple-gate FinFET devices by appropriate consideration of the width of the device. Thus, in reality, the top gate of a FinFET device contributes only to the effective width and current drive of the device without affecting its mathematical description. Therefore, in this chapter, the mathematical formulation of the current flow in FinFET devices is presented considering FinFETs as DG-MOSFETs.

It is found that properly designed FinFETs with appropriate target values of channel length L and fin thickness t_{fin} offer short-channel effects (SCE) immunity. In order to suppress SCEs in FinFETs, the requirement for device architecture is $t_{fin} \ll L$ [11–13]. This scaling relation between t_{fin} and L can easily be understood from the basic principle of semiconductor physics. Thus, to establish the gate scaling rule, let us consider an ideal symmetric DG-FinFET structure as shown in Figure 5.3.

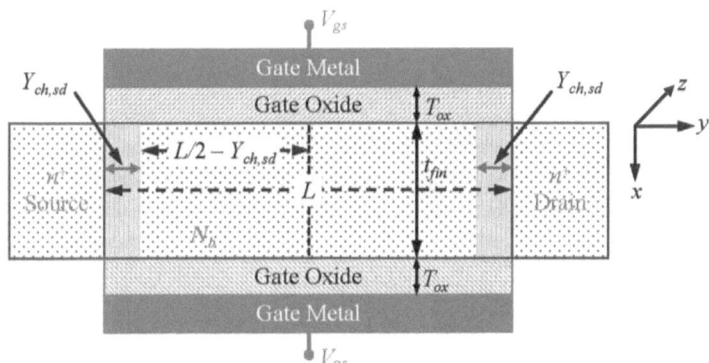

FIGURE 5.3 A typical symmetric n-channel DG-FinFET device structure: t_{fin}, T_{ox}, and N_b are the fin thickness, gate oxide thickness, and body doping concentration, respectively; $Y_{ch,sd}$ is the depletion width in the y-direction along the channel length due to the applied drain bias V_{ds} (not shown).

In order to ensure a complete gate control of the channel, it is required that t_{fin} is completely depleted by gate bias V_{gs} only so that the fin depletion width ($X_{ch,g}$) satisfies the relation

$$X_{ch,g} \geq \frac{t_{fin}}{2} \tag{5.1}$$

where:
$X_{ch,g}$ is the thickness of the depletion region due to each gate

Again, in order to prevent the source-drain punchthrough, the width ($Y_{ch,d}$) of the lateral channel depletion region due to V_{ds} at each end of the channel must be such that the neutral channel length ($L/2 - Y_{ch,sd}$) in the y-direction along the channel must satisfy

$$\frac{L}{2} - Y_{ch,sd} > X_{ch,g} \tag{5.2}$$

Using the limiting condition, $X_{ch,g} = t_{fin}/2$ from Equation 5.1, we get from Equation 5.2

$$\frac{L}{2} - Y_{ch,sd} > \frac{t_{fin}}{2}. \tag{5.3}$$

Now, expressing the depletion width in silicon $Y_{ch,d}$ in terms of the equivalent gate dielectric thickness [Equation 3.158], and re-arranging Equation 5.3, we get the condition for *scaling* FinFET device structure

$$\frac{t_{fin}}{2} + \frac{K_{si}}{K_{ox}} T_{ox} \ll \frac{L}{2}. \tag{5.4}$$

where:
K_{si} is the dielectric constant of silicon fin
K_{ox} is the dielectric constant of gate oxide
T_{ox} is the gate oxide thickness

Typically, $Y_{ch,d} \ll t_{fin}/2$, therefore, the scaling rule for FinFETs can safely be defined by

$$\frac{t_{fin}}{2} \ll \frac{L}{2}. \tag{5.5}$$

Thus, if the fin is sufficiently thin with a thickness t_{fin} smaller than L, then SCEs are suppressed and subthreshold swing (S) is expected to be near its ideal value of about 60 mV/decade (at room temperature) [13]. Thus, the new device architecture results in a new *scaling rule* given in Equation 5.5, that is, L can be scaled by maintaining the condition $t_{fin} < L$, relaxing the constraints on scaling the gate dielectric thickness and body doping.

5.3   FinFET DEVICE OPERATION

Figures 5.1 and 5.4 show that a bulk-FinFET is a four-terminal device with *gate g*, *source s*, *drain d*, and *substrate* or *body b* perpendicular to the surface (not shown) with applied biases V_g, V_s, V_d, and V_b, respectively, and L and W are the channel length and channel width of the device, respectively. The devices may be fabricated as symmetrical DG-FinFETs with same gate oxide thickness for the top and bottom gates or asymmetric DG-FinFETs with different gate oxide thicknesses for the top and bottom gates. The body terminal offers more flexibility of devices at circuit operation. The device may be operated as a common gate configuration with all gates tied together with a common gate voltage or independently operate to modulate the channel inversion layers for device-specific applications.

Theoretically, a FinFET device has three modes of operation such as *accumulation*, *depletion*, and *inversion* similar to a multigate MOS capacitor system discussed in Chapter 3. Therefore, the theory developed for multigate MOS capacitors can be directly extended to FinFETs by considering the channel potential due to the lateral electric field from the source to drain terminals of the structure shown in Figure 5.4.

In conventional common gate FinFET device operation, the source is used as the reference terminal with bias $V_s = 0$ and a drain voltage V_{ds} with reference to the source is applied to the drain so that the source-drain *pn*-junctions are reverse biased. Under this biasing condition, the body or substrate current, $I_b = 0$ and the gate current, $I_g = 0$. The applied gate voltage with reference to source bias V_{gs} controls the surface carrier densities. A certain value of V_{gs}, defined as the threshold voltage (V_{th}), is required to create the channel *inversion layer*; where V_{th} is determined by the properties of the structure and the integrated circuit (IC) technology. Thus, with reference to source potential,

- For $V_{gs} < V_{th}$, the FinFET structure consists of two back-to-back *pn*-junctions and only leakage currents (~ I_o of source-drain *pn*-junctions) flow from the source to drain of the device, that is, $I_{ds} \sim 0$
- For $V_{gs} > V_{th}$, an inversion layer exists, that is, a conducting channel exists from the drain to source of the device and a drain current I_{ds} will flow

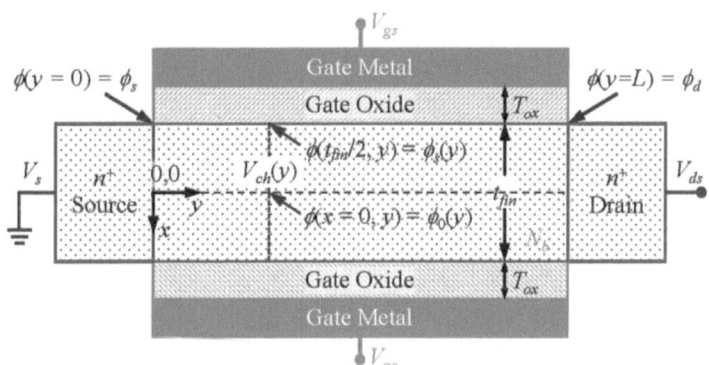

FIGURE 5.4   2D cross-section of an *n*-channel DG-FinFET showing the biasing conditions and co-ordinate system; *x* and *y* represent the spaces along the fin thickness and gate length of the device, respectively.

In the case of independent gate FinFETs, the two gates can independently modulate the inversion layer to offer more flexibility of devices at circuit operation. In common multigate bulk-FinFET device operation, V_{bs} is applied to reverse bias the source/drain pn-junctions.

For mathematical formulation describing the performance of FinFETs, all multitigate MOS capacitor equations derived in Chapter 3 are valid for large L and large H_{fin} devices with appropriate consideration of the lateral electric field E_y due to the applied drain bias V_{ds} as shown in Figure 5.4. Thus, to derive the drain current I_{ds} expression for FinFETs, let us consider the source potential $V_s = 0$ as the reference voltage. Then due to the applied V_{ds}, the surface potential ϕ_s is a function of location y along the channel such that $\phi_s = \phi_s(y)$. Therefore, a channel potential $V_{ch}(y)$ exists along the channel from the source to drain such that

$$V_{ch}(y) = \begin{cases} V_{sb}; & \text{at } y = 0 \\ V_{sb} + V_{ds}; & \text{at } y = L. \end{cases} \tag{5.6}$$

where:

V_{sb} is the source voltage with reference to body bias, if any (for bulk-FinFETs)

In order to derive mathematical expressions to model FinFET device performance, let us consider an n-channel DG-FinFET device with lightly doped substrate with concentration N_b as shown in Figure 5.4. For the sake of simplicity, we assume a large geometry device so that the effects of geometry on device performance can be neglected. We will derive a generalized large geometry FinFET drain current expression using several simplifying assumptions.

5.4 DRAIN CURRENT FORMULATION

In general, the static and dynamic characteristics of a semiconductor device under the influence of external fields can be described by the following three sets of coupled differential equations.

1. The Poisson's equation for electrostatic potential ϕ given by [Equation 2.72]

$$\nabla^2 \phi = -\frac{\rho}{K_{si}\varepsilon_0} \tag{5.7}$$

where:
   ρ is the charge density
   K_{si} is the dielectric constant of silicon
   ε_0 is the permittivity of free space

2. The *current density* equations for electron current density (J_n) and hole current density (J_p) [Equations 2.76 and 2.77]

$$J_n = q\mu_n nE + qD_n\nabla n \qquad \text{(electrons)}$$

$$J_p = q\mu_p pE - qD_p\nabla p \qquad \text{(holes)} \tag{5.8}$$

The expressions in Equation 5.8 under the non-equilibrium condition are represented by

$$J_n = -qn\mu_n\nabla\phi_n \quad \text{(electrons)}$$
$$J_p = -qp\mu_p\nabla\phi_p \quad \text{(holes)}$$

(5.9)

where:

n and p represent electron and hole concentrations, respectively
q is the electronic charge
E is the electric field
ϕ_n and ϕ_p are the electron and hole quasi-Fermi potentials, respectively, in non-equilibrium condition
μ_n and μ_p are the electron and hole mobilities, respectively

The total current density (J) flowing through the device is given by $J = J_n + J_p$.
3. The *current continuity* equations for electrons [Equation 2.81] and holes [Equation 2.80]

$$\frac{\partial n}{\partial t} = \frac{1}{q}\nabla.J_n - (G_n - R_n) \quad \text{(electrons)}$$
$$\frac{\partial p}{\partial t} = -\frac{1}{q}\nabla.J_p + (G_p - R_p) \quad \text{(holes)}$$

(5.10)

where:

G_n and G_p are the generation rates for electrons and holes, respectively
R_n and R_p represent the recombination rates for electrons and holes, respectively

Generally, mathematical formulation for FinFET device performance is a 3D problem, however, assuming large geometry devices, we can treat a large geometry FinFET device as a 2D problem in the x and y directions only as shown in Figure 5.4. Even as a 2D problem, the mathematical expressions are fairly complex and can only be solved exactly using numerical device simulation tools [14–17] and techniques [18]. However, in order to obtain a simplified analytical expression for I_{ds}, we make a number of valid simplifying assumptions to derive basic device theory to mathematically describe the behavior of FinFET devices.

In the following sections, we make a number of valid simplifying assumptions to develop a generalized expression for drain current I_{ds} of large geometry FinFETs with a uniformly doped fin body.

Assumption 1: First of all, we assume that the variation of the electric field E_y in the y-direction along the channel is much less than the corresponding variation of the electric field E_x in the x-direction into the substrate. Then we have

$$\frac{\partial E_y}{\partial y} \ll \frac{\partial E_x}{\partial x}; \quad \therefore \frac{\partial^2\phi}{\partial y^2} \ll \frac{\partial^2\phi}{\partial x^2}$$

(5.11)

Equation 5.11 is referred to as the *gradual channel approximation* (GCA) [19]. Therefore, like a multigate MOS capacitor, we solve for ϕ in the x-direction along the thickness of the channel only to obtain the total charge Q_s in the semiconductor. For DG-FinFETs in *inversion* ($\phi_B < \phi_s < 2\phi_B$), the total charge $Q_s = Q_s(y)$ due to the channel potential $V_{ch}(y)$ can be derived from the multigate MOS capacitor theory [Equation 3.53]. Thus, we assume GCA so that we need to solve only one-dimensional (1D) Poisson's equation described in Equation 2.72 and is given by

$$\frac{d^2\phi}{dx^2} = -\frac{\rho(x)}{\varepsilon_{si}} \qquad (5.12)$$

where:

$\rho(x)$ is the net charge density at any point x along the fin thickness t_{fin}

$\varepsilon_{si} = K_{si}\varepsilon_0$ is the permittivity of silicon fin

2D numerical analysis shows that the GCA is valid for most of the channel length except near the drain end of the channel region. Near the drain end of the channel, the longitudinal electric field E_y is comparable to the transverse electric field E_x even for long-channel devices and GCA breaks down [12,20]. In spite of its failure near the drain end, the GCA is used as it reduces the system to a 1D current flow problem. The fact that we have to solve only a 1D Poisson's equation means that the charge expressions developed in Chapter 3 for a multigate MOS capacitor system could be used for a DG-FinFET device, with the modification that charge and potential will now be position-dependent in the y-direction.

Assumption 2: Assume that only minority carriers contribute to I_{ds}; e.g., for an nFinFET device, the hole current can be neglected. Since the fin is undoped or lightly doped, the number of minority carriers in inversion is much higher than the majority carriers. In nFinFETs, the majority carrier holes are created by impact ionization and become important in describing the device characteristics in the avalanche or breakdown regime. However, in the normal range of operation of the FinFET devices, the drain current does not include the breakdown regime and therefore, the assumption that the current in FinFETs is due to the minority carriers is valid under the normal biasing conditions, for example, for an n-channel FinFET, hereafter referred to as the nFinFET, $V_{ds} \geq 0$ and $V_{bs} \leq 0$. Thus, for drain current calculation, we consider only the minority carrier current density J_n for nFinFET devices.

Assumption 3: Assume there are no generation and recombination of carriers, that is, for an nFinFET device $R_n = 0 = G_n$. Then considering only the static characteristics of the device, the continuity Equation 5.10 becomes

$$\nabla \cdot J_n = 0 \qquad (5.13)$$

This implies that the total drain current I_{ds} is a constant at any point along the channel of the device.

Assumption 4: Assume that the current flows in the y-direction along the channel only, that is, $d\phi_n/dx = 0$. Thus, the electron quasi-Fermi potential ϕ_n is a constant in the x-direction. Then from Equation 5.8, the electron current density is given by

$$J_n(y) = -qn(x, y)\mu_n(x, y)\frac{\partial\phi_n}{\partial y} \tag{5.14}$$

Since the cross-sectional area of the channel in which the current flows is the channel width $W \cong 2H_{fin}$ times the channel length L, therefore, integrating Equation 5.14 across the fin thickness x and width z, we get I_{ds} at any point y in the channel as

$$I_{ds}(y) = -W\int_{-t_{fin}/2}^{+t_{fin}/2}\left[qn(x, y)\mu_n(x, y)\frac{\partial\phi_n}{\partial y}\right]dx = \text{constant} \tag{5.15}$$

where:

μ_n is the surface mobility of the channel electrons in nFinFET devices and is often referred to as the surface mobility μ_s in order to distinguish it from the bulk mobility deep into the substrate described in Section 2.2.5.1.

In the rest of the discussion, we will replace μ_n by μ_s to represent the surface mobility of the inversion carriers in FinFET devices.

In device operation, the application of source and drain voltages relative to the substrate results in a lowering of the quasi-Fermi level E_{fn} (or potential ϕ_n) at the source end of the device by an amount qV_{gs}, and the drain end of the device by an amount $(qV_{gs} + qV_{ds})$, relative to equilibrium Fermi level E_f in the body. This difference in ϕ_n between the source and drain drives the electrons down the channel. Thus, the channel potential $V_{ch}(y)$ at any point y in the channel is given by

$$V_{ch}(y) = \phi_n(y) - \phi_n\big|_{\text{source}} \tag{5.16}$$

where:

$\phi_n\big|_{\text{source}}$ is the quasi-Fermi potential at the source end of the device

Again, from Equation 5.6, $V_{ch}(0) = 0$ at the source end of the channel ($V_{sb} = 0$) and $V_{ch}(L) = V_{ds}$ at the drain end of the channel. Thus, compared to the case of an MOS capacitor, the quasi-Fermi potential is lowered by an amount $V_{ch}(y)$ at the surface region of a FinFET device. As a result, the surface electron concentration (n_s) is lowered by a factor $\exp(-V_{ch}(y)/v_{kT})$. Then following the derivation of minority carrier density given by Equation 3.50 for an MOS capacitor, we can write the minority carrier surface electron concentration at any point y in the channel of a FinFET device as

$$n(x, y) = N_b\exp\left[\frac{\phi(x, y) - 2\phi_B - V_{ch}(y)}{v_{kT}}\right]; \tag{5.17}$$

where:

ϕ_B is the bulk potential given by Equation 3.5

$v_{kT} (= kT/q)$ is the *thermal voltage* at the ambient temperature T with k and q representing the Boltzmann constant and electronic charge, respectively

Thus, from Equation 5.17, we find that the minority carrier electron concentration for n-channel FinFETs changes due to the applied bias; however, the majority carrier hole concentration does not change with applied bias and therefore, following MOS capacitor Equation 3.47, we can write for the majority carrier concentration in MOSFETs as $p = N_b \exp\left[(-\phi(x,y))/v_{kT}\right]$.

Now, substituting for ϕ_n from Equation 5.16 into Equation 5.15, the drain current expression in FinFET devices can be written as

$$I_{ds}(y) = -W\frac{dV_{ch}(y)}{dy}\int_{-t_{fin}/2}^{+t_{fin}/2} qn(x,y)\mu_s(x,y)\,dx \qquad (5.18)$$

Assumption 5: For the simplicity of long-channel I_{ds} calculation, we assume μ_s is a constant at some average gate and drain electric field though μ_s depends on both E_x and E_y, as discussed in Chapter 6. With this assumption, we can write Equation 5.18 as

$$I_{ds}(y) = -W\mu_s\frac{dV_{ch}(y)}{dy}\int_{-t_{fin}/2}^{+t_{fin}/2} qn(x,y)\,dx \qquad (5.19)$$

Now, the total charge density Q_i of mobile minority carriers is given by

$$Q_i(y) = -q\int_{-t_{fin}/2}^{+t_{fin}/2} n(x,y)\,dx \qquad (5.20)$$

Then substituting Equation 5.20 into Equation 5.19, we get the general expression for $I_{ds}(y)$ as

$$I_{ds}(y) = W\mu_s Q_i(y)\frac{dV_{ch}}{dy} \qquad (5.21)$$

Again, assuming GCA is valid along the entire length of the channel, we get the expression for I_{ds} after integrating Equation 5.21 along the channel length from $y = 0$ to $y = L$ as

$$I_{ds} = \begin{cases} \left(\dfrac{W}{L}\right)\mu_s\displaystyle\int_0^{V_{ds}} Q_i(y)dV_{ch}; & \text{(surface potential-based } I_{ds}) \\[4mm] \left(\dfrac{W}{L}\right)\mu_s\displaystyle\int_{Q_{is}}^{Q_{id}} Q_i(y)dV_{ch}; & \text{(inversion charge-based } I_{ds}) \end{cases} \qquad (5.22)$$

where:
Q_{is} is the inversion charge density at the source end of the device at $y = 0$
Q_{id} is the inversion charge density at the drain end of the device at $y = L$

In Equation 5.22, we have used $V_{ch}(y = 0) = V_{sb} = 0$ at $y = 0$ since $V_s = 0$ and we assume $V_b = 0$. Equation 5.22 represents the general expressions for I_{ds} flowing through a FinFET device. In order to calculate I_{ds}, we need to calculate the mobile inversion charge density $Q_i(y)$ in the channel region. In the following section, we will derive an expression for I_{ds} by computing $Q_i(y)$.

5.4.1 Derivation of Electrostatic Potential

In order to derive an expression for I_{ds} from Equation 5.22, first of all, we solve for potential $\phi(x,y)$ in the ultrathin-body fin of FinFET devices to obtain surface potential. To solve for $\phi(x,y)$, let us consider a symmetric n-channel FinFET device as shown in Figure 5.4 with undoped (or lightly p-type doped) fin so that the bulk term N_b is negligibly small. Therefore, considering only the mobile charge (electrons) term in Equation 3.54, Poisson's equation can be expressed as

$$\frac{d^2\phi(x,y)}{\partial x^2} = \frac{qn_i}{\varepsilon_{si}} \exp\left(\frac{\phi(x,y) - \phi_B - V_{ch}(y)}{v_{kT}}\right) \tag{5.23}$$

where:
n_i = intrinsic carrier concentration
$\phi(x,y)$ = electrostatic potential
$V_{ch}(y)$ = channel potential due to drain bias along the y-direction
ϕ_B is the bulk potential defined in Equation 3.43

Here, we consider an nFinFET device with $\phi(x,y)/v_{kT} \gg 1$, that is, the band bending is downward (Section 3.2.5). Since the current flows predominantly from the source to the drain along the y-direction, the gradient of the electron quasi-Fermi level is also in the y-direction. This justifies the gradual channel approximation so that $V_{ch}(y)$ is a constant in the x-direction. Then Equation 5.23 can be solved for surface potential $\phi(x)$. Now, following the procedure used in Section 3.4.2, we can express Equation 5.23 as

$$\frac{d}{dx}\left(\frac{d\phi(x,y)}{dx}\right)^2 = \frac{2qn_i}{\varepsilon_{si}}\left[e^{\frac{\phi(x,y)-\phi_B-V_{ch}(y)}{v_{kT}}}\right]\frac{d\phi(x,y)}{dx} \tag{5.24}$$

Now, we integrate Equation 5.24 from the center point, $x = 0$ to a point (x,y) to find the potential distribution along the thickness of the structure shown in Figure 5.4. Since for a symmetric DG-FinFET structure the vertical component of the electric field, $E(x) = 0 = -d\phi/dx$ at the center, $x = 0$; where $\phi(x = 0,y) = \phi_0(y)$, we use the following boundary conditions for integration of Equation 5.24.

$$\begin{cases} \phi(x,y) = \phi_0(y), \dfrac{d\phi(x,y)}{dx} = 0; & \text{at } x = 0 \\[3mm] \phi(x,y) = \phi(x), \dfrac{d\phi(x,y)}{dx} = \dfrac{d\phi(x,y)}{dx}; & \text{at } x = x \end{cases} \tag{5.25}$$

Then, using the boundary conditions given by Equation 5.25, we can express Equation 5.24 as

$$
\int_{0}^{\frac{d\phi(x,y)}{dx}} d\left(\frac{d\phi(x,y)}{dx}\right)^2 = \frac{2qn_i}{\varepsilon_{si}} \int_{\phi_0(y)}^{\phi(x,y)} e^{\frac{\phi(x,y)-\phi_B-V_{ch}(y)}{v_{kT}}} d\phi(x,y) \tag{5.26}
$$

After integrating Equation 5.26 and using $n_i \exp\left(-\phi_B / v_{kT}\right) = n_i^2 / N_b$ from Equation 3.49, we get

$$
\left(\frac{d\phi(x,y)}{dx}\right)^2 = \frac{2q}{\varepsilon_{si}} v_{kT} \frac{n_i^2}{N_b}\left(e^{\frac{\phi(x,y)-V_{ch}(y)}{v_{kT}}} - e^{\frac{\phi_0(y)-V_{ch}(y)}{v_{kT}}}\right) \tag{5.27}
$$

Or,

$$
\frac{d\phi(x,y)}{dx} = \pm\sqrt{\frac{2v_{kT}q}{\varepsilon_{si}} \frac{n_i^2}{N_b}\left(e^{\frac{\phi(x,y)-V_{ch}(y)}{v_{kT}}} - e^{\frac{\phi_0(y)-V_{ch}(y)}{v_{kT}}}\right)} \tag{5.28}
$$

where:
  the "+" sign is for $x > 0$
  the "−" sign is for $x < 0$

As discussed in Chapter 3, after the second integration of Equation 5.28, we obtain an expression for $\phi(x,y)$. In order to perform integration on Equation 5.28, let us consider the general mathematical solution of Poison's Equation 5.23. The differential Equation 5.23 has two mathematical solutions: one is a trigonometric function (proportional to $1/\cos^2\xi$) and the other is a hyperbolic function [21,22]. Considering the trigonometric solution similar to that used in Section 3.4.2.2 given by

$$
e^{\frac{\phi(x,y)}{v_{kT}}} = e^{\frac{\phi_0(y)}{v_{kT}}} \sec^2\xi \tag{5.29}
$$

where:
  ξ is a function of position x

Then substituting Equation 5.29 into Equation 5.28, we can show after simplification

$$
\frac{d\phi(x,y)}{dx} = \pm\sqrt{\frac{2v_{kT}q}{\varepsilon_{si}} \frac{n_i^2}{N_b}\left(e^{\frac{\phi_0(y)-V_{ch}(y)}{v_{kT}}}\sec^2\xi - e^{\frac{\phi_0(y)-V_{ch}(y)}{v_{kT}}}\right)}
$$

$$
= \pm\sqrt{\frac{2v_{kT}q}{\varepsilon_{si}} \frac{n_i^2}{N_b} \cdot e^{\frac{\phi_0(y)-V_{ch}(y)}{v_{kT}}}\left(\sec^2\xi - 1\right)} \tag{5.30}
$$

Therefore, $\dfrac{d\phi(x,y)}{dx} = \pm\sqrt{\dfrac{2v_{kT}q}{\varepsilon_{si}} \dfrac{n_i^2}{N_b} \cdot e^{\frac{\phi_0(y)-V_{ch}(y)}{v_{kT}}}\tan^2\xi}$

where:

$$\tan^2 \xi = \left(\sec^2 \xi - 1\right)$$

We can also obtain an expression for $d\phi(x,y)/dx$ by differentiating (trigonometric solution) Equation 5.29 with respect to x, to get

$$\frac{d}{dx}\left(e^{\frac{\phi(x,y)}{v_{kT}}}\right) = \frac{d}{dx}\left(e^{\frac{\phi_0(y)}{v_{kT}}}\sec^2 \xi\right)$$

(5.31)

or, $\frac{d}{d\phi}\left(e^{\frac{\phi(x,y)}{v_{kT}}}\right)\frac{d\phi(x,y)}{dx} = \frac{d}{d\xi}\left(e^{\frac{\phi_0(y)}{v_{kT}}}\sec^2 \xi\right)\frac{d\xi}{dx}$

or, $\frac{e^{\frac{\phi(x,y)}{v_{kT}}}}{v_{kT}}\frac{d\phi(x,y)}{dx} = e^{\frac{\phi_0(y)}{v_{kT}}}2\sec \xi\left(\sec \xi \tan \xi\right)\frac{d\xi}{dx}$ (5.32)

Then substituting for $e^{\frac{\phi(x,y)}{v_{kT}}} = e^{\frac{\phi_0(y)}{v_{kT}}}\sec^2 \xi$ from Equation 5.29 into the left-hand side of Equation 5.32, we get after simplification

$$\frac{d\phi(x,y)}{dx} = 2v_{kT}\tan \xi \frac{d\xi}{dx}$$

(5.33)

Now, the expressions for $d\phi(x,y)/dx$ in Equations 5.30 and 5.33 are obtained from the solution of the same Poisson's Equation 5.23 and therefore, are equal. Then equating Equations 5.30 and 5.33, we get, after simplification

$$2v_{kT}\tan \xi \frac{d\xi}{dx} = \pm\sqrt{\frac{2v_{kT}q}{\varepsilon_{si}}\frac{n_i^2}{N_b}e^{\frac{\phi_0(y)-V_{ch}(y)}{v_{kT}}}\tan^2 \xi}$$

(5.34)

or,

$$\frac{d\xi}{dx} = \pm\sqrt{\frac{q}{2v_{kT}\varepsilon_{si}}\frac{n_i^2}{N_b}e^{\frac{\phi_0(y)-V_{ch}(y)}{v_{kT}}}}$$

(5.35)

Now, we integrate Equation 5.35 from $(x = 0,y)$ to any point (x,y); since $\xi = f(x)$, therefore, at $x = 0$, $\xi = 0$; and at $x = x$; $\xi = \xi$. Therefore,

$$\int_0^\xi d\xi = \sqrt{\frac{q}{2v_{kT}\varepsilon_{si}}\frac{n_i^2}{N_b}e^{\frac{\phi_0(y)-V_{ch}(y)}{v_{kT}}}}\int_0^x dx$$

(5.36)

or,

$$\xi = \sqrt{\frac{q}{2v_{kT}\varepsilon_{si}}\frac{n_i^2}{N_b}e^{\frac{\phi_0(y)-V_{ch}(y)}{v_{kT}}}} \cdot x$$

(5.37)

Also, we can derive another expression for ξ from the general trigonometric solution, Equation 5.29 as shown below

$$e^{\frac{\phi(x,y)}{v_{kT}}} = \frac{e^{\frac{\phi_0(y)}{v_{kT}}}}{\cos^2 \xi} \tag{5.38}$$

$$\text{or,} \quad \cos \xi == \frac{1}{\sqrt{e^{\frac{\phi(x,y)-\phi_0(y)}{v_{kT}}}}} \tag{5.39}$$

$$\xi = \cos^{-1} \left(\frac{1}{\sqrt{e^{\frac{\phi(x,y)-\phi_0(y)}{v_{kT}}}}} \right) \tag{5.40}$$

Thus, we can eliminate ξ by equating Equation 5.37 and Equation 5.40, to get

$$\cos^{-1} \left(\frac{1}{\sqrt{e^{\frac{\phi(x,y)-\phi_0(y)}{v_{kT}}}}} \right) = \sqrt{\frac{q}{2v_{kT}\varepsilon_{si}} \frac{n_i^2}{N_b} e^{\frac{\phi_0(y)-V_{ch}(y)}{v_{kT}}}} \cdot x$$

$$\text{or,} \quad \frac{1}{\sqrt{e^{\frac{\phi(x,y)-\phi_0(y)}{v_{kT}}}}} = \cos \left(\sqrt{\frac{q}{2v_{kT}\varepsilon_{si}} \frac{n_i^2}{N_b} e^{\frac{\phi_0(y)-V_{ch}(y)}{v_{kT}}}} \cdot x \right)$$

$$\tag{5.41}$$

$$\text{or,} \quad e^{-\frac{\phi(x,y)-\phi_0(y)}{2v_{kT}}} = \cos \left(\sqrt{\frac{q}{2v_{kT}\varepsilon_{si}} \frac{n_i^2}{N_b} e^{\frac{\phi_0(y)-V_{ch}(y)}{v_{kT}}}} \cdot x \right)$$

$$\text{or,} \quad -\frac{\phi(x,y)-\phi_0(y)}{2v_{kT}} = \ln \left[\cos \left(\sqrt{\frac{q}{2v_{kT}\varepsilon_{si}} \frac{n_i^2}{N_b} e^{\frac{\phi_0(y)-V_{ch}(y)}{v_{kT}}}} \cdot x \right) \right]$$

After simplification of Equation 5.41, we get the expression for $\phi(x,y)$ as

$$\phi(x,y) = \phi_0(y) - 2v_{kT} \ln \left[\cos \left(\sqrt{\frac{q}{2\varepsilon_{si}v_{kT}} \frac{n_i^2}{N_b} \exp \left(\frac{\phi_0(y)-V_{ch}(y)}{v_{kT}} \right)} \cdot x \right) \right] \tag{5.42}$$

Equation 5.42 is valid for the entire range $-t_{fin}/2 \le x \le t_{fin}/2$. Note that the right-hand side of Equation 5.42 is always positive. Then the surface potential $\phi_s \equiv \phi(x = \pm t_{fin}/2)$ is given by

$$\phi\left(x = \frac{t_{fin}}{2}, y\right) \equiv \phi_s = \phi_0(y) - 2v_{kT} \ln\left[\cos\left(\sqrt{\frac{q}{2\varepsilon_{si}v_{kT}} \frac{n_i^2}{N_b} \exp\left(\frac{\phi_0 - V_{ch}(y)}{v_{kT}}\right)} \cdot \frac{t_{fin}}{2}\right)\right]$$

(5.43)

Following the procedure described in Section 3.4.2.2 and using the boundary condition Equation 3.100, we can express surface potential ϕ_s in terms of the applied gate bias V_{gs}. Then we solve coupled ϕ_s-equations to determine ϕ_s and ϕ_0 for every bias point and calculate current-voltage characteristics of FinFET devices. However, for the simplicity of mathematical formulation, let us derive a surface potential function as described in Section 3.4.2.4 to calculate surface potential using a single continuous equation.

In order to derive the surface potential function, we define a transformation variable $\beta(V_{ch}(y))$ as the argument of the cosine function in $\phi(x = t_{si}/2, y)$ in Equation 5.43 to get

$$\sqrt{\frac{q}{2\varepsilon_{si}v_{kT}} \frac{n_i^2}{N_b} \exp\left(\frac{\phi_0(y) - V_{ch}(y)}{v_{kT}}\right)} \cdot \frac{t_{fin}}{2} \equiv \beta$$

(5.44)

$$\text{or,} \quad \sqrt{\frac{q}{2\varepsilon_{si}v_{kT}} \frac{n_i^2}{N_b} \exp\left(\frac{\phi_0(y) - V_{ch}(y)}{v_{kT}}\right)} = \frac{2\beta}{t_{fin}}$$

Now, let us define

$$a \equiv \sqrt{\frac{q}{2\varepsilon_{si}v_{kT}} \frac{n_i^2}{N_b}};$$

(5.45)

$$b \equiv \sqrt{\frac{q}{2\varepsilon_{si}v_{kT}} \frac{n_i^2}{N_b} \exp\left(\frac{\phi_0(y) - V_{ch}(y)}{v_{kT}}\right)} = \frac{2\beta}{t_{fin}}$$

Then from Equation 5.45, we can show

$$a \exp\left(\frac{\phi_0(y) - V_{ch}(y)}{2v_{kT}}\right) = b = \frac{2\beta}{t_{fin}}$$

(5.46)

$$\text{or,} \quad \exp\left(\frac{\phi_0(y) - V_{ch}(y)}{2v_{kT}}\right) = \frac{b}{a} = \frac{1}{a} \frac{2\beta}{t_{fin}}$$

From the second expression of Equation 5.46, we get

$$\phi_0(y) = V_{ch}(y) + 2v_{kT} \ln\left(\frac{b}{a}\right) = V_{ch}(y) + 2v_{kT} \ln\left(\frac{1}{a} \frac{2\beta}{t_{fin}}\right)$$

(5.47)

Now, substituting for the argument of cosine function (β) from Equation 5.45 and $\phi_0(y)$ from Equation 5.47, we can express Equation 5.42 for $\phi(x, y)$, in terms of the transformation parameter β as

$$\phi(x,y) = \phi_0(y) - 2v_{kT} \ln\left[\cos\left(\sqrt{\frac{q}{2\varepsilon_{si}v_{kT}}\frac{n_i^2}{N_b}}\exp\left(\frac{\phi_0(y) - V_{ch}(y)}{v_{kT}}\right) \cdot x\right)\right]$$

$$= V_{ch}(y) + 2v_{kT} \ln\left(\frac{1}{a}\frac{2\beta}{t_{fin}}\right) - 2v_{kT}\ln\left[\cos\left(\frac{2\beta}{t_{fin}} \cdot x\right)\right]$$

(5.48)

$$= V_{ch}(y) - 2v_{kT}\left\{-\ln\left(\frac{1}{a}\frac{2\beta}{t_{fin}}\right) + \ln\left[\cos\left(\frac{2\beta}{t_{fin}} \cdot x\right)\right]\right\}$$

$$= V_{ch}(y) - 2v_{kT}\ln\left\{a \cdot \frac{t_{fin}}{2\beta}\cos\left(\frac{2\beta}{t_{fin}} \cdot x\right)\right\}$$

Then substituting for the expression for the parameter a from Equation 5.45, we can express the potential $\phi(x,y)$ at any point (x,y) of the ultrathin-body fin in terms of the transformation variable β as

$$\phi(x,y) = V_{ch}(y) - 2v_{kT}\ln\left[\frac{t_{fin}}{2\beta}\sqrt{\frac{q}{2\varepsilon_{si}v_{kT}}\frac{n_i^2}{N_b}}\cos\left(\frac{2\beta}{t_{fin}}x\right)\right]$$

(5.49)

Rearranging Equation 5.49, we can show

$$\phi(x,y) = V_{ch}(y) - 2v_{kT}\left[-\ln\beta - \ln\left(\frac{2}{t_{fin}}\sqrt{\frac{2\varepsilon_{si}v_{kT}}{q}\frac{N_b}{n_i^2}}\right) + \ln\left\{\cos\left(\frac{2\beta}{t_{fin}}x\right)\right\}\right]$$

(5.50)

From Equation 5.50, we can show that the surface potential, at $x = \pm t_{fin}/2$ (Si/SiO$_2$ interface) is given by

$$\phi\left(x = \pm\frac{t_{fin}}{2}, y\right) = V_{ch}(y) - 2v_{kT}\left[-\ln\beta - \ln\left(\frac{2}{t_{fin}}\sqrt{\frac{2\varepsilon_{si}v_{kT}}{qn_i}\frac{N_b}{n_i^2}}\right) + \ln\left\{\cos\left(\beta\right)\right\}\right]$$

(5.51)

Also, from Equation 5.50, the potential gradient with respect to x is given by

$$\frac{d\phi(x,y)}{dx} = 2v_{kT}\left[-\frac{-\sin\left(2\beta x / t_{fin}\right)}{\cos\left(2\beta x / t_{fin}\right)} \cdot \frac{2\beta}{t_{fin}}\right]$$

(5.52)

$$\text{or,} \quad \frac{d\phi(x,y)}{dx} = 2v_{kT}\frac{2}{t_{fin}}\beta\tan\left(\frac{2\beta x}{t_{fin}}\right)$$

where:

β is a constant (of x) to be determined from the boundary condition at the Si/SiO$_2$ interface at $x = \pm t_{fin}/2$

In order to derive an expression for β relating the gate voltage V_{gs} and surface potential ϕ_s, we use the boundary condition given by Equation 3.100, we get

$$\varepsilon_{ox}\frac{V_{gs} - V_{fb} - \phi(x = \pm t_{fin}/2)}{T_{ox}} = \pm\varepsilon_{si}\frac{d\phi}{dx}\bigg|_{x=\pm t_{fin}/2} \quad (5.53)$$

where:

ε_{ox} is the permittivity of gate oxide

V_{fb} = flat band voltage of the silicon body $\cong \Phi_{ms}$ = workfunction difference between metal gate and fin channel

T_{ox} = gate oxide thickness

Now, from Equation 5.52, we can show

$$\frac{d\phi}{dx}\bigg|_{x=\pm t_{fin}/2} = 2v_{kT}\frac{2}{t_{fin}}\beta\tan\beta \quad (5.54)$$

Then substituting for $\phi(x = \pm t_{fin}/2)$ from Equation 5.51 and $d\phi(x = \pm t_{fin}/2)/dx$ from Equation 5.54 into Equation 5.53, we get

$$\frac{\varepsilon_{ox}}{T_{ox}}\left[V_{gs} - V_{fb} - V_{ch}(y) + 2v_{kT}\left\{-\ln\beta - \ln\left(\frac{2}{t_{fin}}\sqrt{\frac{2\varepsilon_{si}v_{kT}}{q}\frac{N_b}{n_i^2}}\right) + \ln(\cos\beta)\right\}\right] = \varepsilon_{si}\left(2v_{kT}\frac{2}{t_{fin}}\beta\tan\beta\right)$$

or,
$$\frac{V_{gs} - V_{fb} - V_{ch}(y)}{2v_{kT}} - \ln\beta - \ln\left(\frac{2}{t_{fin}}\sqrt{\frac{2\varepsilon_{si}v_{kT}}{q}\frac{N_b}{n_i^2}}\right) + \ln(\cos\beta) = \frac{2\varepsilon_{si}T_{ox}}{\varepsilon_{ox}t_{fin}}\beta\tan\beta$$

$$(5.55)$$

Considering undoped or lightly doped substrate with $n_i^2/N_b = n_i$ [Equation 3.49], we can further simplify Equation 5.55 to get

$$\frac{V_{gs} - V_{fb} - V_{ch}(y)}{2v_{kT}} - \ln\left(\frac{2}{t_{fin}}\sqrt{\frac{2\varepsilon_{si}v_{kT}}{qn_i}}\right) = \ln\beta - \ln(\cos\beta) + \frac{2\varepsilon_{si}T_{ox}}{\varepsilon_{ox}t_{fin}}\beta\tan\beta \quad (5.56)$$

Again, from Equation 5.44, we find that β is a function of the channel potential $V_{ch}(y)$. Thus, for a given value of V_{gs}, the parameter β can be obtained from Equation 5.56. Both V_{gs} and β vary from source to drain and therefore, depend on the position y along the direction of the current flow.

Now, for the simplicity of mathematical presentation, we define

$$\frac{\varepsilon_{si} T_{ox}}{\varepsilon_{ox} t_{fin}} \equiv r \tag{5.57}$$

where:

r is a structural parameter of a FinFET device

Then Equation 5.56, can be expressed as

$$\frac{V_{gs} - V_{fb} - V_{ch}(y)}{2v_{kT}} - \ln\left(\frac{2}{t_{fin}}\sqrt{\frac{2\varepsilon_{si}v_{kT}}{qn_i}}\right) = \ln\beta - \ln(\cos\beta) + 2r\beta\tan\beta \equiv f(\beta) \tag{5.58}$$

We will use Equation 5.58 to formulate drain current in FinFET devices under different biasing conditions.

5.4.2 CONTINUOUS DRAIN CURRENT EQUATION FOR SYMMETRIC DG-FinFETs

In order to derive an expression for continuous drain current I_{ds} equation for FinFETs, the fundamental dependence of $V_{ch}(y)$ and $\beta(y)$ is determined by the current continuity condition as described in Section 5.4. Therefore, assuming GCA is valid along the entire length of the channel, we get after integrating Equation 5.22 along the channel length from $y = 0$ to $y = L$

$$I_{ds} = \mu_s\left(\frac{W}{L}\right)\int_0^{V_{ds}} Q_i(y)dV_{ch}(y) \tag{5.59}$$

Now, changing the variable from V_{ch} to β, Equation 5.59 can be expressed as

$$I_{ds} = \mu_s\left(\frac{W}{L}\right)\int_{\beta_s}^{\beta_d} Q_i(\beta)\frac{dV_{ch}}{d\beta}d\beta \tag{5.60}$$

where:

β_s is the value of β when $V_{ch}(y) = 0$ at the source end
β_d is the value of β when $V_{ch}(y) = V_{ds}$ at the drain end

For any given V_{gs}, both β_s and β_d are determined from Equation 5.44. Thus, in order to derive I_{ds} from Equation 5.60, we need to determine the expression for the mobile or inversion charge $Q_i(y)$ in the fin body of FinFETs. We know that the total charge Q_s in silicon is given by

$$Q_s = Q_i + Q_b \tag{5.61}$$

For an undoped or lightly doped body, $Q_i \gg Q_b$, therefore, from Gauss' law [Equation 3.25], we can show

$$Q_i \cong Q_s = 2\varepsilon_{si} E_s = 2\varepsilon_{si} \left(\frac{d\phi(x)}{dx} \right) \Bigg|_{x=t_{fin}/2} \tag{5.62}$$

where:

    factor "2" on the right-hand side is used to account for the two gates of the DG-FinFET device

    E_s is the surface electric field $= d\phi/dx$ at $x = t_{fin}/2$

Now, to derive an expression for I_{ds} from Equation 5.60, let us find $Q_i(\beta)$ from Equation 5.62. Then substituting for $d\phi/dx$ at $x = t_{fin}/2$ from Equation 5.54 into Equation 5.62, we get

$$Q_i(\beta) = \frac{4\varepsilon_{si}}{t_{fin}} \left(2v_{kT} \right) \beta \tan \beta \tag{5.63}$$

Now, substituting Equation 5.63 into Equation 5.60, we get

$$I_{ds} = \mu_s \left(\frac{W}{L} \right) \frac{4\varepsilon_{si}}{t_{fin}} \left(2v_{kT} \right) \int_{\beta_s}^{\beta_d} \beta \tan \beta \frac{dV_{ch}}{d\beta} d\beta \tag{5.64}$$

Then in order to evaluate Equation 5.64, let us find $dV_{ch}/d\beta$ from Equation 5.58 which is given after rearranging as

$$V_{ch}(y) = V_{gs} - V_{fb} - 2v_{kT} \left[\ln \beta - \ln(\cos \beta) + 2r\beta \tan \beta + \ln \left(\frac{2}{t_{fin}} \sqrt{\frac{2\varepsilon_{si} v_{kT}}{qn_i}} \right) \right] \tag{5.65}$$

Thus, we get

$$\frac{dV_{ch}}{d\beta} = -2v_{kT} \left[\frac{1}{\beta} + \tan \beta + 2r \frac{d}{d\beta} \left(\beta \tan \beta \right) \right] \tag{5.66}$$

Now, substituting for $dV_{ch}/d\beta$ from Equation 5.66 into Equation 5.64, we get

$$I_{ds} = -\mu_s \left(\frac{W}{L} \right) \frac{4\varepsilon_{si}}{t_{fin}} \left(2v_{kT} \right)^2 \int_{\beta_s}^{\beta_d} \beta \tan \beta \left[\frac{1}{\beta} + \tan \beta + 2r \frac{d}{d\beta} \left(\beta \tan \beta \right) \right] d\beta$$

$$\text{or,} \quad I_{ds} = \mu_s \left(\frac{W}{L} \right) \frac{4\varepsilon_{si}}{t_{fin}} \left(2v_{kT} \right)^2 \int_{\beta_d}^{\beta_s} \left[\tan \beta + \beta \tan^2 \beta + 2r \left(\beta \tan \beta \frac{d}{d\beta} \left(\beta \tan \beta \right) \right) \right] d\beta$$

$$\tag{5.67}$$

To derive the final expression for I_{ds}, we use the following integral solutions

$$\int (\tan \beta)\,d\beta = -\ln(\cos \beta)$$

$$\int (\beta \tan^2 \beta)\,d\beta = \beta \tan \beta + \ln(\cos \beta) - \frac{\beta^2}{2} \tag{5.68}$$

$$\int (\beta \tan \beta)\frac{d}{d\beta}(\beta \tan \beta)\,d\beta = \frac{1}{2}\int \frac{d}{d\beta}(\beta \tan \beta)^2\,d\beta = \frac{1}{2}\beta^2 \tan^2 \beta$$

Substituting the integral solutions from Equation 5.68 into Equation 5.67, we get the expression for continuous I_{ds} after simplification as

$$I_{ds} = \mu_s \left(\frac{W}{L}\right)\frac{4\varepsilon_{si}}{t_{fin}}(2v_{kT})^2 \left[\beta \tan \beta - \frac{\beta^2}{2} + r\beta^2 \tan^2 \beta\right]_{\beta_d}^{\beta_s} \tag{5.69}$$

It is to be noted that Equation 5.69 is a continuous DG-FinFET drive current expression applicable for all regions of device operation from accumulation to strong inversion and must be solved numerically. Thus, I_{ds} can be computed from solution of coupled Equations 5.69 and 5.58. For the simplicity of I_{ds} computation, let us define term within the square brackets of Equation 5.69 as

$$g_r(\beta) = \beta \tan \beta - \frac{\beta^2}{2} + r\beta^2 \tan^2 \beta \tag{5.70}$$

Then Equation 5.69 can be expression as

$$I_{ds} = \mu_s \left(\frac{W}{L}\right)\frac{4\varepsilon_{si}}{t_{fin}}(2v_{kT})^2 \left[g_r(\beta_s) - g_r(\beta_d)\right] \tag{5.71}$$

In order to compute I_{ds}, β is determined from Equation 5.58 and is defined as

$$f_r(\beta) = \ln \beta - \ln(\cos \beta) + 2r\beta \tan \beta \tag{5.72}$$

Therefore, from Equation 5.58, we can write

$$f_r(\beta) = \frac{V_{gs} - V_{fb} - V_{ch}(y)}{2v_{kT}} - \ln\left(\frac{2}{t_{fin}}\sqrt{\frac{2\varepsilon_{si}v_{kT}}{qn_i}}\right) \tag{5.73}$$

Thus, for any given set of values of V_{gs} and V_{ds}, the parameters β_s and β_d can be computed from the surface potential function given by Equations 5.72 and 5.73 and then used in Equation 5.70 to calculate functions $g_r(\beta_s)$ and $g_r(\beta_r)$ to compute I_{ds} from Equation 5.71. Since the argument of cosine function [Equation 5.44] is β, therefore, $0 < \beta < \pi/2$. Then for any given values of V_{gs} and V_{ds}, β_s and β_d are computed from Equation 5.73 as described below.

At the source end of a FinFET device: $V_{ch}(y) = V_s = 0$ and $\beta = \beta_s$; then from Equation 5.73

$$f_r(\beta_s) = \frac{1}{2v_{kT}}\left[V_{gs} - \left\{V_{fb} + 2v_{kT}\ln\left(\frac{2}{t_{fin}}\sqrt{\frac{2\varepsilon_{si}v_{kT}}{qn_i}}\right)\right\}\right] \qquad (5.74)$$

Now, if we define

$$V_0 = V_{fb} + 2v_{kT}\ln\left(\frac{2}{t_{fin}}\sqrt{\frac{2\varepsilon_{si}v_{kT}}{qn_i}}\right) \qquad (5.75)$$

Then we get

$$f_r(\beta_s) = \frac{\left(V_{gs} - V_0\right)}{2v_{kT}} \qquad (5.76)$$

Similarly, at the drain end of the device: $V_{ch}(y) = V_{ds}$ and $\beta = \beta_d$ so that

$$f_r(\beta_d) = \frac{\left(V_{gs} - V_0 - V_{ds}\right)}{2v_{kT}} \qquad (5.77)$$

Thus, using Equations 5.76 and 5.77, I_{ds} can be easily computed from Equation 5.71 [2]. It is to be noted that Equation 5.69 is a continuous DG-FinFET drive current expression that describes I_{ds} for all regions of device operations: (1) linear; (2) saturation; and (3) subthreshold and is obtained computationally as shown in Figures 5.5 and 5.6.

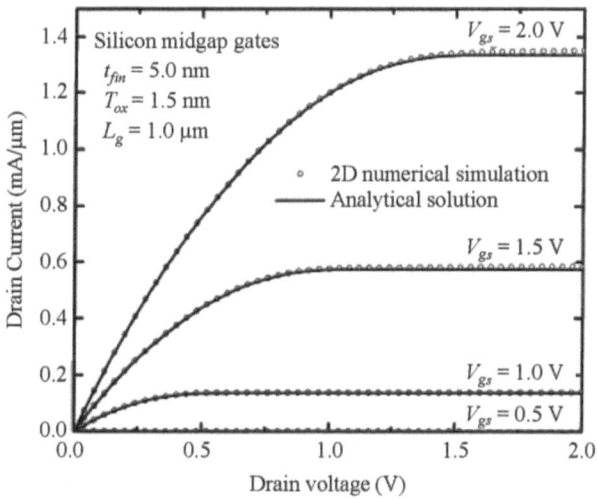

FIGURE 5.5 $I_{ds} - V_{ds}$ characteristics calculated from the analytic model (solid curves), compared with the 2D numerical simulation results (open circles). A constant mobility of 300 cm^{-2} V^{-1} sec^{-1} is used in both calculations [2].

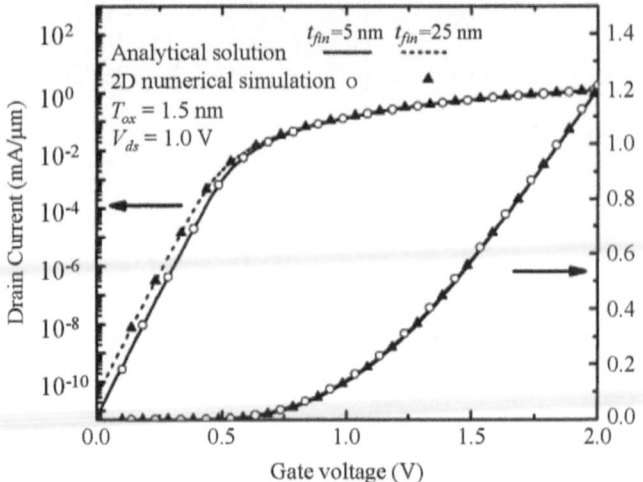

FIGURE 5.6 $I_{ds} - V_{gs}$ characteristics obtained from the analytic model for two different values of t_{fin} (solid and dashed curves), compared with the 2D numerical simulation results (symbols). The same currents are plotted on both logarithmic (left) and linear (right) scales. A constant mobility of 300 cm^{-2} V^{-1} sec^{-1} is used in calculations [2].

For the simplicity of calculation of FinFET device characteristics in the different regions of device operation, we can derive expressions for each region from the continuous model given in Equation 5.71 as described in Section 5.4.3.

5.4.3 REGIONAL DRAIN CURRENT FORMULATION FOR SYMMETRIC DG-FinFETs

From the continuous drain current Equation 5.71 we can derive separate expressions for the linear, saturation, and subthreshold regime of FinFET operation under an applied biasing condition. In this section, we will make several simplifying assumptions to derive I_{ds} expressions for each operating regime of FinFETs in terms of a defined threshold voltage of the devices.

5.4.3.1 Threshold Voltage Formulation

In order to derive an expression for V_{th}, we write Equation 5.58 as

$$\frac{1}{2v_{kT}}\left[V_{gs} - V_{ch}(y) - \left\{V_{fb} + 2v_{kT}\ln\left(\frac{2}{t_{fin}}\sqrt{\frac{2\varepsilon_{si}v_{kT}}{qn_i}}\right)\right\}\right] = \ln\beta - \ln(\cos\beta) + 2r\beta\tan\beta$$

$$(5.78)$$

The second term within the braces on the left-hand side of Equation 5.78 is the expression for V_0 given in Equation 5.75; therefore, Equation 5.78 can be expressed as

$$\frac{V_{gs} - V_0 - V_{ch}(y)}{2v_{kT}} = \ln \beta - \ln(\cos \beta) + 2r\beta \tan \beta \qquad (5.79)$$

At threshold condition, $V_{gs} = V_{th}$: we have, $V_{ch}(y) = V_s = 0$ at the source end of the device. Then from Equation 5.79, we get

$$\frac{V_{gs} - V_0}{2v_{kT}} = \ln \beta - \ln(\cos \beta) + 2r\beta \tan \beta \qquad (5.80)$$

Also, at threshold condition: $\beta \sim \pi/2$ (upper bound); thus, the $2r\beta \tan \beta$ term on the right-hand side of Equation 5.80 is dominant; therefore, at threshold condition, we can write

$$\frac{V_{gs} - V_0}{2v_{kT}} \cong 2r\beta \tan \beta = 2r\beta \frac{\sin \beta}{\cos \beta} \qquad (5.81)$$

Since the value of $\sin \beta \approx 1$ when $\beta \approx \pi/2$ at $V_g = V_{th}$, we can simplify Equation 5.81 as

$$\frac{V_{gs} - V_0}{4rv_{kT}} = \frac{\beta}{\cos \beta} \qquad (5.82)$$

Now, below V_{th} or subthreshold region, $2r\beta \tan \beta \sim 0$; therefore, we get from Equation 5.80

$$V_{gs} - V_0 \approx 2v_{kT} \left[\ln \beta - \ln(\cos \beta) \right]$$

$$\text{or,} \quad V_{gs} - V_0 = 2v_{kT} \ln \left(\frac{\beta}{\cos \beta} \right) \qquad (5.83)$$

Note that Equation 5.82 represents a point (V_{gs}) in the $I_{ds} - V_{gs}$ plot at threshold condition and Equation 5.83 represents V_{gs} at $I_{ds} \approx 0$ on the same plot; therefore, combining Equations 5.82 and 5.83, we get the expressions for extrapolated V_{th} as

$$V_{th} = V_0 + 2v_{kT} \ln \left(\frac{V_{gs} - V_0}{4rv_{kT}} \right) \qquad (5.84)$$

In reality, the second term, $2v_{kT} \ln \left(\dfrac{V_{gs} - V_0}{4rv_{kT}} \right)$ on the right-hand side of Equation 5.84, is a second-order effect (~ 50 mV) [1]. It is worth noting that both r [Equation 5.57] and V_0 [Equation 5.75] contain t_{fin} in their respective denominator; thus, canceling the effect of body thickness in V_{th} and therefore, V_{th} is independent of t_{fin}.

5.4.3.2 Linear Region I_{ds} Equation

In the linear or above threshold voltage region, $I_{ds} > 0$; therefore, $f_r(\beta_s), f_r(\beta_d) \gg 1$ and $\beta_s = \beta_d \sim \pi/2$. Therefore, from Equation 5.72, we can show

$$f_r(\beta) \cong 2r\beta \tan \beta \qquad (5.85)$$

And, from Equation 5.70, we get

$$g_r(\beta) \cong r\beta^2 \tan^2 \beta$$

$$\text{or,} \quad g_r(\beta) \cong \frac{1}{4r}(2r\beta \tan \beta)^2 \tag{5.86}$$

Now, combining Equations 5.85 and 5.86, we get

$$g_r(\beta) \cong \frac{1}{4r}\left(f_r(\beta)\right)^2 \tag{5.87}$$

Again, at V_{th} condition, we can express Equations 5.76 and 5.77 by using V_{th} for V_0 as

$$f_r(\beta_s) = \frac{\left(V_{gs} - V_{th}\right)}{2v_{kT}}$$

$$f_r(\beta_d) = \frac{\left(V_g - V_{th} - V_{ds}\right)}{2v_{kT}} \tag{5.88}$$

Then for any applied values of V_{gs} and V_{ds}, $g_r(\beta_s)$ and $g_r(\beta_d)$ can be obtained from Equation 5.87 using the corresponding expressions for $f_r(\beta_s)$ and $f_r(\beta_d)$ given in Equation 5.88. Therefore, from Equations 5.87 and 5.88, we get

$$g_r(\beta_s) = \frac{1}{4r}\left(\frac{V_g - V_{th}}{2v_{kT}}\right)^2$$

$$g_r(\beta_d) = \frac{1}{4r}\left(\frac{V_g - V_{th} - V_{ds}}{2v_{kT}}\right)^2 \tag{5.89}$$

Now, substituting the expressions for $g_r(\beta_s)$ and $g_r(\beta_d)$ from Equation 5.89 into Equation 5.71, we get

$$I_{ds} = \mu_s\left(\frac{W}{L}\right)\frac{4\varepsilon_{si}}{t_{fin}}(2v_{kT})^2\left[\frac{1}{4r}\left\{\left(\frac{V_{gs} - V_{th}}{2v_{kT}}\right)^2 - \left(\frac{V_{gs} - V_{th} - V_{ds}}{2v_{kT}}\right)^2\right\}\right]$$

$$= \mu_s\left(\frac{W}{L}\right)\frac{4\varepsilon_{si}}{t_{fin}}(2v_{kT})^2\frac{1}{4r(2v_{kT})^2}\left[\left(V_{gs} - V_{th}\right)^2 - \left(V_{gs} - V_{th} - V_{ds}\right)^2\right] \tag{5.90}$$

$$= \mu_s\left(\frac{W}{L}\right)\frac{\varepsilon_{si}}{t_{fin}}\frac{\varepsilon_{ox}t_{fin}}{\varepsilon_{si}T_{ox}}\left[\left(V_{gs} - V_{th}\right)^2 - \left(V_{gs} - V_{th} - V_{ds}\right)^2\right]$$

In Equation 5.90, we have used $r = \dfrac{\varepsilon_{si}T_{ox}}{\varepsilon_{ox}t_{fin}}$ from Equation 5.57. Then using $C_{ox} = \varepsilon_{ox}/T_{ox}$, we get the linear region I_{ds} equation after simplification of Equation 5.90 ($V_{gs} > V_{th}$) at low V_{ds} as

$$I_{ds} = 2\mu_s \left(\frac{W}{L}\right) C_{ox} \left[\left(V_{gs} - V_{th} - \frac{V_{ds}}{2}\right) V_{ds}\right] \tag{5.91}$$

Equation 5.91 shows that the linear region current expression is similar to that of the planar CMOS drive current with a factor "2" due to two sidewall gates [12]. The linear region current is valid for $V_{ds} < V_{dsat}$ and is given by $V_{dsat} = (V_{gs} - V_{th})$ obtained at the condition: $\left.\dfrac{dI_{ds}}{dV_{gs}}\right|_{V_{ds}=V_{dsat}} = 0.$

5.4.3.3 Saturation Region I_{ds} Equation

In the saturation region, we have $I_{ds} > 0$; therefore, at the source end, $\beta_s \sim \pi/2$ and $f_r(\beta_s) \gg 1$. Therefore, similar to linear region formulation [Equation 5.89], we get

$$g_r(\beta_s) = \frac{1}{4r}\left(\frac{V_g - V_{th}}{2v_{kT}}\right)^2 \tag{5.92}$$

However, in the saturation region, the channel region is pinched-off at the drain end as shown in Figure 5.7, and therefore, GCA does not hold. Due to this depleted channel region, the silicon body at the drain end is below threshold. Thus, in the saturation region, $f_r(\beta_d) \ll -1$ and $\beta_d \ll 1$. Then, from Equation 5.72, we get

$$f_r(\beta_d) \cong \ln \beta_d \tag{5.93}$$

And, from Equation 5.70, we get

$$g_r(\beta_d) \cong -\frac{\beta_d^2}{2} \tag{5.94}$$

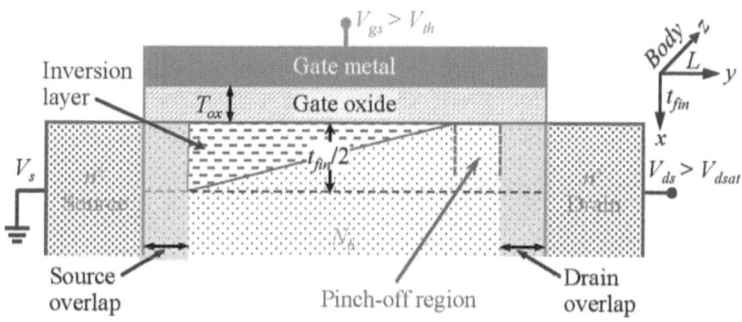

FIGURE 5.7 Schematic diagram of a FinFET device in the saturation regime showing channel pinch-off at the drain end; top gate is shown for illustration only.

Now, equating the expressions for $f_r(\beta_d)$ from Equations 5.77 and 5.93, we get

$$f_r(\beta_d) = \frac{\left(V_{gs} - V_0 - V_{ds}\right)}{2v_{kT}} = \ln \beta_d$$

$$\text{or,} \quad \beta_d = e^{\frac{\left(V_{gs} - V_0 - V_{ds}\right)}{2v_{kT}}}$$

(5.95)

Then substituting β_d from Equation 5.95 into Equation 5.94, we get

$$g_r(\beta_d) \cong -\frac{\beta_d^2}{2} = -\frac{1}{2} e^{\frac{\left(V_{gs} - V_0 - V_{ds}\right)}{v_{kT}}}$$

(5.96)

Finally, by substituting the expressions for $g_r(\beta_s)$ and $g_r(\beta_d)$ from Equations 5.92 and 5.96, respectively, into Equation 5.71, we obtain the saturation region I_{ds} as

$$I_{ds} = \mu_s \left(\frac{W}{L}\right) \frac{4\varepsilon_{si}}{t_{fin}} \left(2v_{kT}\right)^2 \left[\frac{1}{4r}\left\{\left(\frac{V_{gs} - V_{th}}{2v_{kT}}\right)^2 - \frac{1}{2} e^{\frac{\left(V_{gs} - V_0 - V_{ds}\right)}{v_{kT}}}\right\}\right]$$

$$= \mu_s \left(\frac{W}{L}\right) \frac{4\varepsilon_{si}}{t_{fin}} \left(2v_{kT}\right)^2 \frac{1}{4r\left(2v_{kT}\right)^2} \left[\left(V_{gs} - V_{th}\right)^2 - 8rv_{kT}^2 e^{\frac{\left(V_{gs} - V_0 - V_{ds}\right)}{v_{kT}}}\right]$$

(5.97)

Now, using $r = \dfrac{\varepsilon_{si} T_{ox}}{\varepsilon_{ox} t_{fin}}$ from Equation 5.57 outside the square brackets and $C_{ox} = \varepsilon_{ox}/T_{ox}$, we get the saturation region I_{ds} equation after simplification of Equation 5.97 as

$$I_{ds} = \mu_s \left(\frac{W}{L}\right) C_{ox} \left[\left(V_{gs} - V_{th}\right)^2 - 8rv_{kT}^2 e^{\frac{\left(V_{gs} - V_0 - V_{ds}\right)}{v_{kT}}}\right]$$

(5.98)

Note that the saturation I_{ds} expression Equation 5.98 obtained from the continuous current formulation approaches the saturation value with an exponentially decreasing term with increasing V_{ds}, in contrast to the piece-wise I_{ds} expressions derived for conventional MOSFETs in which the current is made constant in the saturation regime [12].

5.4.3.4 Subthreshold Conduction

In the subthreshold region, we have $I_{ds} \sim 0$; therefore, both β_s, $\beta_d \ll 1$ so that from Equation 5.72, we get both β_s, β_d denoted as $\beta_{d/s}$

$$f_r(\beta_{d/s}) \cong \ln \beta_{d/s}$$

(5.99)

And, from Equation 5.70, we get

$$g_r\left(\beta_{d/s}\right) \cong -\frac{\beta_{d/s}^2}{2} \tag{5.100}$$

Now, from Equations 5.76 and 5.99 we get $f_r(\beta_s)$ for source end, $V_s = 0$

$$f_r(\beta_s) = \frac{\left(V_{gs} - V_0\right)}{2v_{kT}} \cong \ln\beta_s \tag{5.101}$$

$$\text{or,} \quad \beta_s^2 = e^{\frac{\left(V_g - V_0\right)}{v_{kT}}}$$

Then, combining Equations 5.100 and 5.101, we get

$$g_r\left(\beta_s\right) \cong -\frac{\beta_s^2}{2} = -\frac{1}{2}e^{\frac{\left(V_{gs} - V_0\right)}{v_{kT}}} \tag{5.102}$$

Similarly, we can show $g_r(\beta_d)$ for drain end, $V_{ch}(y) = V_{ds}$

$$g_r\left(\beta_d\right) \cong -\frac{\beta_d^2}{2} = -\frac{1}{2}e^{\frac{\left(V_{gs} - V_0 - V_{ds}\right)}{v_{kT}}} \tag{5.103}$$

Now, to present subthreshold region I_{ds} in terms of V_{gs} and V_{ds}, we eliminate the parameter V_0 using Equation 5.75 given by

$$V_0 = V_{fb} + 2v_{kT}\ln\left(\frac{2}{t_{fin}}\sqrt{\frac{2\varepsilon_{si}v_{kT}}{qn_i}}\right)$$

$$\text{or,} \quad \frac{V_0 - V_{fb}}{2v_{kT}} = \ln\left(\frac{2}{t_{fin}}\sqrt{\frac{2\varepsilon_{si}v_{kT}}{qn_i}}\right)$$

$$\text{or,} \quad e^{\frac{V_0 - V_{fb}}{2v_{kT}}} = \frac{2}{t_{fin}}\sqrt{\frac{2\varepsilon_{si}v_{kT}}{qn_i}} \tag{5.104}$$

$$\text{or,} \quad e^{\frac{-V_0}{v_{kT}}} = \frac{t_{fin}^2 qn_i}{8\varepsilon_{si}v_{kT}}e^{\frac{-V_{fb}}{v_{kT}}}$$

Then substituting for $\exp(-V_0/v_{kT})$ from Equation 5.104 into Equations 5.102 and 5.103, we get the expressions for $g_r(\beta_s)$ and $g_r(\beta_d)$ after simplification as

$$g_r\left(\beta_s\right) = -\frac{t_{fin}^2 qn_i}{16\varepsilon_{si}v_{kT}}e^{\frac{\left(V_{gs} - V_{fb}\right)}{v_{kT}}} \tag{5.105}$$

$$g_r\left(\beta_d\right) = -\frac{t_{fin}^2 qn_i}{16\varepsilon_{si}v_{kT}}e^{\frac{\left(V_{gs} - V_{fb} - V_{ds}\right)}{v_{kT}}} \tag{5.106}$$

Finally, substituting the expressions for $g_r(\beta_s)$ and $g_r(\beta_d)$ from Equations 5.105 and 5.106 into Equation 5.71, we get

$$I_{ds} = -\mu_s \left(\frac{W}{L}\right)\frac{4\varepsilon_{si}}{t_{fin}}(2v_{kT})^2 \left[\frac{t_{fin}^2 qn_i}{16\varepsilon_{si}v_{kT}}e^{\frac{(V_{gs}-V_{fb})}{v_{kT}}} - \frac{t_{fin}^2 qn_i}{16\varepsilon_{si}v_{kT}}e^{\frac{(V_{gs}-V_{fb}-V_{ds})}{v_{kT}}}\right]$$

$$\text{or,} \quad I_{ds} = -\mu_s \left(\frac{W}{L}\right)\frac{4\varepsilon_{si}}{t_{fin}}(2v_{kT})^2 \frac{t_{fin}^2 qn_i}{16\varepsilon_{si}v_{kT}}e^{\frac{(V_{gs}-V_{fb})}{v_{kT}}}\left(1-e^{-\frac{V_{ds}}{v_{kT}}}\right)$$

$$(5.107)$$

After simplification of Equation 5.107 (dropping the −ve sign which indicates the direction of current flow), we get the drain current in the subthreshold region operation of FinFETs as

$$I_{ds} = \mu_s \left(\frac{W}{L}\right)kTt_{fin}n_i e^{\frac{(V_{gs}-V_{fb})}{v_{kT}}}\left(1-e^{-\frac{V_{ds}}{v_{kT}}}\right) \qquad (5.108)$$

We note from Equation 5.108:

1. The subthreshold region current I_{ds} is similar to that derived using diffusion current expression $J_{diff} = -qD_n(dn/dx)$ [12]
2. The subthreshold region I_{ds} is directly proportional to t_{fin}, that is, decreases with decreasing body thickness
3. The subthreshold region I_{ds} is independent of C_{ox} – a manifestation of the volume inversion [23]
4. In contrast to subthreshold current, linear region and saturation currents are proportional to C_{ox} and independent t_{fin}

Thus, from the continuous current expression given by Equation 5.71, we have derived the regional current expressions for FinFET devices given by Equations 5.91, 5.98, and 5.108. Then combining these expressions, we have a complete set of piecewise I_{ds} equations given by

$$I_{ds} = \begin{cases} \mu_s\left(\frac{W}{L}\right)kTt_{fin}n_i e^{\frac{(V_{gs}-V_{fb})}{v_{kT}}}\left(1-e^{-\frac{V_{ds}}{v_{kT}}}\right); & V_{gs} > V_{fb} \text{ and } V_{ds} < V_{dsat} \\[3mm] 2\mu_s\left(\frac{W}{L}\right)C_{ox}\left[\left(V_{gs}-V_{th}-\frac{V_{ds}}{2}\right)V_{ds}\right]; & V_{gs} > V_{fb} \text{ and } V_{ds} < V_{dsat} \\[3mm] \mu_s\left(\frac{W}{L}\right)C_{ox}\left[\left(V_{gs}-V_{th}\right)^2 - 8rv_{kT}^2 e^{\frac{(V_{gs}-V_0-V_{ds})}{v_{kT}}}\right]; & V_{gs} > V_{fb} \text{ and } V_{ds} > V_{dsat} \end{cases}$$

$$(5.109)$$

Equation 5.109 offers an alternative solution to analyze FinFET device performance without the complex numerical computation of the continuous I_{ds} expression given by Equation 5.71.

Subthreshold slope: An important characteristic of the subthreshold region operation is the gate voltage swing of the device from its off-state to on-state. This gate voltage is also called the *subthreshold swing S*, or *SS*, or *S-factor*. It is the inverse of the slope of $I_{ds} - V_{gs}$ plot and is defined as the change in the gate voltage V_{gs} required to change the subthreshold current I_{ds} by one decade. Thus, S is a measure of the on-off characteristics of FinFET devices.

If we take two points (I_{ds1}, V_{gs1}) and (I_{ds2}, V_{gs2}) in the subthreshold region shown in Figure 5.8, then by definition $(V_{gs2} - V_{gs1})$ required to changing the ratio of (I_{ds2}/I_{ds1}) by one decade or 10 can be defined as

$$S \equiv \frac{V_{gs2} - V_{gs1}}{\log I_{ds2} - \log I_{ds1}} = \frac{dV_{gs}}{d(\log I_{ds})} = 2.3 \frac{dV_{gs}}{d(\ln I_{ds})} \qquad (5.110)$$

In Equation 5.110 we have used $(\ln I_{ds}) = 2.3 \times (\log I_{ds})$ for the conversion from the logarithm base "10" to natural logarithm base "e." In reality, S varies with I_{ds} in the subthreshold region, however, this variation is negligible over one decade of current so that S can be considered as a gate swing per decade of current change. Therefore, from Equation 5.108, we get

$$\ln I_{ds} = \ln \left(\frac{\mu_s W k T t_{fin} n_i}{L} \right) + \frac{V_{gs} - V_{fb}}{v_{kT}} + \ln \left(1 - e^{-\frac{V_{ds}}{v_{kT}}} \right) \qquad (5,111)$$

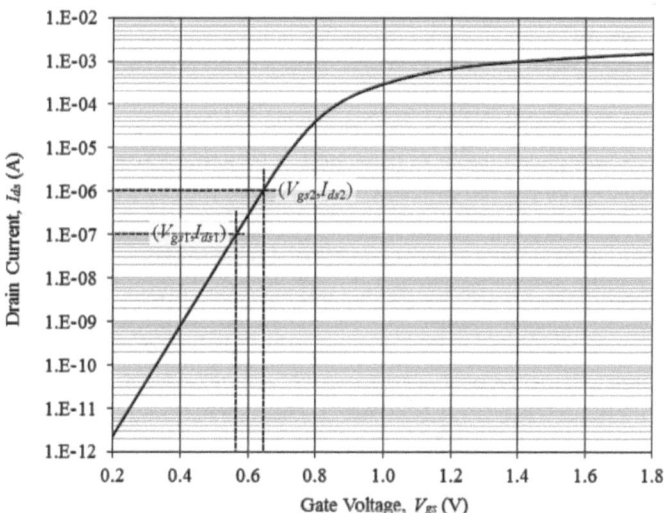

FIGURE 5.8 $\log(I_{ds})$ versus V_{gs} characteristics of a typical FinFET device to calculate *S-factor*; the ratio of two data points, (V_{gs1}, I_{ds1}) and (V_{gs2}, I_{ds2}), shown in the subthreshold current is one decade.

Now, taking the derivative of Equation 5.111 with respect to V_{gs}, we get

$$d\left(\ln I_{ds}\right) = \frac{dV_{gs}}{v_{kT}}$$

(5,112)

$$\text{or,} \quad \frac{dV_{gs}}{d\left(\ln I_{ds}\right)} = v_{kT}$$

Then, substituting Equation 5.112 in Equation 5.110, we get

$$S = 2.3 v_{kT}$$

(5.113)

Since at room temperature ($T \sim 300$ K), $v_{kT} \cong 25.9$ mV, therefore, Equation 5.113 shows that the theoretical minimum value of subthreshold swing is given by

$$S = 2.3 v_{kT} \cong 60 \text{ mV/decade}$$

(5.114)

Thus, the minimum attainable S for a FinFET device is approximately 60 mV per decade of drain current at room temperature. Since I_{ds} in the subthreshold region is independent of C_{ox} and depletion capacitance, S does not depend on an ideality factor as in planar-MOSFETs [12].

5.5 SUMMARY

This chapter presented the basic mathematical formulations of large geometry DG-FinFET devices. First of all, the basic features of FinFETs are overviewed and it was shown that a DG-FinFET structure can be represented by a DG-MOSFET structure. Thus, the DG-MOSFET structure is used to analyze the performance of DG-FinFET devices. After the structure definition, a set of simplifying assumptions are made to derive a continuous drain current expression for long-channel devices applicable to all regions of device operation. This continuous drain current expression must be solved computationally to generate $I - V$ characteristics of FinFET devices. However, for intuitive analysis of device performance, the drain current expressions for linear, saturation, and subthreshold regions are derived in terms of threshold voltage from the continuous drain current expression. Furthermore, a number of simplified assumptions are also used to derive an expression for threshold voltage V_{th} of long-channel devices. Thus, the regional current equations can be used to estimate the basic design parameters including V_{th}, I_{on}, I_{off}, and S of FinFETs for VLSI circuits. Finally, the subthreshold characteristics of DG-FinFET devices are discussed and it is shown that the ideal subthreshold slope of 60 mV/decade can be achieved in FinFET devices.

REFERENCES

1. Y. Taur, "An analytical solution to a double-gate MOSFET with undoped body," *IEEE Electron Device Letters*, 21(5), pp. 245–247, 2000.
2. Y. Taur, X. Liang, W. Wang, and H. Lu, "A continuous, analytic drain current model for DG MOSFETs," *IEEE Electron Device Letters*, 25(2), pp. 107–109, 2004.

3. J. Sallese, F. Krummenacher, F. Pregaldiny, *et al.*, "A design oriented charge-based current model for symmetric DG MOSFET and its correlation with EKV formalism," *Solid-State Electronics*, 49(3), pp. 485–489, 2005.

4. A. Ortiz-Conde, F. Garcia Sanchez, and J. Muci, "Rigorous analytic solution for the drain current of undoped symmetric dual-gate MOSFETs," *Solid-State Electronics*, 49(4), pp. 640–647, 2005.

5. M.V. Dunga, C.-H. Lin, X. Xi, *et al.*, "Modeling advanced FET technology in a compact model," *IEEE Transactions on Electron Devices*, 53(9), pp. 1971–1978, 2006.

6. O. Moldovan, A. Cerdeira, D. Jimenez, et al., "Compact model for highly-doped double-gate SOI MOSFETs targeting baseband analog applications," *Solid-State Electronics*, 51(5), pp. 655–661, 2007.

7. G. Smit, A. Scholten, G. Curatola, *et al.*, "PSP-based scalable compact FinFET model," *Proceedings NSTI Nanotechnologies*, 3, pp. 520–525, 2007.

8. M.V. Dunga, "Nanoscale CMOS modeling," Ph.D. dissertation, Electrical Engineering and Computer Science, University of California Berkeley, Berkeley, CA, 2008.

9. F. Lime, and B. Iniguez, "A quasi-two-dimensional compact drain-current model for undoped symmetric double-gate MOSFETs including short-channel effects," *IEEE Transactions on Electron Devices*, 55(6), pp. 1441–1448, 2008.

10. F. Liu, L. Zhang, J. Zhang, *et al.*, "Effects of body doping on threshold voltage and channel potential of symmetric DG MOSFETs with continuous solution from accumulation to strong-inversion regions," *Semiconductor Science and Technology*, 24(8), p. 2009.

11. Y.S. Chauhan, D.D. Lu, S. Venugopalan, *et al.*, *FinFET Modeling for IC Simulation and Design: Using the BSIM-CMG Standard*, Academic Press, San Diego, CA, 2015.

12. S.K. Saha, *Compact Models for Integrated Circuit Design: Conventional Transistors and Beyond*, CRC Press, Taylor & Francis Group, Boca Raton, FL, 2015.

13. C. Auth, C. Allen, A. Blattner, *et al.*, "A 22-nm-high performance and low-power CMOS technology featuring fully-depleted tri-gate transistors, self-aligned contacts and high density MIM capacitors." In: *Symposium on VLS Technology*, pp. 131–132, 2012.

14. ATLAS *User's Manual: Device Simulation Software*, Silvaco. Inc., Santa Clara, CA, 2013.

15. *Taurus MEDICI Manuals*, Synopsys, Inc., Mountain View, CA, 2007.

16. *MINIMOS User's Guide*, Institut für Mikroelektronik, Technische University, Vienna, Austria, 2013.

17. *Sentaurus TCAD Manuals*, Synopsys, Inc., Mountain View, CA, 2012.

18. S.K. Saha, "Introduction to technology computer aided design." In: *Technology Computer Aided Design: Simulation for VLSI MOSFET*, C.K. Sarkar, (ed.), CRC Press, Boca Raton, FL, 2013.

19. H.C. Pao and C.T. Sah, "Effects of diffusion current on characteristics of metal-oxide (insulator)-semiconductor transistors," *Solid-State Electronics*, 9(10), pp. 927–937, 1966.

20. N. Arora, *MOSFET Models for VLSI Circuit Simulation: Theory and Practice*, Springer–Verlag, Vienna, 1993.

21. J. He, X. Xi, C.-H. Lin *et al.*, "A non-charge-sheet analytical theory for undoped symmetric double-gate MOSFETs from the exact solution of Poisson's equation using SPP approach," *Proceedings of the NSTI Nanotech*, 2, pp. 124–127, 2004.

22. N. Pandey and Y.S. Chauhan, Private communications, December, 2018.

23. F. Balestra, S. Cristoloveanu, M. Benachir, J. Brini, and T. Elewa, "Double-gate silicon-on-insulator transistor with volume inversion: A new device with greatly enhanced performance," *IEEE Electron Devices Letters*, 8(9), pp. 410–412, 1987.

6 Small Geometry FinFETs
Physical Effects on Device Performance

6.1 INTRODUCTION

In Chapter 5, we derived the continuous as well as piece-wise drain current I_{ds} expressions for fin field-effect transistors (FinFETs) to characterize the electrostatic performance of ideal long-channel devices. However, in real devices, several physical phenomena critically affect the performance of short-channel devices as the channel length rapidly approaches to its ultimate physical limit [1]. In general, these physical effects, commonly referred to as the *short-channel effects* (SCEs), are less severe for short-channel FinFET devices due to better electrostatic control on the channel by the multiple gates [2,3]. However, the physical phenomena for SCEs in FinFETs must be understood for an accurate characterization of the performance of short-channel FinFET devices [4]. The physical phenomena affecting the performance of short-channel FinFET devices include threshold voltage (V_{th}) degradation, also known as V_{th} roll-off, subthreshold swing (S) degradation [5–8], channel length modulation (CLM) [9], drain-induced barrier lowering (DIBL) [10], mobility degradation due to vertical electric field [11], velocity saturation due to high lateral electric field [12], quantum mechanical (QM) effects [13–15], and parasitic elements due to structural effects [16]. Thus, these physical phenomena in small geometry FinFET devices must be appropriately considered for accurate characterization of the performance of real FinFET devices. In this chapter, some of these critical physical phenomena affecting the electrostatic behavior of short-channel FinFET devices are described.

6.2 SHORT-CHANNEL EFFECTS ON THRESHOLD VOLTAGE

Short-channel effects originate from two-dimensional (2D) electrostatics caused by the close proximity of the drain to source region. Due to this close proximity of the drain to source, the drain significantly affects the potential barrier at the source causing degradation of device performance through V_{th} roll-off and S-degradation. Though there are several approaches to characterize SCEs [2,17–20], the approach assuming a parabolic potential function perpendicular to the silicon/insulator interface to solve 2D Poisson's equation is shown to be effective in predicting SCEs in multiple-gate or multigate metal-oxide-semiconductor field-effect transistors (MOSFETs) [2,19]. In this approach, the electrostatic integrity of a device is characterized by a parameter called the *natural length* or *scale length* (λ) which defines the spread of the drain field in the channel causing the SCEs in field-effect transistors (FETs). The following section presents the mathematical formulation to derive an expression for λ of multigate MOSFET devices.

6.2.1 FORMULATION OF NATURAL LENGTH

The natural length or scale length is the characteristic field penetration length that defines the amount of SCE in a FET and predicts the variation in the extent of the drain field penetrating into the silicon body. It is a function of physical parameters such as fin thickness t_{fin} and gate oxide thickness T_{ox}.

For the simplicity of mathematical steps in deriving an expression for λ, let us consider a 2D double-gate (DG) FinFET structure with co-ordinate system (x,y) as shown in Figure 6.1, with $x = 0$ at the Si/SiO$_2$ interface and $x = t_{fin}/2$ at the center of the body of thickness t_{fin} and the potential, $\phi(0,y) = \phi_s(y)$ (surface potential) at $x=0$ and $\phi(t_{fin}/2,y) = \phi_c(y)$ (center potential) at $x = t_{fin}/2$. Since in the subthreshold regime, the inversion charge Q_i is negligibly small compared to bulk charge Q_b, therefore, the potential $\phi(x,y)$ at any point in the silicon body of a DG-FinFET can be obtained by solving 2D Poisson's equation considering only the majority carrier term in Equation 3.54 and is given by

$$\frac{d^2\phi(x,y)}{dx^2} + \frac{d^2\phi(x,y)}{dy^2} = \frac{qN_b}{\varepsilon_{si}} \tag{6.1}$$

where:

N_b is the channel doping concentration

q is the electronic charge

ε_{si} = permittivity of silicon

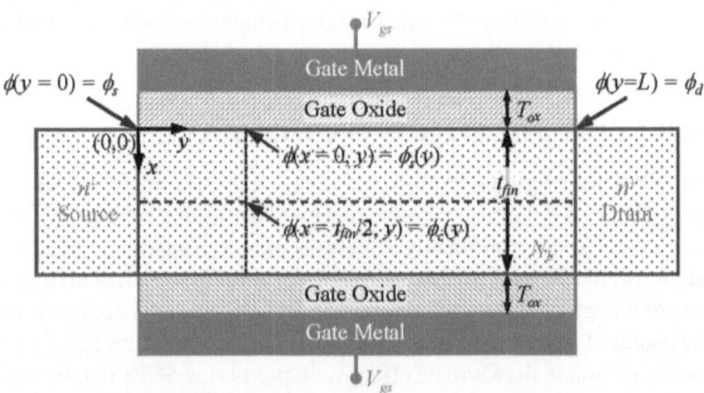

FIGURE 6.1   2D cross-section of a symmetric common DG-FinFET device structure: x-axis is along the thickness t_{fin} and y-axis is parallel to the channel along the length of the device.

Now, assuming a parabolic potential distribution in silicon along the thickness in the *x*-direction, we can write the solution of Equation 6.1 as

$$\phi(x, y) = a_0(y) + a_1(y)x + a_2(y)x^2 \tag{6.2}$$

where:

$a_0(y)$, $a_1(y)$, and $a_2(y)$ are the coefficients of the parabolic potential function

The coefficients $a_0(y)$, $a_1(y)$, and $a_2(y)$ are determined using the following boundary conditions (BCs) with reference to Figure 6.1.

BC-1: At the Si/SiO$_2$ interface, $x = 0$ and $\phi(0,y) = \phi_s(y)$ is the surface potential at the top-gate

BC-2: At the Si/SiO$_2$ interface, $x = 0$, we can write from Equation 3.100

$$\varepsilon_{ox} \frac{V_{gs} - V_{fb} - \phi(x = 0, y)}{T_{ox}} = -\varepsilon_{si} \frac{d\phi}{dx}\bigg|_{x=0} \tag{6.3}$$

$$\text{or,} \quad \frac{d\phi}{dx}\bigg|_{x=0} = \frac{\varepsilon_{ox}}{\varepsilon_{si}} \frac{\phi_s(y) - V_{gst}}{T_{ox}}$$

where:

ε_{ox} is the permittivity of oxide

$\phi(x = 0,y) = \phi_s(y)$, from BC-1

T_{ox} is the gate oxide thickness

V_{gst} is the gate overdrive voltage defined as $(V_{gs} - V_{fb})$, with V_{gs} and V_{fb} as the gate-to-source voltage and flat band voltage, respectively

BC-3: At the center of the fin channel, $x = t_{fin}/2$, $\phi(x = t_{fin}/2, y) \equiv \phi_c(y)$, and

$$\frac{d\phi(x)}{dx}\bigg|_{x=t_{fin}/2} = 0$$

Now, applying the boundary condition, BC-1, we get from Equation 6.2

$$\phi(x = 0, y) = a_0(y) = \phi_s(y) \tag{6.4}$$

Applying boundary condition BC-2 Equation 6.3, we get from Equation 6.2 after simplification

$$\frac{\varepsilon_{ox}}{\varepsilon_{si}} \frac{\phi_s(y) - V_{gst}}{T_{ox}} = a_1(y) \tag{6.5}$$

From BC-3, we get from Equation 6.2

$$\frac{d\phi}{dx}\bigg|_{x=t_{fin}/2} = 0 = a_1(y) + t_{fin}a_2(y) \tag{6.6}$$

Simplifying Equation 6.6, we get

$$a_2(y) = -\frac{1}{t_{fin}} a_1(y) \tag{6.7}$$

Then substituting $a_1(y)$ from Equation 6.5 into Equation 6.7, we get

$$a_2(y) = -\frac{\varepsilon_{ox}}{\varepsilon_{si} t_{fin}} \frac{\phi_s(y) - V_{gst}}{T_{ox}} \tag{6.8}$$

Thus, from the boundary conditions BC-1, BC-2, and BC-3, we get the coefficients of the parabolic potential function [Equation 6.2] given by Equations 6.4, 6.5, and 6.8 as

$$a_0(y) = \phi_s(y)$$

$$a_1(y) = \frac{\varepsilon_{ox}}{\varepsilon_{si}} \frac{\phi_s(y) - V_{gst}}{T_{ox}} \tag{6.9}$$

$$a_2(y) = -\frac{\varepsilon_{ox}}{\varepsilon_{si} t_{fin}} \frac{\phi_s(y) - V_{gst}}{T_{ox}}$$

Now, substituting the coefficients from Equation 6.9 into Equation 6.2, we get the parabolic potential at any point (x,y) of the silicon fin as

$$\phi(x, y) = \phi_s(y) + \frac{\varepsilon_{ox}}{\varepsilon_{si}} \frac{\phi_s(y) - V_{gst}}{T_{ox}} x - \frac{\varepsilon_{ox}}{\varepsilon_{si} t_{fin}} \frac{\phi_s(y) - V_{gst}}{T_{ox}} x^2 \tag{6.10}$$

Then separating the terms containing ϕ_s, we can write Equation 6.10 as

$$\phi(x, y) = \phi_s(y) + \frac{\varepsilon_{ox}}{\varepsilon_{si} T_{ox}} \phi_s(y) \cdot x - \frac{\varepsilon_{ox}}{\varepsilon_{si} t_{fin} T_{ox}} \phi_s \cdot x^2 - \frac{\varepsilon_{ox}}{\varepsilon_{si} T_{ox}} V_{gst} \cdot x + \frac{\varepsilon_{ox}}{\varepsilon_{si} t_{fin} T_{ox}} V_{gst} \cdot x^2$$

or,

$$\phi(x, y) = \left(1 + \frac{\varepsilon_{ox}}{\varepsilon_{si} T_{ox}} x - \frac{\varepsilon_{ox}}{\varepsilon_{si} t_{fin} T_{ox}} x^2 \right) \cdot \phi_s(y) - \left(\frac{\varepsilon_{ox}}{\varepsilon_{si} T_{ox}} x - \frac{\varepsilon_{ox}}{\varepsilon_{si} t_{fin} T_{ox}} x^2 \right) \cdot V_{gst} \tag{6.11}$$

Equation 6.11 represents the general expression for the variation of potential as a function of x along the thickness of the silicon fin of a DG-FinFET perpendicular to the direction of current flow.

Now, in a DG-FinFET, the center of the fin is the farthest away from the gate electrodes. Therefore, gate control is the weakest at this point. Thus, the potential at the center $\phi_c(y)$ is the most relevant to SCEs in DG-FinFETs. Therefore, we will derive a relationship between $\phi_c(y)$ and $\phi_s(y)$ from Equation 6.11 by using the BC-3: $x = t_{fin}/2$, $\phi(x = t_{fin}/2, y) \equiv \phi_c(y)$ to get

$$\phi_c(y) = \phi(x,y)\big|_{x=t_{fin}/2}$$

$$= \left(1 + \frac{\varepsilon_{ox}}{\varepsilon_{si}T_{ox}}\frac{t_{fin}}{2} - \frac{\varepsilon_{ox}}{\varepsilon_{si}t_{fin}T_{ox}}\frac{t_{fin}^2}{4}\right)\cdot\phi_s(y) - \left(\frac{\varepsilon_{ox}}{\varepsilon_{si}T_{ox}}\frac{t_{fin}}{2} - \frac{\varepsilon_{ox}}{\varepsilon_{si}t_{fin}T_{ox}}\frac{t_{fin}^2}{4}\right)\cdot V_{gst}$$

$$= \left(1 + \frac{\varepsilon_{ox}t_{fin}}{2\varepsilon_{si}T_{ox}} - \frac{\varepsilon_{ox}t_{fin}}{4\varepsilon_{si}T_{ox}}\right)\cdot\phi_s(y) - \left(\frac{\varepsilon_{ox}t_{fin}}{2\varepsilon_{si}T_{ox}} - \frac{\varepsilon_{ox}t_{fin}}{4\varepsilon_{si}T_{ox}}\right)\cdot V_{gst}$$

or, $\phi_c(y) = \left(1 + \dfrac{\varepsilon_{ox}t_{fin}}{4\varepsilon_{si}T_{ox}}\right)\phi_s(y) - \dfrac{\varepsilon_{ox}t_{fin}}{4\varepsilon_{si}T_{ox}}V_{gst}$

$$(6.12)$$

From Equation 6.12, we get the expression for $\phi_s(y)$ as

$$\phi_s(y) = \frac{1}{\left(1 + \dfrac{\varepsilon_{ox}t_{fin}}{4\varepsilon_{si}T_{ox}}\right)}\left(\phi_c(y) + \frac{\varepsilon_{ox}t_{fin}}{4\varepsilon_{si}T_{ox}}V_{gst}\right) \qquad (6.13)$$

Now, substituting the expression for $\phi_s(y)$ from Equation 6.13 into Equation 6.11, we get the potential distribution in the silicon fin at any point (x,y) as

$$\phi(x,y) = \left(1 + \frac{\varepsilon_{ox}}{\varepsilon_{si}T_{ox}}x - \frac{\varepsilon_{ox}}{\varepsilon_{si}t_{fin}T_{ox}}x^2\right)\cdot\frac{1}{\left(1 + \dfrac{\varepsilon_{ox}t_{fin}}{4\varepsilon_{si}T_{ox}}\right)}\left(\phi_c(y) + \frac{\varepsilon_{ox}t_{fin}}{4\varepsilon_{si}T_{ox}}V_{gst}\right)$$

$$- \left(\frac{\varepsilon_{ox}}{\varepsilon_{si}T_{ox}}x - \frac{\varepsilon_{ox}}{\varepsilon_{si}t_{fin}T_{ox}}x^2\right)\cdot V_{gst}$$

$$(6.14)$$

We can use Equation 6.14 in Poisson's Equation 6.1 to determine the penetration of the drain field in the silicon fin under the gate of a device. In order to solve Equation 6.1, we determine the second derivative of the potential given in Equation 6.14 with respect to x and y as follows.

$$\frac{d^2\phi(x,y)}{dx^2} = \frac{-2\varepsilon_{ox}}{\varepsilon_{si}t_{fin}T_{ox}}\frac{1}{\left(1+\frac{\varepsilon_{ox}t_{fin}}{4\varepsilon_{si}T_{ox}}\right)}\phi_c(y) - \frac{2\varepsilon_{ox}}{\varepsilon_{si}t_{fin}T_{ox}}\frac{\frac{\varepsilon_{ox}t_{fin}}{4\varepsilon_{si}T_{ox}}}{\left(1+\frac{\varepsilon_{ox}t_{fin}}{4\varepsilon_{si}T_{ox}}\right)}V_{gst} + \frac{2\varepsilon_{ox}}{\varepsilon_{si}t_{fin}T_{ox}}V_{gst}$$

$$= \frac{-2\varepsilon_{ox}}{\varepsilon_{si}t_{fin}T_{ox}}\frac{1}{\left(1+\frac{\varepsilon_{ox}t_{fin}}{4\varepsilon_{si}T_{ox}}\right)}\phi_c(y) - \left[\frac{2\varepsilon_{ox}}{\varepsilon_{si}t_{fin}T_{ox}}\left\{\frac{1}{\left(1+\frac{\varepsilon_{ox}t_{fin}}{4\varepsilon_{si}T_{ox}}\right)}\frac{\varepsilon_{ox}t_{fin}}{4\varepsilon_{si}T_{ox}} - 1\right\}\right]\cdot V_{gst}$$

$$= \frac{-2\varepsilon_{ox}}{\varepsilon_{si}t_{fin}T_{ox}}\frac{1}{\left(1+\frac{\varepsilon_{ox}t_{fin}}{4\varepsilon_{si}T_{ox}}\right)}\phi_c(y) - \left[\frac{2\varepsilon_{ox}}{\varepsilon_{si}t_{fin}T_{ox}}\left\{\frac{\frac{\varepsilon_{ox}t_{fin}}{4\varepsilon_{si}T_{ox}} - \left(1+\frac{\varepsilon_{ox}t_{fin}}{4\varepsilon_{si}T_{ox}}\right)}{\left(1+\frac{\varepsilon_{ox}t_{fin}}{4\varepsilon_{si}T_{ox}}\right)}\right\}\right]\cdot V_{gst}$$

$$= \frac{-2\varepsilon_{ox}}{\varepsilon_{si}t_{fin}T_{ox}}\frac{1}{\left(1+\frac{\varepsilon_{ox}t_{fin}}{4\varepsilon_{si}T_{ox}}\right)}\phi_c(y) - \left[\frac{2\varepsilon_{ox}}{\varepsilon_{si}t_{fin}T_{ox}}\frac{-V_{gst}}{\left(1+\frac{\varepsilon_{ox}t_{fin}}{4\varepsilon_{si}T_{ox}}\right)}\right].$$

$$(6.15)$$

Thus, after simplification, we get the x-derivative of Poisson's equation as

$$\frac{d^2\phi(x,y)}{dx^2} = \frac{2\varepsilon_{ox}}{\varepsilon_{si}t_{fin}T_{ox}}\frac{1}{\left(1+\frac{\varepsilon_{ox}t_{fin}}{4\varepsilon_{si}T_{ox}}\right)}\left(V_{gst} - \phi_c(y)\right) \qquad (6.16)$$

Again, from Equation 6.14, the y-derivative of Poisson's equation along the channel is given by

$$\frac{d^2\phi(x,y)}{dy^2} = \left(1+\frac{\varepsilon_{ox}}{\varepsilon_{si}T_{ox}}x - \frac{\varepsilon_{ox}}{\varepsilon_{si}t_{fin}T_{ox}}x^2\right)\cdot\frac{1}{\left(1+\frac{\varepsilon_{ox}t_{fin}}{4\varepsilon_{si}T_{ox}}\right)}\frac{d^2\phi_c(y)}{dy^2} \qquad (6.17)$$

Now, setting $x = t_{fin}/2$ in Equation 6.17, along the center of the channel, we get the y-component of Poisson's equation along the center of the device as

$$\frac{d^2\phi(x,y)}{dy^2} = \frac{d^2\phi(x,y)}{dy^2}\bigg|_{x=t_{fin}/2} = \frac{d^2\phi_c(y)}{dy^2} \qquad (6.18)$$

Then substituting Equations 6.16 and 6.18 into Equation 6.1, we can express Poisson's equation in terms of the center-potential $\phi_c(y)$ as

$$\frac{2\varepsilon_{ox}}{\varepsilon_{si}t_{fin}T_{ox}}\frac{1}{\left(1+\dfrac{\varepsilon_{ox}t_{fin}}{4\varepsilon_{si}T_{ox}}\right)}\left(V_{gst}-\phi_c(y)\right)+\frac{d^2\phi_c(y)}{dy^2}=\frac{qN_b}{\varepsilon_{si}}$$

(6.19)

$$\text{or,} \quad \frac{d^2\phi_c(y)}{dy^2}+\frac{V_{gst}-\phi_c(y)}{\dfrac{\varepsilon_{si}}{2\varepsilon_{ox}}\left(1+\dfrac{\varepsilon_{ox}t_{fin}}{4\varepsilon_{si}T_{ox}}\right)t_{fin}T_{ox}}-\frac{qN_b}{\varepsilon_{si}}=0$$

Now, if we define

$$\lambda\equiv\sqrt{\frac{\varepsilon_{si}}{2\varepsilon_{ox}}\left(1+\frac{\varepsilon_{ox}t_{fin}}{4\varepsilon_{si}T_{ox}}\right)t_{fin}T_{ox}}$$

(6.20)

Then we can express Equation 6.19 after rearranging as

$$\frac{d^2\phi_c(y)}{dy^2}+\frac{V_{gst}-\phi_c(y)}{\lambda^2}-\frac{qN_b}{\varepsilon_{si}}=0$$

(6.21)

Equation 6.21 describes the variation of the center potential along the channel length of a device with a technology-dependent parameter λ, referred to as the *natural length* of a device given by Equation 6.20.

Again, Equation 6.21 can be written as

$$\lambda^2\frac{d^2\phi_c(y)}{dy^2}=\phi_c(y)-\left(V_{gst}-\frac{qN_b}{\varepsilon_{si}}\lambda^2\right)$$

(6.22)

For a long-channel device where the drain bias does not have a significant influence on source potential, $\dfrac{d^2\phi_c(y)}{dy^2}=\dfrac{dE_c(y)}{dy}\cong 0$; and we can safely assume that the center potential $\phi_c(y)$ at the source-end of a long-channel device is given by the potential at the source ϕ_{csL} of the device. Thus, from Equation 6.22, we get an expression for source potential for the long-channel devices as

$$\phi_{csL}=\left(V_{gst}-\frac{qN_b}{\varepsilon_{si}}\lambda^2\right)$$

(6.23)

where:

ϕ_{csL} is equivalent to the center potential for the long-channel devices

The *characteristics* or *natural length* λ defined in Equation 6.20 represent the spread of drain electric potential in the y-direction along the length of the device. As observed from Equation 6.20, λ depends on gate oxide thickness T_{ox} and silicon fin thickness t_{fin}. Therefore, appropriate choice of T_{ox} and t_{fin} can reduce the influence of the drain electric field in the channel region. The natural length can be used to estimate the maximum silicon fin thickness and width that can be used in order to avoid SCEs. Numerical device simulation data show that for an effective control of SCEs, the channel length L must be at least five times larger than the natural length λ given in Equation 6.20 [21]. Thus, the natural length scale can be effectively used as a design guideline [22]. In the following section, we will discuss the SCEs in FinFETs using the concept of the natural length.

6.2.2 CHANNEL POTENTIAL

In order to characterize SCEs in FinFETs, we use 2D Poisson's Equation 6.1 to determine the distribution of electrostatic potential along the channel at high electric field at the drain end of the device. Thus, let us consider an infinitesimal rectangular box of width dy and height $t_{fin}/2$ as shown in Figure 6.2.

Now, applying Gauss' law [Equation 2.71] in two dimensions (x,y) on the rectangular box near the drain end of the channel as shown in Figure 6.2, we get

$$\left[\frac{E_y(y)-E_y(y+dy)}{dy}\right]+\left[\frac{E_x(0)-E_x(t_{fin}/2)}{t_{fin}/2}\right]=\frac{qN_b}{\varepsilon_{si}} \quad (6.24)$$

Multiplying both sides of Equation 6.24 by $t_{fin}/2$, we get

$$\frac{t_{fin}}{2}\left[\frac{E_y(y)-E_y(y+dy)}{dy}\right]+\left[E_x(0)-E_x(t_{fin}/2)\right]=\frac{qN_b}{2\varepsilon_{si}}t_{fin} \quad (6.25)$$

$$\text{or,} \quad \frac{t_{fin}}{2}\frac{dE_y}{dy}+\left[E_x(0)-E_x(t_{fin}/2)\right]=\frac{qN_b}{2\varepsilon_{si}}t_{fin}$$

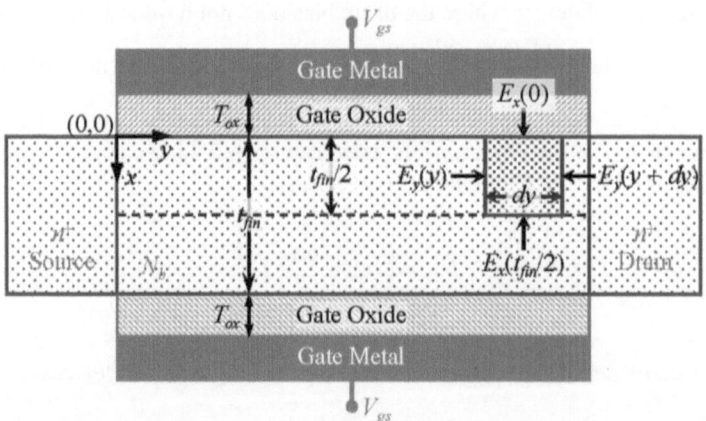

FIGURE 6.2 2D cross-section of a symmetric common DG-FinFET illustrating the velocity saturation region from the top gate only; x-axis is along the thickness t_{fin} and y-axis is parallel to the channel along the length of the device; dy is a small elemental length at the drain end of the channel.

We know $E_x = -d\phi/dx$ and $E_y = -d\phi/dy$, therefore we can express Equation 6.25 as

$$\frac{t_{fin}}{2}\frac{d^2\phi_s(y)}{dy^2} + \left[-\frac{d\phi(x)}{dx}\bigg|_{x=0} + \frac{d\phi(x)}{dx}\bigg|_{x=t_{fin}/2} \right] = \frac{qN_b}{2\varepsilon_{si}}t_{fin} \tag{6.26}$$

Now, substituting the expressions for $d\phi/dx$ from Equation 6.3 (BC-2) into Equation 6.26, we get

$$\frac{t_{fin}}{2}\frac{d^2\phi_s(y)}{dy^2} - \frac{\varepsilon_{ox}}{\varepsilon_{si}}\frac{\phi_s(y)-V_{gst}}{T_{ox}} = \frac{qN_b}{2\varepsilon_{si}}t_{fin}$$

$$\left(\frac{\varepsilon_{si}}{2\varepsilon_{ox}}t_{fin}T_{ox}\right)\frac{d^2\phi_s(y)}{dy^2} = \phi_s(y) - \left(V_{gst} - \frac{qN_b}{2C_{ox}}t_{fin}\right) \tag{6.27}$$

$$\lambda^2\frac{d^2\phi_s(y)}{dy^2} = \phi_s(y) - \left(V_{gst} - \frac{qN_b}{2C_{ox}}t_{fin}\right)$$

where:

$C_{ox} = \varepsilon_{ox}/T_{ox}$

$\lambda = \sqrt{\dfrac{\varepsilon_{si}}{2\varepsilon_{ox}}t_{fin}T_{ox}}$ is the characteristics length

Note that the expression for λ in Equation 6.27 is different from that obtained in Equation 6.20. This is due to the fact that Equation 6.27 is derived using pseudo-2D analysis of potential using a linear variation of electric field whereas Equation 6.20 is obtained using a parabolic potential distribution. In reality, for DG-FinFET devices the expression for λ given in Equation 6.20 is used. Then substituting the expression for long-channel center potential ϕ_{csL} from Equation 6.23 into Equation 6.27, we get after simplification

$$\frac{d^2\phi_s(y)}{dy^2} = \frac{\phi_s(y)}{\lambda^2} - \phi_{csL} \tag{6.28}$$

Now, we can solve Equation 6.28 for the surface potential $\phi_s(y)$. We know that the surface potentials at the source and drain ends are $\phi_{s0} = V_s + V_{bi}$ and $\phi_{sL} = V_{ds} + V_{bi}$, respectively; where V_{bi} is the built-in potential of the source-drain to body pn-junction described in Section 2.3.2. Then equating the expression for $\phi_s(y)$ obtained from the solution of Equation 6.28 to that given by Equation 6.13, we can show the expression for the center potential [23] as

$$\phi_c(y) = \phi_{csL} + \left(V_{bi} - \phi_{csL}\right)\frac{\sinh\left(\dfrac{L-y}{\lambda}\right)}{\sinh\left(\dfrac{L}{\lambda}\right)} + \left(V_{bi} - V_{ds} - \phi_{csL}\right)\frac{\sinh\left(\dfrac{y}{\lambda}\right)}{\sinh\left(\dfrac{L}{\lambda}\right)} \tag{6.29}$$

Then from Equation 6.29, we can show a general expression for the shift in the threshold voltage ΔV_{th} due to SCE with reference to long-channel devices as

$$\Delta V_{th} = -\frac{2\left(V_{bi} - \phi_{csL}\right) + V_{ds}}{2\cosh\left(\dfrac{L}{2\lambda}\right) - 2} \tag{6.30}$$

Equation 6.30 is used to predict the channel length dependence (V_{th} roll-off) as well as V_{ds} dependence (DIBL) phenomena of SCEs as described in the following sections.

6.2.3 THRESHOLD VOLTAGE ROLL-OFF

The threshold voltage roll-off is defined as the decrease in V_{th} with the decrease in the channel length. We know that V_{th} is the minimum gate voltage to induce a conducting channel from the source to drain initiating current flow in FinFET devices with negligible drain voltage, $V_{ds} \approx 0$. Under this condition, the center potential ϕ_{csL} can be taken as the surface potential at the source end ϕ_{st}. Then by substituting $V_{ds} \approx 0$, we get the expression for ΔV_{th} roll-off due to SCE from Equation 6.30 as

$$\Delta V_{th,SCE} = -\frac{\left(V_{bi} - \phi_{st}\right)}{\cosh\left(\dfrac{L}{2\lambda}\right) - 1} \tag{6.31}$$

Now, ϕ_{csL} can be considered as the surface potential, ϕ_{st} ($\approx E_g/2$) at the source end in the subthreshold region as discussed in Chapter 3. Therefore, from Equation 6.31, we can determine the V_{th} roll-off due to SCE.

6.2.4 DIBL EFFECT ON THRESHOLD VOLTAGE

We discussed in Section 1.2.2 that the drain voltage reduces the barrier height V_{bi} at the source end of the device, thus increasing the carrier injection from the source to channel. This phenomenon is referred to as the drain-induced barrier lowering. The decrease in the barrier height results in a decrease in V_{th} as V_{ds} increases. Thus, at the high drain bias condition $V_{ds} \gg 2(V_{bi} - \phi_{csL})$. Then from Equation 6.30, we can write an expression for DIBL as

$$\Delta V_{th,DIBL} = -\frac{V_{ds}}{2\left[\cosh\left(\dfrac{L}{2\lambda}\right) - 1\right]} \tag{6.32}$$

6.3 QUANTUM MECHANICAL EFFECTS

In Chapter 1 we discussed that the structural dimensions of multigate FinFETs are rapidly approaching the values near five-nanometer regime [1]. Under this condition, the electrons in the "channel" (for an n-channel device) form either a 2D electron gas (2DEG) for a DG-FET or a one-dimensional (1D) electron gas (1DEG) for a triple- or quadruple-gate FET. For a DG-FinFET device with ultrathin-body of fin thickness t_{fin} shown in Figure 6.1, the electrons are free to move in the y-direction along the length as well as z-direction along the fin height H_{fin} (perpendicular to the surface) of the device; however, the electrons are confined in the t_{fin}-direction. On the other hand, in a thin-body and narrow triple- or quadruple-gate device, electrons are free to move in the y-direction along the flow of current) only and are confined in the x and z directions. This results in the formation of energy sub-bands and electron distributions in the silicon film that can be significantly different from that predicted by classical theory. In particular, the inversion layers are not localized at the surface of the silicon fin but are instead found at a depth away from the Si/SiO$_2$ interface of the fin as described in Section 3.5, giving rise to *volume inversion* [13]. This confinement of electrons within the ultrathin-body is called the *inversion layer quantization* or *quantum mechanical* (QM) effect in FinFETs. The QM effect affects the performance of single as well as multigate FinFET devices through degradation in oxide thickness and consequently threshold voltage (V_{th}) [24] and mobility enhancement [25].

6.3.1   VOLUME INVERSION

The volume inversion is a phenomenon that describes the confinement of the inversion carrier concentration at the center of the ultrathin-body multigate FinFET instead of near the Si/SiO$_2$ interface as predicted by classical device physics. Volume inversion was first reported in 1987 [13], observed on gate-all-around (GAA) MOSFETs in 1990 [14], measured on DG devices in 1994 [26,27], and confirmed by numerous follow-on studies [28–34]. Volume inversion is a QM effect and is predicted by a self-consistent solution of Schrödinger [35] and Poisson equations.

Volume inversion in multigate-FETs can be solved using Poisson-Schrödinger solver. When a multigate FinFET operates in the volume inversion regime, the electrons form a low-dimensional electron gas (a 2DEG for a DG-FETs and 1DEG for a triple-gate, Π-gate, Ω-gate, or surrounding gate FET). As a result, sub-bands are formed. The j-th electron wave function and the corresponding energy level E_j can be found by a self-consistent solution of the Schrödinger equation given by

$$\left(-\frac{\hbar^2}{2m^*}\nabla^2 - q\phi \right)\Psi_j = E_j\Psi_j \tag{6.33}$$

and Poisson's equation given by

$$\nabla^2\phi = -\frac{q}{\varepsilon_{si}}\left[p - n + N_d - N_a \right] \tag{6.34}$$

where:
   m^* is the effective mass of electrons
   $\hbar$ is the reduced Planck's constant
   p is the concentration of holes
   n is the concentration of electrons
   N_a is the acceptor concentration in the fin channel
   N_d is the donor concentration in the fin channel

Thus, using the effective mass approximation to solve the Schrödinger equation, the electron concentration is given by

$$n = \sum_j \left[\left(\Psi_j \times \Psi_j^* \right) \times \int_{E_j}^{\infty} \rho_j(E) f_{FD}(E) dE \right] \qquad (6.35)$$

where:

$\rho_j(E)$ is the density of states as a function of the energy

$f_{FD}(E)$ is the Fermi–Dirac distribution function

In a 2DEG, the density of states is a constant (independent of energy), and in a 1DEG, it is a function of $(E - E_j)^{-1/2}$ [36,37]. It is to be noted that Equation 6.33 is anisotropic and is equal to

$$\left[-\frac{\hbar^2}{2} \left\{ \frac{\partial}{\partial x} \left(\frac{1}{m_x^*} \frac{\partial}{\partial x} \right) + \frac{\partial}{\partial y} \left(\frac{1}{m_y^*} \frac{\partial}{\partial y} \right) + \frac{\partial}{\partial z} \left(\frac{1}{m_z^*} \frac{\partial}{\partial z} \right) \right\} - q\phi \right] \Psi_j = E_j \Psi_j \qquad (6.36)$$

In Equation 6.36, the effective masses m_x^*, m_y^*, and m_z^* correspond to different valleys and depend on the crystal orientation [38]. Figure 6.3 shows the electron concentration in DG-FinFET for different silicon fin thickness t_{fin} and gate voltage $V_{gs} > V_{th}$ [39]. A high electron concentration is observed at the center of the silicon fin due to volume inversion [40]. The more appropriate QM analysis shows peaks of the electron concentration distributions at a distance of the order of a few nanometers away from the surfaces as shown in Figure 6.3. If the semiconductor film is thinner than

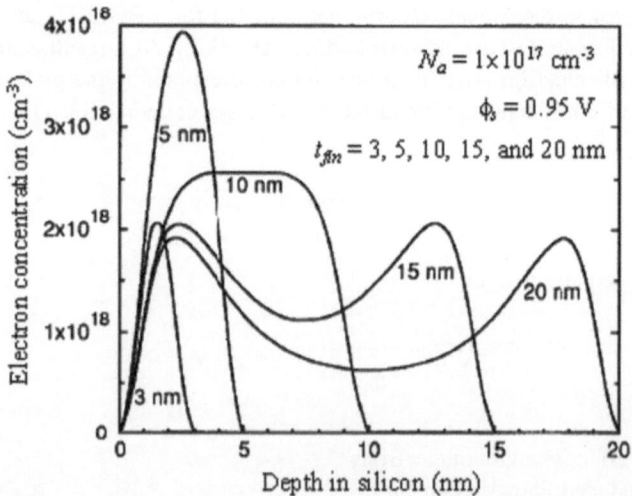

FIGURE 6.3 Quantum mechanical effect on fin thickness of DG-FinFET: plots showing the influence of the fin thickness on the distribution of electron concentration due to quantum mechanical effect obtained at a constant value of the surface potential [39].

10 nm, $N_b = 1 \times 10^{17}$ cm^{-3}, and $\phi_s = 0.95$ V, then the two surface regions overlap so strongly that the peak electron concentration is located at the middle of the semi-conductor layer as shown for $t_{fin} = 5$ nm devices and the distribution of electron concentration is totally different from that obtained by classical physics [39]. A direct consequence of the volume inversion in FETs is the increase in the inversion carrier mobility of the devices [25].

6.3.2 QM EFFECT ON MOBILITY

The carriers in volume inversion experience less interface scattering than the carriers in a surface inversion layer. As a result, an increase in the mobility and trans-conductance is observed in DG-FET devices. Furthermore, the phonon scattering rate is lower in DG devices than in SG transistors. However, in thick fins, there is no interaction between the front and back channels and there is no volume inversion as shown for $t_{fin} > 10$ nm devices in Figure 6.3. As a result, there is no difference in mobility between the DG-FinFETs and bulk-MOSFETs. If the film gets thinner, volume inversion appears, and mobility is increased because of the reduced Si/SiO$_2$ interface scattering.

In thicker films, the inversion carriers are concentrated near the interfaces, but in thinner films, most of the carriers are concentrated near the center of the silicon film, further away from the interface scattering centers as shown in Figure 6.3 which increases their mobility. In very thin silicon films, however, the inversion carriers in the volume inversion layer do experience surface scattering because of their physical proximity to the interfaces, and mobility drops with any decrease in film thickness [41,42].

6.3.3 QM EFFECT ON THRESHOLD VOLTAGE

According to classical device physics, V_{th} of a fully depleted SOI-MOSFET with uniformly doped ultrathin-body, t_{fin} decreases as body thickness decreases [43]. This is due to the reduction of depletion charge $Q_b = qN_b t_{fin}$ with the decrease in t_{fin}, and for any value of t_{fin} below 10 nm, Q_b is negligibly small compared to the inversion charge Q_i. However, due to QM effect, the inversion carrier concentration n decreases as discussed in Section 3.5 [44]. Therefore, two non-classical contributions to V_{th} have to be taken into account. First of all, the concentration of inversion carriers must be higher than that obtained by classical physics in order to reach the threshold condition. Thus, the potential ϕ in the ultrathin-body must be higher than the potential $2\phi_B$. Secondly, since the minimum energy of the sub-bands (and therefore, the conduction band) increases with decreasing fin thickness due to the splitting of the conduction band, an increase in the gate voltage V_{gs} is required to reach the same number of inversion carrier concentration as obtained by classical physics. This apparent bandgap widening also causes V_{th} to increase [39,45–47].

QM effects in FinFETs at the nanometer node cause a shift in V_{th} to a higher value due to the higher electron ground energy and the electron concentration peaks away from the surface in contrast to the classical solution in which the carrier density peaks at the surface [48]. The numerical simulation data from the solution of Poisson and Schrödinger equations show a shift in V_{th} and the degradation of the slope of $I_{ds} - V_{gs}$ curve, hence the gate capacitance C_g as a result of QM effects. Therefore, as the thicknesses of the fin and gate oxide of DG-FinFETs decrease to the sub-10 nm regime, the QM effects become significant and must be taken into account for accurate prediction of device performance.

As discussed in Section 3.5, since the energy barrier at the Si/SiO$_2$ interface is very high (~3.1 eV), the silicon film can be approximated to be an infinite potential well. Also, we discussed in Section 5.4.3.4, the potential in the subthreshold region of DG-FinFETs is essentially flat throughout the silicon fin regardless of the oxide thickness due to volume inversion [13,49]. In particular in the case of a DG-FinFET device, neglecting any potential variation in the silicon, Schrödinger Equation 6.33, can be solved to find the minimum energy of the first conduction sub-band. The general solution to Equation 6.33 is given by

$$E_n = \frac{\hbar^2}{2m^*}\left(\frac{\pi}{t_{fin}}\right)^2 n^2 \tag{6.37}$$

where:
$n = 1, 2, 3, ..., n$ represents the sub-band energy levels
$m^* = $ effective mass along the confinement direction

Now, from Equations 6.33 and 6.37, the energy of the lowest sub-band, $n = 1$ is given by

$$E \equiv E_1 = E_{c0} + \frac{\pi^2\hbar^2}{2m^*t_{fin}^2} \tag{6.38}$$

where:
E_{c0} is the classical, "three-dimensional" minimum energy of the conduction band
Thus, the increase of channel potential due to QM effects is given by

$$\Delta\phi_s^{QM} = \frac{E - E_{c0}}{q} = \frac{\pi^2\hbar^2}{2qm^*t_{fin}^2} \tag{6.39}$$

Now, we can derive an expression for V_{th} considering a single-gate device using the basic capacitance relationship given by

$$C_{inv} \cong C_{ox} + C_b \tag{6.40}$$

where:
$C_{inv} = $ inversion layer capacitance
$C_{ox} = $ gate oxide capacitance
$C_b = $ depletion layer capacitance of the silicon body

For an undoped or lightly doped fully depleted device, $C_{inv} \gg C_b$; therefore, we can express Equation 6.40 as

$$C_{inv} \cong C_{ox} \tag{6.41}$$

Under the same condition and in the subthreshold operation, the inversion charge Q_i in a DG device is given by [Equation 3.50]

$$Q_i = -t_{fin} q n_i \exp\left(\frac{\phi_s}{v_{kT}}\right) \tag{6.42}$$

Therefore, the capacitance in a DG device is given by

$$C_{inv} = -\frac{1}{2}\frac{dQ_i}{d\phi} = \frac{1}{2v_{kT}}\left[t_{fin} q n_i \exp\left(\frac{\phi_s}{v_{kT}}\right)\right]$$

$$\text{or,} \quad C_{inv} = -\frac{1}{2v_{kT}} Q_i \tag{6.43}$$

where:
  ϕ_s is the surface potential in the fin channel
  "1/2" factor is due to two surfaces of the DG device

Thus, from Equation 6.43, we get

$$Q_i = -2v_{kT} C_{ox} \tag{6.44}$$

Then substituting the expression for Q_i from Equation 6.42 into Equation 6.44, we get

$$-q n_i t_{fin} \exp\left(\frac{\phi_s}{v_{kT}}\right) = -2v_{kT} C_{ox}$$

$$\text{or,} \quad \phi_s = v_{kT} \ln\left[\frac{2C_{ox} v_{kT}}{q n_i t_{fin}}\right] \tag{6.45}$$

Now, from Equation 3.30, we know that the gate voltage of a FinFET device is given by

$$V_{gs} = V_{fb} + \phi_s - \frac{Q_s}{C_{ox}} \tag{6.46}$$

where:
  V_{fb} is the flat band voltage, to compensate for the workfunction difference between the gate and silicon body

Then substituting the expression for ϕ_s from Equation 6.45 into Equation 6.46 and considering lightly doped n-channel FinFET (nFinFET) device with $Q_s \cong -Q_i$, we get the expression for V_{gs} at the threshold condition as

$$V_{gs} \equiv V_{th} = V_{fb} + v_{kT} \ln\left[\frac{2C_{ox}v_{kT}}{qn_i t_{fin}}\right] + \frac{Q_i}{C_{ox}}$$

$$\therefore \quad V_{th} = V_{fb} + v_{kT} \ln\left[\frac{2C_{ox}v_{kT}}{qn_i t_{fin}}\right] - 2v_{kT} \tag{6.47}$$

In Equation 6.47, we have used $Q_i/C_{ox} = -2v_{kT}$ from Equation 6.44. Since $2v_{kT}$ is negligibly small compared to the first and second terms on the right-hand side of Equation 6.47, we can safely write the expression for V_{th} without QM effect as

$$V_{th} = V_{fb} + v_{kT} \ln\left[\frac{2C_{ox}v_{kT}}{qn_i t_{fin}}\right] \tag{6.48}$$

Now, adding the increase in the channel potential ϕ_s^{QM} due to QM effect from Equation 6.39, we find the expression for V_{th} with QM effect as

$$V_{th} = V_{fb} + v_{kT} \ln\left[\frac{2C_{ox}v_{kT}}{qn_i t_{si}}\right] + \frac{\pi^2 \hbar^2}{2qm^* t_{fin}^2} \tag{6.49}$$

The first term in Equation 6.49 is due to the workfunction difference between the gate and silicon body. The second term of the expression represents the potential ϕ_s in the channel, which is inversely proportional to the silicon film thickness t_{fin}. In very thin fin body, ϕ_s can be significantly larger than $2\phi_B$. As a result, the inversion carrier concentration at threshold condition can be much larger in an ultrathin-body device than a thicker one [25].

Although the above mathematical formulation is based on the solution of one-dimensional Poisson and Schrödinger equations for electrons, the final expressions are equally applicable to holes, because QM effects of holes can also be approximated particles-in-a-potential-well problem.

6.3.4   QM Effect on Drain Current

From Figure 6.3, we observed the shift in the carrier concentration away from the Si/SiO$_2$ interface due to QM confinement of carriers. As a result, the inversion layer thickness increases and the gate capacitance is degraded. The increased inversion layer thickness can be computed by numerical device simulation. Thus, from the numerical simulation data, an empirical expression for the increase in the inversion layer thickness can be shown as [48]

$$\delta t_{inv} = \left(\frac{7\varepsilon_{si}\hbar^2}{qm^* Q_i}\right)^{1/3} \tag{6.50}$$

This increase in δt_{inv} increases the equivalent oxide thickness (EOT) of gate dielectric. It is to be noted that the structural parameter of DG-MOSFETs, $r \equiv \dfrac{\varepsilon_{si} T_{ox}}{\varepsilon_{ox} t_{fin}}$, defined in Equation 5.57 is a function of the T_{ox}, hence the boundary condition for β in Equation 5.58. In order to establish an appropriate QM correction, the boundary condition in Equation 5.58 must be reformulated using EOT due to QM effects

$$\frac{V_{gs} - V_{fb} - V_{ch}(y)}{2v_{kT}} - \ln\left(\frac{2}{t_{fin}}\sqrt{\frac{2\varepsilon_{si}v_{kT}}{qn_i}}\right) = \ln\beta - \ln\left(\cos\beta\right) + 2r^{QM}\beta\tan\beta \qquad (6.51)$$

$$\text{Or, } f_r^{QM}(\beta) = \ln\beta - \ln\left(\cos\beta\right) + 2r^{QM}\beta\tan\beta \qquad (6.52)$$

where:

$$r^{QM} \equiv \frac{\varepsilon_{si}\left(T_{ox} + \delta t_{inv}\right)}{\varepsilon_{ox}t_{fin}} \qquad (6.53)$$

Now, applying the current continuity, the drain current expression given by Equation 5.71 can be written with QM effect as

$$I_{ds}^{QM} = \mu_s\left(\frac{W}{L}\right)\frac{4\varepsilon_{si}}{t_{fin}}\left(2v_{kT}\right)^2\left[g_r^{QM}(\beta_s) - g_r^{QM}(\beta_d)\right] \qquad (6.54)$$

where:

$$f_r^{QM}(\beta) = \ln\beta - \ln\left(\cos\beta\right) + 2r\beta\tan\beta + 2\frac{\varepsilon_{si}\delta t_{inv}}{\varepsilon_{ox}t_{fin}}\beta\tan\beta \qquad (6.55)$$

$$g_r^{QM}(\beta) = \beta\tan\beta - \frac{\beta^2}{2} + \beta\tan^2\beta + r\beta^2\tan^2\beta + \frac{\varepsilon_{si}\delta t_{inv}}{\varepsilon_{ox}t_{fin}}\beta^2\tan^2\beta \qquad (6.56)$$

Due to the increase in the effective oxide thickness (EOT) by the non-zero inversion layer thickness, the drain current obtained by Equation 6.54 is degraded compared to that given by Equation 5.71, depending on the value of oxide thickness T_{ox} and QM effects decided by the electron effective mass m^* along the confinement direction. It should be noted that each confinement direction is associated with its specific effective mass m^* as described in Equation 6.36.

6.4 SURFACE MOBILITY

The accurate prediction of FinFET drain current depends on the accuracy of inversion layer mobility model. In Chapter 5, we have assumed a constant surface mobility μ_s for deriving the expression for FinFET drain current I_{ds}. This assumption is not valid under high electric field operation of the devices. As the vertical electric field E_x and lateral electric field E_y increase with increasing gate voltage V_{gs} and drain voltage V_{ds}, respectively, the inversion carriers suffer increased scattering. Therefore, μ_s strongly depends on E_x and E_y. Thus, in the following section, we will derive an expression for the surface mobility to use in computer-aided design (CAD) of very large scale integrated (VLSI) circuits using small geometry FinFETs.

Now, let us determine the surface mobility by considering only the effect of E_x along the thickness of a FinFET device, that is, $V_{ds} \approx 0$. For the simplicity of I_{ds} formulation, let us define an *effective mobility* as the average mobility of carriers given by

$$\mu_{eff} = \frac{\displaystyle\int_0^{X_{inv}} \mu_s(x,y).n(x,y)dx}{\displaystyle\int_0^{X_{inv}} n(x,y)dx} \tag{6.57}$$

Using the above definition of mobility, we can write the general expression for I_{ds} [Equation 5.22] as

$$I_{ds} = \frac{W}{L}\mu_{eff}\int_0^{V_{ds}} Q_i dV \tag{6.58}$$

In reality, μ_{eff} is highly reduced by large vertical electric field due to the high applied V_{gs}. The vertical electric field E_x pulls the inversion layer electrons in DG-nMOSFETs towards the surfaces, causing higher surface scattering as well as coulomb scattering due to the interaction of electrons with oxide charges (Q_f, N_{it}) as discussed in Chapter 3. Since the electric field varies vertically through the inversion layer, the average field in the inversion layer is given by

$$E_{eff} = \frac{E_{x1} + E_{x2}}{2} \tag{6.59}$$

where:
E_{x1} is the vertical electric field at the Si/SiO$_2$ interface
E_{x2} is the vertical electric field at the channel/depletion layer interface as shown in Figure 6.4

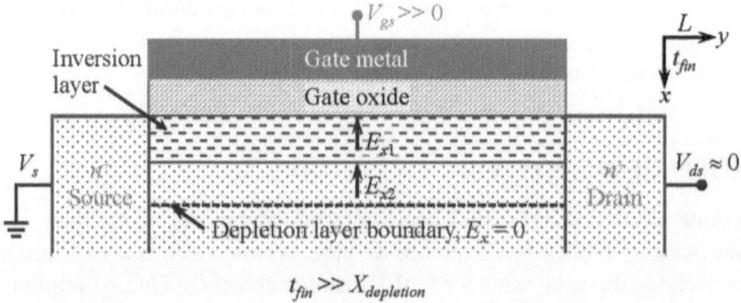

FIGURE 6.4 Effective vertical electric field on FinFET inversion carriers due to the high applied gate bias V_{gs}: E_{x1} is the vertical electric field at the Si/SiO$_2$ interface and E_{x2} is the vertical electric field at the channel depletion layer boundary.

Now, from Gauss' law we can show that

$$E_{x1} - E_{x2} = \frac{Q_i}{\varepsilon_{si}},$$

and, (6.60)

$$E_{x2} = \frac{Q_b}{\varepsilon_{si}}.$$

Substituting for E_{x1} and E_{x2} from Equation 6.60 into Equation 6.59, we can show

$$E_{eff} = \frac{1}{\varepsilon_{si}}\left(\frac{1}{2}Q_i + Q_b\right) \qquad (6.61)$$

Thus, the general expression for the effective electric field representing both the electrons and holes, is given by

$$E_{eff} = \frac{1}{\varepsilon_{si}}\left(\eta Q_i + Q_b\right) \qquad (6.62)$$

where:
$\eta = 1/2$ for electrons and $\eta = 1/3$ for holes [50–52]

The measured μ_{eff} versus E_{eff} plots show a *universal behavior* independent of doping concentration at high effective vertical electrical fields and dependence on the substrate doping concentration and interface charge at low effective vertical electric fields as shown in Figure 6.5(a) [51].

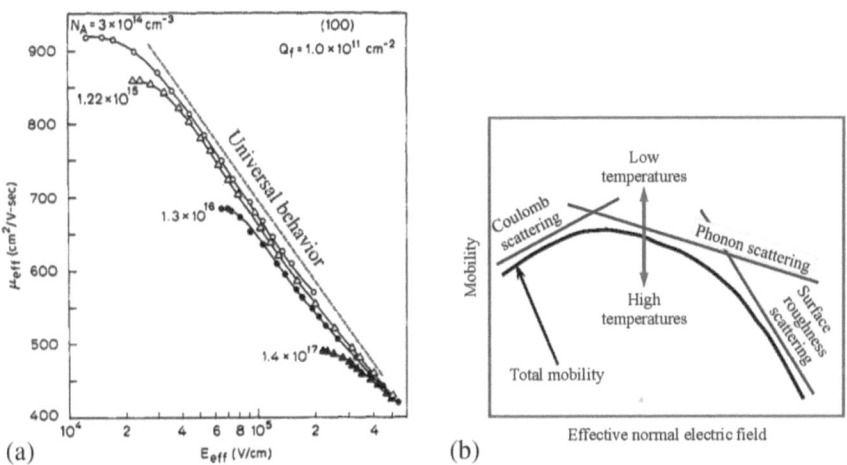

(a) (b)

FIGURE 6.5 Low-field mobility of inversion carriers in MOSFETs: (a) Universal mobility behavior of inversion layer electrons in *n*MOSFETs experimental data from [51]; (b) physical mechanisms showing the dependence of the inversion layer mobility on the effective vertical electric field.

The observed data for bulk-MOSFETs shown in Figure 6.5(a) are valid for FinFET devices. The experimentally observed universal mobility behavior shown in Figure 6.5 is due to the relative contributions of different scattering mechanisms [53,54] set by the strength of the electrical fields perpendicular to the gate as shown in Figure 6.4. From Figure 6.5(b), μ_{eff} is determined by the *coulomb* scattering of the ionized impurities and oxide charges, *phonon* scattering due to thermal vibration, and *surface roughness* scattering at the Si/SiO$_2$ interface. At high vertical electric fields, surface roughness scattering dominates as the carrier confinement is close to the channel/gate-dielectric interface resulting in a decrease of μ_{eff} with the increase of E_{eff} as observed in Figure 6.5(a).

The deviation from the *universal behavior* observed in Figure 6.5(a), particularly in the heavily doped substrates at low effective electric fields, is due to the ionized impurity scattering, coulomb scattering, and phonon scattering. At low effective vertical electric fields, Q_i is low and $\ll Q_b$. As a result, the ionized impurity scattering and coulomb scattering by ionized impurities and oxide charges become dominant scattering mechanisms in the depletion region of a FinFET device and μ_{eff} becomes a strong function of channel doping concentration as observed experimentally. As the effective electric field increases, the phonon scattering due to lattice vibration becomes important. Thus, phonon scattering is weakly dependent on vertical electric fields and has the strongest temperature dependence on μ_{eff} as shown in Figure 6.5(b).

The above physical analysis describes the behavior of μ_{eff} versus E_{eff}. However, we need to develop an effective mobility expression that can be used in drain current calculation to account for the vertical field effects on device performance. In order to develop μ_{eff} expression for circuit CAD, we consider the inversion charge $Q_i = Q_{ia}$ as the average of the inversion charge density at the source and drain end of the channel Q_{is} and Q_{id}, respectively, so that

$$Q_{ia} = \frac{Q_{is} + Q_{id}}{2} \tag{6.63}$$

Then the effective electric field given in Equation 6.62 can be written as

$$E_{eff} = \frac{1}{\varepsilon_{si}} \left(Q_b + \eta Q_{ia} \right) \tag{6.64}$$

If we normalize the charges by C_{ox} ($= \varepsilon_{ox}/T_{ox}$) for unit voltage, then we can show

$$E_{eff} = \frac{\varepsilon_{ox}}{\varepsilon_{si}} \cdot \frac{1}{T_{ox}} \left(q_b + \eta \cdot q_{ia} \right)$$

$$\text{or,} \quad E_{eff} = \left(\frac{q_b + \eta \cdot q_{ia}}{T_{ox} \left(\varepsilon_{si} / \varepsilon_{ox} \right)} \right) \tag{6.65}$$

where:
q_{ia} is the normalized average inversion charge in the fin
q_b is the normalized bulk charge in the fin

Now, we know that the unified formulation of effective mobility is given by the empirical relation [23,55,56]

$$\mu_{eff} = \frac{\mu_0}{\left(1 + \dfrac{E_{eff}}{E_0}\right)^{\nu}} \tag{6.66}$$

where:
μ_0 is concentration-dependent surface mobility
E_0 is the critical electric field
ν is a constant $\ll 1$

Since the parameter $\nu \ll 1$, we can use Taylor's series expansion of the denominator and neglect the higher order terms to obtain

$$\left(1 + \frac{E_{eff}}{E_0}\right)^{\nu} = 1 + \nu \frac{E_{eff}}{E_0} + \frac{\nu(\nu-1)}{2!}\left(\frac{E_{eff}}{E_0}\right)^2 + \cdots \tag{6.67}$$

Then substituting for E_{eff} from Equation 6.65 to the right-hand side of Equation (5.54), we get

$$\left(1 + \frac{E_{eff}}{E_0}\right)^{\nu} \cong 1 + \frac{\nu}{E_0}\left(E_{eff}\right) + \frac{\nu(\nu-1)}{2E_0^2}\left(\frac{q_b + \eta \cdot q_{ia}}{T_{ox}\left(K_{si}/K_{ox}\right)}\right)^2 + \cdots$$

$$= 1 + \frac{\nu}{E_0}\left(E_{eff}\right) + \frac{\nu(\nu-1)}{2E_0^2}\left(\frac{q_b}{T_{ox}\left(K_{si}/K_{ox}\right)}\right)^2\left(1 + \frac{\eta \cdot q_{ia}}{q_b}\right)^2 + \cdots \tag{6.68}$$

For the simplicity of computation of the surface mobility, we can show after simplification of Equation 6.68

$$\left(1 + \frac{E_{eff}}{E_0}\right)^{\nu} \cong 1 + \mu_a\left(E_{eff}\right)^{eu} + \frac{\mu_d}{\left[\dfrac{1}{2}\left(1 + \dfrac{q_{ia}}{q_b}\right)\right]^{ucs}} \tag{6.69}$$

where:
μ_a is a technology-dependent parameter that describes mobility degradation
μ_d is a technology-dependent parameter that describes the second order effect on mobility
eu is a technology-dependent parameter that describes the primary effect of E_{eff} on the inversion layer mobility
ucs is a technology-dependent parameter that describes the secondary effect of E_{eff} on the inversion layer mobility

The technology-dependent parameter set $\{\mu_a, \mu_d, eu, ucs\}$ are extracted from the measured I_{ds} *versus* V_{gs} characteristics of FinFET devices at low drain bias V_{ds}. Finally, combining Equations 6.64 and 6.69, the low field mobility of the inversion carriers in FinFETs can be shown as [57]

$$\mu_{eff} = \frac{\mu_0}{1 + \mu_a \left(E_{eff}\right)^{eu} + \mu_d \left[\frac{1}{2}\left(1 + \frac{q_{ia}}{q_b}\right)\right]^{-ucs}} \quad (6.70)$$

In order to account for the body bias V_{bs} dependence of mobility for FinFETs on bulk substrate, a term $U_c V_{bs}$ is introduced in the denominator of Equation 6.70 so that

$$\mu_{eff} = \frac{\mu_0}{1 + \left(\mu_a + \mu_c V_{bs}\right)\left(E_{eff}\right)^{eu} + \mu_d \left[\left(1 + \eta \frac{q_{ia}}{q_b}\right)\right]^{-ucs}} \quad (6.71)$$

The mobility Equations 6.70 and 6.71 have been derived assuming strong inversion condition. In the strong inversion regime, the inversion carrier mobility is a function of V_{gs}. In the subthreshold region, the accuracy of the mobility is not critical since Q_{inv} varies with V_{gs} and cannot be modeled accurately. Therefore, in the subthreshold regime, the mobility is usually modeled as a constant concentration-dependent mobility.

It should be pointed out that the mobility expressions given above account only for the influence of the vertical electrical field due to V_{gs} at low lateral electric field and are often referred to as the *low-field mobility* model. The influence of the lateral electric field due to applied V_{ds} on device performance is modeled in drain current calculation by considering the *velocity saturation* in FinFET devices under high lateral electric field.

6.5 HIGH FIELD EFFECTS

In the presence of a high electric field, carriers gain energy from the field as they drift along, thereby accompanying a number of physical phenomena which significantly change the characteristics of FinFET devices. These phenomena are primarily identified to be the velocity saturation and channel length modulation.

6.5.1 VELOCITY SATURATION

The high lateral electric field along the channel due to the applied V_{ds} significantly affects device performance. As we observe from Figure 6.6, the drift velocity v_d for electrons saturates near $E \approx 10^4$ V cm^{-1}. As a result, the relation $v_d = \mu E$ does not hold at high electric field. Since the average electric field for short-channel devices $> 10^4$ V cm^{-1}, therefore, small geometry FinFET devices operate at $v_d = v_{sat} \cong 10^7$ cm sec^{-1}, that is, at saturation velocity of electrons.

We discussed earlier that the inversion carrier mobility is also affected by high lateral electric field. Therefore, we must account for the high lateral electric field

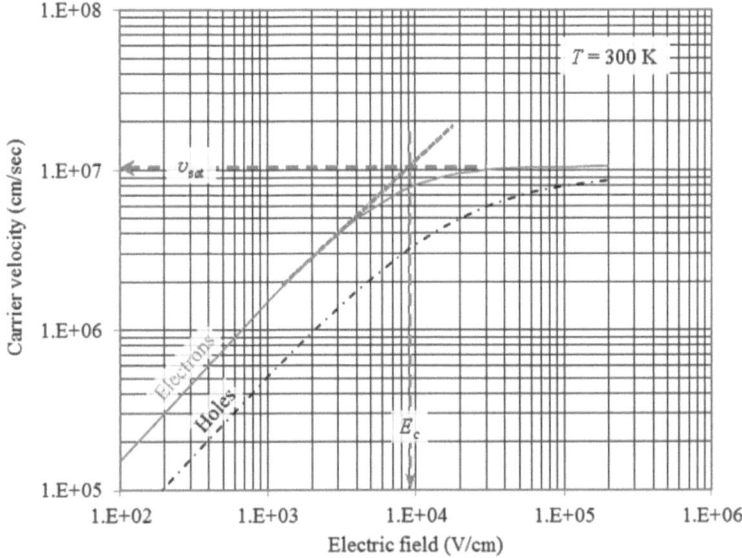

FIGURE 6.6 Drift velocity of electrons and holes as a function of electric field; the electron velocity in silicon saturates at an electric field near 1×10^4 V cm^{-1}.

effect in the expression for I_{ds} derived from Chapter 5. We also discussed that at high electric field along the channel, the FinFET devices operate at a drift velocity, $v_d = v_{sat}$ [58]. Then with reference to Figure 6.6, we can assume a piece-wise linear relation for v_d versus E plot as shown in Figure 6.7 to account for velocity saturation in I_{ds}. Thus, with reference to Figure 6.7, the relation between the drift velocity $v(y)$ and electric field $E(y)$, at any point y in the channel can be written as [58]

$$v(y) = \begin{cases} \dfrac{\mu_{eff} E(y)}{1 + \dfrac{E(y)}{E_c}}, & (E(y) < E_c) \\ \\ v_{sat}, & (E(y) > E_c) \end{cases} \tag{6.72}$$

where:

E_c is the lateral electric field at which carriers are velocity saturated, that is, at $E_y = E_c$, $v_d = v_{sat}$

As shown in Figure 6.7, we assume that v_d saturates abruptly at a critical lateral electric field E_c along the channel.

Now, from Equation 6.72, we can show

$$v_{sat} = \frac{\mu_{eff} E_c}{2}$$

$$\text{or,} \quad E_c = \frac{2 v_{sat}}{\mu_{eff}} \tag{6.73}$$

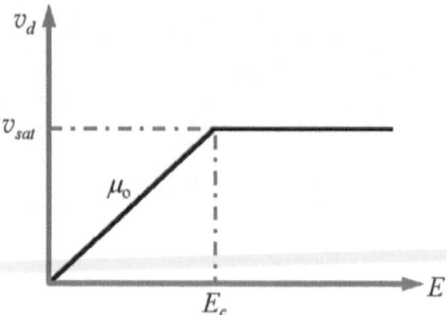

FIGURE 6.7 Piece-wise linear mobility behavior of inversion layer electrons due to high electric field along the channel length L of FinFETs: v_{sat} and μ_0, are the saturation velocity and concentration-dependent mobility of inversion carriers, respectively; and E_c is the critical electric field at which carrier velocity saturates.

We use Equation 6.72 to derive drain current expression to account for the high lateral electric field along the channel due to V_{ds}.

We know that the current density at any point y along the channel in the y-direction of an nFinFET device is given by $J_n(y) = -nqv(y) = -Q_iv(y)$, where n, q, $v(y)$ are the inversion carrier density, electronic charge, and drift velocity of inversion layer electrons, respectively; and $Q_i = nq$ is the inversion carrier charge per unit area. Then we can write the general expression for drain current as

$$I_{ds} = I(y) = WQ_iv(y) \tag{6.74}$$

where:

W is the width of the device

Then substituting for $v(y)$ from Equation 6.72 in Equation 6.74 we get for $E_y < E_c$

$$I_{ds} = WQ_i(y)\frac{\mu_{eff}E(y)}{1+\dfrac{E(y)}{E_c}} \tag{6.75}$$

After simplification, we can show from Equation 6.75

$$E(y) = \frac{I_{ds}}{W\mu_{eff}Q_i(y)-\dfrac{I_{ds}}{E_c}} = \frac{dV(y)}{dy} \tag{6.76}$$

$$\text{or, } \quad I_{ds}dy = \left(W\mu_{eff}Q_i(y)-\frac{I_{ds}}{E_c}\right)dV(y)$$

Integrating Equation 6.76 from ($y = 0$, $V(y) = 0$) to ($y = L_{eff}$, $V(y) = V_{ds}$) and after simplification, we get the linear region ($V_{ds} < V_{dsat}$) current as

$$I_{ds} = \frac{W}{L_{eff}\left(1 + \dfrac{V_{ds}}{E_c L_{eff}}\right)} \mu Q_i(y) V_{ds} \tag{6.77}$$

From Equation 6.77 note that the effect of high lateral electric field is the apparent increase in L_{eff} at higher V_{ds}, thus decreasing the linear region current. Thus, the drain current in DG-FinFETs with velocity saturation is given by

$$I_{ds} = \frac{I_{ds0}}{\left[1 + \left(\dfrac{V_{ds}}{E_c L_{eff}}\right)^{\alpha}\right]^{\frac{1}{\alpha}}} \tag{6.78}$$

where:
  I_{ds0} is the drain current without velocity saturation given by Equation 5.71
  L_{eff} is the effective channel length of the device
  α is an empirical parameter used for accurate prediction of velocity saturation effect in drain current

6.5.2 CHANNEL LENGTH MODULATION

When sufficiently high drain bias is applied, the lateral electric field becomes comparable to the vertical electric field, so that the gradual-channel approximation described in Section 5.4 becomes invalid and carriers are no longer confined to the surface channel. As the drain voltage increases beyond the saturation voltage V_{dsat}, the saturation point where the surface channel collapses or is pinched-off begins to move slightly towards the source as shown in Figure 6.8. In other words, the channel length effectively shrinks to be $L - \Delta L$. This causes non-zero conductance g_{ds} in the saturation region.

A pseudo-2D model based on the application of Gauss' law for the bulk-CMOS can also be used to analyze the channel length modulation of DG-FinFETs. Thus,

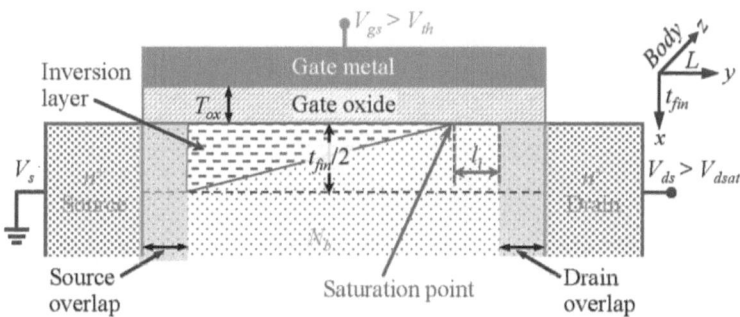

FIGURE 6.8 Schematic diagram of a FinFET device in the saturation regime showing channel pinch-off length $l_i \equiv \Delta L$ at the drain-end; here, the top gate is shown for illustration only.

the procedure described in Section 6.2.2 can be used to derive an expression for ΔL. Then using the pseudo-2D analysis, the channel electric field E can be expressed as [43,59]

$$E(y) = -\frac{dV}{dy} = \sqrt{\frac{\left(V(y) - V_{dsat}\right)^2}{l_i^2} + E_c^2} \qquad (6.79)$$

where:

E_c represents the channel electric field at which the carriers reach velocity saturation (at any point $y = 0$ and $E = E_c$) and is about 2×10^4 V cm^{-1} for electrons

V_{dsat} is the drain voltage at E_c

l_i is effective length of the channel pinch region

The expression for l_i is derived in Section 6.2.2 from pseudo-2D analysis is given by

$$l_i = \sqrt{\frac{\varepsilon_{si}}{2\varepsilon_{ox}} t_{fin} T_{ox}} \qquad (6.80)$$

Now, Equation 6.79 can be written as

$$-\frac{dV}{dy} = \frac{1}{l_i}\sqrt{\left(V(y) - V_{dsat}\right)^2 + \left(E_c l_i\right)^2}$$

$$\text{or,} \quad -\frac{dy}{l_i} = \frac{dV}{\sqrt{\left(V(y) - V_{dsat}\right)^2 + \left(E_c l_i\right)^2}} \qquad (6.81)$$

When $y = L - \Delta L$, $V(y) = V_{dsat}$; and $y = L$, $V(y) = V_{ds}$, then Equation 6.81 can be integrated as

$$\int_{L-\Delta L}^{L} \frac{dy}{l_i} = -\int_{V_{dsat}}^{V_{ds}} \frac{dV}{\sqrt{\left(V(y) - V_{dsat}\right)^2 + \left(E_c l_i\right)^2}} \qquad (6.82)$$

In order to solve Equation 6.82, we use the following integral solution

$$\int \frac{du}{\sqrt{u^2 + a^2}} = \ln\left(u + \sqrt{u^2 + a^2}\right) \qquad (6.83)$$

where:

$$u \equiv V(y) - V_{dsat}$$

$$a \equiv E_c l_i, \quad \text{is a constant} \qquad (6.84)$$

Using the integral solution in Equation 6.83, we get from Equation 6.82

$$-\frac{\Delta L}{l_i} = \ln\left[\left(V(y) - V_{dsat}\right) + \sqrt{\left(V(y) - V_{dsat}\right)^2 + \left(E_c l_i\right)^2}\right]_{V_{dsat}}^{V_{ds}} \qquad (6.85)$$

Now, using the boundary conditions, we can write Equation 6.85 as

$$-\frac{\Delta L}{l_i} = \left\{ \ln\left[(V_{ds} - V_{dsat}) + \sqrt{(V_{ds} - V_{dsat})^2 + (E_c l_i)^2} \right] - \ln\left[(V_{dsat} - V_{dsat}) + \sqrt{(V_{dsat} - V_{dsat})^2 + (E_c l_i)^2} \right] \right\}$$

(6.86)

After simplification, we can show

$$\Delta L = -\sqrt{\frac{\varepsilon_{si} T_{ox} t_{fin}}{2\varepsilon_{ox}}} \ln\left[\frac{(V_{ds} - V_{dsat}) + \ln\sqrt{(V_{ds} - V_{dsat})^2 + (E_c l_i)^2}}{E_c l_i} \right]$$

(6.87)

We know from I_{ds} expression, $I_{ds} \propto 1/L$, therefore, we can write the expression for the drain current I_{ds} to account for CLM as

$$I_{ds} = \frac{L}{L - \Delta L} I_{ds0}$$

(6.88)

where:

I_{ds0} is the drain current without CLM given by Equations 5.71 and 5.98

6.6 OUTPUT RESISTANCE

In order to formulate the output resistance (R_{out}) of FinFET devices, let us consider $I_{ds} - V_{ds}$ characteristics and the reciprocal of the first order derivative of $I - V$ data (R_{out}) superimposed as shown in Figure 6.9 [60]. Though Figure 6.9 is obtained for

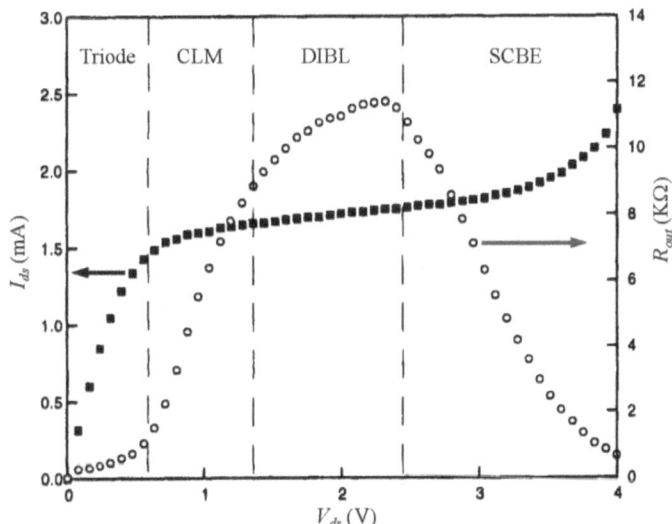

FIGURE 6.9 Drain current I_{ds} and output resistance R_{out} of an nMOSFET device divided into different operating regions based on different physical mechanisms [60].

bulk-MOSFETs, it is applicable to FinFET devices. As shown in Figure 6.9, the R_{out} plot can be divided into four different regions with different operating mechanisms.

From the general behavior of $I - V$ plot and output resistance $R_{out}(V_{ds})$ shown in Figure 6.9, the four separate regions are (1) triode or linear; (2) channel length modulation (CLM); (3) drain-induced barrier lowering (DIBL); and (4) substrate current-induced body effect (SCBE). Three mechanisms CLM, DIBL, and SCBE affect R_{out} in the saturation region; however, each of them dominates in one of the three distinct regions as shown in Figure 6.9.

We know that I_{ds} depends on both V_{gs} and V_{ds} and from Figure 6.9, we find that I_{ds} is weakly dependent on V_{ds} in the saturation region (CLM and DIBL). Since the saturation region I_{ds} depends weakly on V_{ds}, we can use Taylor's series expansion of I_{ds} at $V_{ds} = V_{dsat}$ and neglect the higher order terms to get

$$I_{ds}\left(V_{gs},V_{ds}\right) = I_{ds}\left(V_{gs},V_{dsat}\right) + \frac{dI_{ds}\left(V_{gs},V_{ds}\right)}{dV_{ds}}\left(V_{ds}-V_{dsat}\right)$$

$$= I_{dsat}\left(1 + \frac{1}{I_{dsat}}\frac{dI_{ds}\left(V_{gs},V_{ds}\right)}{dV_{ds}}\left(V_{ds}-V_{dsat}\right)\right) \qquad (6.89)$$

$$= I_{dsat}\left(1 + \frac{V_{ds}-V_{dsat}}{V_A}\right)$$

where:

$$I_{dsat} = I_{ds}\left(V_{gs},V_{dsat}\right)$$

$$V_A = I_{dsat}\left(\frac{dI_{ds}}{dV_{ds}}\right)^{-1}. \qquad (6.90)$$

In Equation 6.90, the expression for I_{dsat} is given by Equation 5.71. In Equation 6.89, V_A is called the *Early voltage* (following the original term used in describing bipolar junction transistor output resistance [44]) and is introduced for the analysis of the output resistance of FinFET devices in the saturation region. In order to determine V_A, we consider the contributions of CLM, DIBL, and SCBE components on output resistance as given by

$$V_{ACLM} = I_{dsat}\left(\frac{dI_{ds}}{dL}\cdot\frac{dL}{dV_{ds}}\right)^{-1} = C_{clm}\cdot\left(V_{ds} - V_{dsat}\right) \qquad (6.91)$$

$$V_{ADIBL} = I_{dsat}\left(\frac{dI_{ds}}{dV_{th}}\cdot\frac{dV_{th}}{dV_{ds}}\right)^{-1} \qquad (6.92)$$

$$V_{SCBE} = I_{dsat}\left(\frac{dI_{ds}}{dI_{sub}}\cdot\frac{dI_{sub}}{dV_{ds}}\right)^{-1} \qquad (6.93)$$

6.7 SUMMARY

This chapter discussed the characteristics of small geometry FinFET devices to account for the physical effects on real device performance. For accurate characterization of small geometry devices, the different structural and physical effects are discussed. First of all, the scale length of the devices that describes the influence of drain field penetration into the channel and hence short-channel effect is formulated. Then the mathematical formulations for V_{th} roll-off, DIBL, quantum mechanical effects, low-field mobility, velocity saturation, and channel length modulation are described. Finally, the output resistance and its effect on drain current are presented.

REFERENCES

1. F.-L. Yang, D.-H. Lee, H.-Y. Chen, *et al.*, "5nm-gate nanowire FinFET." In: *Symposium on VLSI Technology*, pp. 196–197, 2004.
2. K. Suzuki, T. Tanaka, Y. Tosaka, H. Horie, Y. Arimoto, "Scaling theory for double-gate SOI MOSFETs," *IEEE Transactions on Electron Devices*, 40(12), pp. 2326–2329, 1993.
3. H.-S.P. Wong, D.J. Frank, and P.M. Solomon, "Device design considerations for double-gate, ground plane and single-gate ultra-thin SOI MOSFETs at the 25nm channel length consideration." In: *IEEE International Electron Devices Meeting Technical Digest*, pp. 407–410, 1998.
4. J. Kedzierski, D.M. Fried, E.J. Nowak, *et al.*, "High-performance symmetric-gate and CMOS-compatible Vt asymmetric-gate FinFET devices." In: *IEEE International Electron Devices Meeting Technical Digest*, pp. 437–440, 2001.
5. C. Duvvury, "A guide to short channel effects in MOSFETS," *IEEE Circuit and Devices Magazine*, 2(6), pp. 6–10, 1986.
6. C.Y. Lu and J.M. Sung, "Reverse short channel effects on threshold voltage in submicron salicide devices," *IEEE Electron Device Letters*, 10(10), pp. 446–448, 1989.
7. E.H. Li, K.M. Hong, Y.C. Cheng, and K.Y. Chan, "The narrow channel effect in MOSFET with semi-recessed oxide structures," *IEEE Transactions on Electron Devices*, 37(3), pp. 692–701, 1990.
8. L.A. Akers, "The inverse narrow width effect," *IEEE Electron Device Letters*, 7(7), pp. 419–421, 1986.
9. W. Fichtner and H.W. Potzl, "MOS modeling by analytical approximations-subthreshold current and subthreshold voltage," *International Journal of Electronics*, 46(1), pp. 33–35, 1979.
10. R.R. Troutman, "VLSI limitations from drain-induced barrier lowering," *IEEE Transactions on Electron Devices*, ED-26(4), pp. 461–469, 1979.
11. M.S. Liang, J.Y. Choi, P.-K. Ko, and C. Hu, "Inversion layer capacitance and mobility of very thin oxide MOSFETs," *IEEE Transactions on Electron Devices*, 33(3), pp. 409–413, 1986.
12. C.G. Sodini, P.K. Ko, and J.L. Moll, "The effects of high fields on MOS device and circuit performance," *IEEE Transaction on Electron Devices*, ED-31(10), pp. 1386–1393, 1984.
13. F. Balestra, S. Cristoloveanu, M. Benachir, J. Brini, T. Elewa, "Double-gate silicon-on-insulator transistor with volume inversion: A new device with greatly enhanced performance," *IEEE Electron Devices Letters*, 8(9), pp. 410–412, 1987.
14. J.P. Colinge, M.H. Gao, A. Romano, *et al.*, "Silicon-on-insulator 'gate-all-around' MOS device." In: *IEEE International Electron Devices Meeting Technical Digest*, pp. 595–598, 1990.

15. M.V. Dunga, "Nanoscale CMOS modeling," Ph.D. dissertation, Electrical Engineering and Computer Science, University of California Berkeley, Berkeley, CA, 2008.

16. D. Lu, "Compact models for future generation CMOS," Ph.D. dissertation, Electrical Engineering and Computer Science, University of California Berkeley, Berkeley, CA, 2011.

17. X. Liang and Y. Taur, "A 2-D analytical solution for SCEs in DG MOSFETs," *IEEE Transactions on Electron Devices*, 51(9), pp. 1385–1391, 2004.

18. Q. Chen, E.M. Harrell, and J.D. Meindl, "A physical short-channel threshold voltage model for undoped symmetric double-gate MOSFETs," *IEEE Transactions on Electron Devices*, 50(7), pp. 1631–1637, 2003.

19. K. Suzuki, Y. Tosaka, and T. Sugii, "Analytical threshold voltage model for short channel n+/p+ double-gate SOI MOSFETs," *IEEE Transactions on Electron Devices*, 43(5), pp. 732–738, 1996.

20. K.K. Young, "Short-channel effect in fully depleted SOI MOSFET's," *IEEE Transactions on Electron Devices*, 36(2), pp. 399–402, 1989.

21. W. Xiong, "Multigate MOSFET technology." In: *FinFETs and Other Multi-Gate Transistors*, J.-P. Colinge, (ed.), Springer, Cambridge, MA, 2008.

22. R.-H. Yan, A. Ourmazd, and K.F. Lee, "Scaling the Si MOSFET: From bulk to SOI to bulk," *IEEE Transactions on Electron Devices*, 30(7), pp. 1704–1710, 1992.

23. Z.-H. Liu, C. Hu, J.-H. Huang, *et al.*, "Threshold voltage model for deep-submicron MOSFETs," *IEEE Transactions on Electron Devices*, 40(1), pp. 86–95, 1993.

24. S. Saha, "Effects of inversion layer quantization on channel profile engineering for nMOSFETs with 0.1 μm channel lengths," *Solid-State Electronics*, 42(11), pp. 1985–1991, 1998.

25. J.-P. Colinge, "The SOI MOSFET: From single gate to multigate." In: *FinFETs and Other Multi-Gate Transistors*, J.-P. Colinge, (ed.), Springer, Cambridge, MA, 2008.

26. T. Oussie, "Self-consistent quantum-mechanical calculations in ultrathin silicon-on-insulator structures," *Journal of Applied Physics*, 76(10), pp. 5989–5995, 1994.

27. J.P. Colinge, X. Baie, and V. Bayot, "Evidence of two-dimensional carrier confinement in thin n-channel SOI gate-all-around (GAA) devices," *IEEE Electron Device Letters*, 15(6), pp. 193–195, 1994.

28. S. Cristoloveanu and D.E. Ioannou, "Adjustable confinement of the electron gas in dual-gate silicon-on-insulator mosfet's," *Superlattices and Microstructures*, 8(1), pp. 131–135, 1990.

29. X. Baie and J.P. Colinge, "Two-dimensional confinement effects in gate-all-around (GAA) MOSFETs," *Solid-State Electronics*, 42(4), pp. 499–504, 1998.

30. T. Oussie, T. Oussie, D.K. Maude, *et al.*, "Subband structure and anomalous valley splitting in ultra-thin silicon-on-insulator MOSFET's," *Physica Part B: Condensed Matter*, 249–251(6), pp. 731–734, 1998.

31. G. Baccarani and S. Reggiani, "A compact double-gateMOSFET model comprising quantum-mechanical and nonstatic effects," *IEEE Transactions on Electron Devices*, 46(8), pp. 1656–1666, 1999.

32. L. Ge and J.G. Fossum, "Analytical modeling of quantization and volume inversion in thin Si-film DG-MOSFETs," *IEEE Transactions on Electron Devices*, 49(2), pp. 287–292, 2002.

33. A. Rahman and M.S. Lundstrom, "A compact scattering model for the nanoscale double-gate MOSFET," *IEEE Transactions on Electron Devices*, 49(3), pp. 481–489, 2012.

34. S. Venugopalan, M.A. Karim, S. Salahuddin, *et al.*, "Phenomenological compact model for QM charge centroid in multi gate FETs," *IEEE Transactions on Electron Devices*, 60(4), pp. 480–484, 2013.

35. L.D. Landau and E.M. Lifshitz, *Quantum Mechanics*, Addison-Wesley, Reading, MA, 1990.

36. T. Ando, A.B. Fowler, and F. Stern, "Electronic properties of two-dimensional systems," *Review of Modern Physics*, 54(2), pp. 437–672, 1982.
37. P.N. Butcher, "Theory of electron transport in low-dimensional semiconductor structures." In: *Physics of Low-Dimensional Semiconductor Structures: Physics of Solids and Liquids*, P.N. Butcher, N.H. March, and M.P. Tosi, (eds.), pp. 95–176, Springer, Boston, MA, 1993.
38. X. Shao and Z. Yu, "Nanoscale FinFET simulation: A quasi-3D quantum mechanical model using NEGF," *Solid-State Electronics*, 49(8), pp. 1435–1445, 2005.
39. B. Majkusiac, T. Janik, and J. Walczak, "Semiconductor thickness effects in the double-gate SOI MOSFET," *IEEE Transactions on Electron Devices*, 45(5), pp. 1127–1134, 1998.
40. J.-P. Colinge, J.C. Alderman, W. Xiong, and C.R. Cleavelin, "Quantum-mechanical effects in trigate SOI MOSFETs," *IEEE Transactions on Electron Devices*, 53(5), pp. 1131–1136, 2006.
41. L. Ge, J.G. Fossum, and F. Gamiz, "Mobility Enhancement via volume inversion in double-gate MOSFETs." In: *Proceedings of the IEEE International Conference on SOI*, pp. 153–154, 2003.
42. Ge. Tsutsui, M. Saitoh, T. Saraya, *et al.*, "Mobility enhancement due to volume inversion in [110]-oriented ultra-thin body double-gate nMOSFETs with body thickness less than 5 nm." In: *International Electron Devices Meeting Technical Digest*, pp. 729–732, 2005.
43. H.K. Lim and J.G. Fossum, "Threshold voltage of thin-film Silicon-on-insulator (SOI) MOSFET's," *IEEE Transactions on Electron Devices*, 30(10), pp. 1244–1251, 1983.
44. S.K. Saha, *Compact Models for Integrated Circuit Design: Conventional Transistors and Beyond*, CRC Press, Taylor & Francis Group, Boca Raton, FL, 2015.
45. Y. Omura, S. Horiguchi, M. Tabe, and K. Kishi, "Quantum-mechanical effects on the threshold voltage of ultrathin-SOI nMOSFETs," *IEEE Electron Device Letters*, 14(12), pp. 569–571, 1993.
46. K. Uchida, J. Koga, R. Ohba, *et al.*, "Experimental evidences of quantum-mechanical effects on low-field mobility, gate-channel capacitance, and threshold voltage of ultra-thin body SOI MOSFETs." In: *IEEE International Electron Devices Meeting Technical Digest*, pp. 633–636, 2001.
47. T. Ernst, S. Cristoloveanu, G. Ghibaudo, *et al.*, "Ultimately thin double-gate SOI MOSFETs," *IEEE Transactions on Electron Devices*, 50(3), pp. 830–838, 2003.
48. F. Stern and W.E. Howard, "Properties of semiconductor surface inversion layers in the electric quantum limit," *Physical Review*, 163(3), pp. 816–835, 1967.
49. W. Wei, "Quantum modeling of symmetric double-gate MOSFETs," Ph. D. dissertation, Electrical Engineering (Applied Physics), University of California San Diego, La Jolla, CA, 2007.
50. N.D. Arora and G.Sh. Gildenblat, "A semi-empirical model of the MOSFET inversion layer mobility for low temperature operation," *IEEE Transactions on Electron Devices*, 34(1), pp. 89–93, 1987.
51. S.C. Sun and J.D. Plummer, "Electron mobility in inversion and accumulation layers on thermally grown oxidized silicon surfaces," *IEEE Transactions on Electron Devices*, 27(8), pp. 1497–1508, 1980.
52. D.T. Amm, H. Mingam, P. Delpech, and T.T. D'ouville, "Surface mobility in n+ and p+ doped polysilicon gate PMOS transistors," *IEEE Transactions on Electron Devices*, 36(5), pp. 963–968, 1989.
53. M.S. Liang, J.Y. Choi, P.-K. Ko, and C. Hu, "Inversion layer capacitance and mobility of very thin oxide MOSFETs," *IEEE Transactions on Electron Devices*, 33(3), pp. 409–413, 1986.

54. K. Lee, J.Y. Choi, S.P. Sim, and C.K. Kim, "Physical understanding of low field carrier mobility in silicon inversion layer," *IEEE Transactions on Electron Devices*, 38(8), pp. 1905–1912, 1991.

55. K. Chen, H.C. Wann, J. Dunster, *et al.*, "MOSFET carrier mobility model based on gate oxide thickness, threshold and gate voltages," *Solid-State Electronics*, 39(10), pp. 1515–1518, 1996.

56. Y. Cheng, M.-C. Jeng, Z. Liu, *et al.*, "A physical and scalable BSIM3v3 I-V model for analog/digital circuit simulation," *IEEE Transactions on Electron Devices*, 44(2), pp. 227–287, 1997.

57. Y.S. Chauhan, D.D. Lu, S. Venugopalan, *et al.*, *FinFET Modeling for IC Simulation and Design: Using the BSIM-CMG Standard*, Academic Press, San Diego, CA, 2015.

58. C. Sodini, P.K. Ko, and J. Moll, "The effect of high fields on MOS device and circuit simulation," *Transactions on Electron Devices*, 31(10), pp. 1386–1393, 1984.

59. N.D. Arora, *MOSFET Models for VLSI Circuit Simulation: Theory and Practice*, Springer – Verlag, Vienna, 1993.

60. J.H. Huang, H. Liu, M.C. Jeng, *et al.*, "A physical model for MOSFET output resistance." In: *IEEE International Electron Devices Meeting Technical Digest*, pp. 569–572, 1992.

7 Leakage Currents in FinFETs

7.1 INTRODUCTION

In Chapter 1, we discussed that the fin field-effect transistor (FinFET) device structure offers reduced short-channel effects (SCEs) due to increased gate controllability on the ultrathin fin channel by multiple gates or multigate around the fin. However, as the device dimensions approach the ultimate scaling limit [1], FinFETs become susceptible to leakage currents due to the inherent physical mechanisms during device operation. The major sources of leakage currents in FinFETs include: subthreshold leakage between the drain and source in the weak-inversion region [2]; gate-induced drain leakage (GIDL) and gate-induced source leakage (GISL) between the drain and body terminals and between the source and body terminals, respectively [3–6]; impact ionization or substrate leakage [7–10]; gate oxide leakage between gate to source, drain, and body terminals [11–14]; and source-body and drain-body *pn*-junctions leakage currents in the *on-* as well as *off*-state of the devices [2,15]. The overall drain current I_{ds} with the addition of the source-drain *pn*-junction leakage currents, gate oxide tunneling currents, and impact ionization leakage current becomes noticeably different in the on-state of the devices. Thus, it is critical to characterize different leakage currents in FinFET devices for accurate analysis of the drive current and device performance in very large scale integrated (VLSI) circuits and systems. In this chapter, we discuss the physical mechanisms of different leakage currents in FinFETs and present mathematical formulation for quantitative analysis of these leakage currents in FinFET devices.

7.2 SUBTHRESHOLD LEAKAGE CURRENTS

In the subthreshold or weak-inversion region of FinFET operation, the applied gate bias (V_{gs}) is below the device threshold voltage (V_{th}) to induce a conducting channel from the source to drain.

Thus, in the $|V_{gs}| < |V_{th}|$ operation regime, FinFET structure consists of two back-to-back *pn*-junctions and only leakage current flows from source to drain of the device. In *n*FinFETs (*n*-channel FinFETs), this leakage current is due to the minority carrier electrons in the n^+-source diffusion region with enough energy overcoming the source-channel potential barrier (V_{bi}) diffuse to the drain side of the device. Thus, these electrons generate a non-zero drain current I_{ds} for any applied drain bias, $V_{ds} > 0$ in the subthreshold regime. This leakage current is commonly known as the *weak-inversion* or *subthreshold current* and is the dominant leakage mechanism in advanced FinFET devices. Since the number of these minority carriers increases exponentially with V_{gs} (for $|V_{gs}| < |V_{th}|$) [Figure 5.8], the weak-inversion current increases exponentially as shown in Figure 7.1.

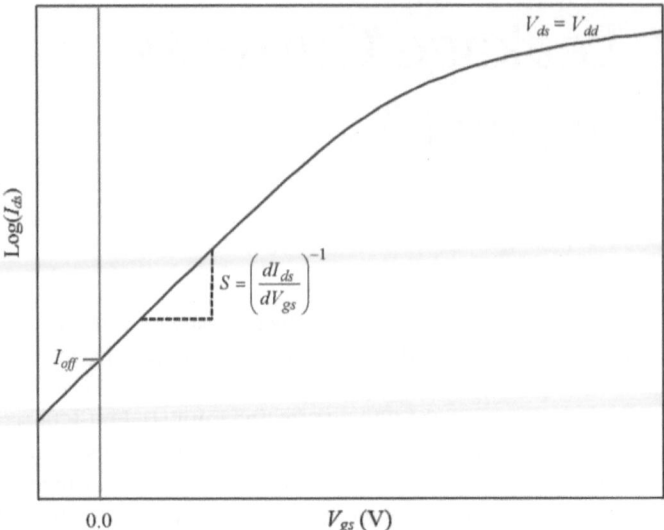

FIGURE 7.1 $I_{ds} - V_{gs}$ characteristics of a typical DG-FinFET device at $V_{ds} = V_{dd}$ (supply voltage); I_{off} and S are the off-state leakage current and subthreshold swing, respectively.

The inverse in the slope $\left(dI_{ds} / dV_{gs}\right)^{-1}$ of the $\log(I_{ds}) - V_{gs}$ characteristics, shown in Figure 7.1, is defined as the *subthreshold swing* (S). Typically, the unit of S is in mV of V_{gs} per decade of I_{ds}. The ideal value of S is about 60 (mV/decade) at room temperature (300 °K) as described in Section 5.4.3.4. This value of S is a fundamental limit at room temperature and represents the fact that the value of V_{gs} must be increased by 60 mV in order to achieve an increase in the value of I_{ds} by a factor of 10 over the potential barrier.

Again, as discussed in Chapter 1, FinFET devices offer great ability to suppress SCE and reduce off-state leakage current (I_{off}) due to a tighter electrostatic control of the channel. However, the close proximity of the source to drain region and their charge sharing with the gate in short-channel devices causes V_{th}-degradation at higher values of drain voltage V_{ds} as discussed in Section 6.2.4. This is due to the fact that the drain is close enough to source so that V_{ds} can effectively lower the source-channel barrier height at the Si/SiO$_2$ interface [2,15,16]. This effect of lowering the barrier at the source-channel interface by drain bias is defined as the *drain-induced barrier lowering* (DIBL). Ideally, the DIBL is defined as [17]

$$\frac{dV_{th}}{dV_{ds}} = \frac{V_{th}\left(V_{ds,high}\right) - V_{th}\left(V_{ds,low}\right)}{V_{ds,high} - V_{ds,low}} \tag{7.1}$$

where:

$V_{th}(V_{ds,high})$ is the value of V_{th} measured at high drain bias, $V_{ds} = V_{dd}$ (= supply voltage)

$V_{th}(V_{ds,low})$ is the value of V_{th} measured at low drain bias, $V_{ds} \leq 50$ mV (e.g., 10 mV)

Since the value of V_{th} decreases with the increase in the value of V_{ds} due to DIBL, therefore, from Equation 7.1 the value of ΔV_{th} due to DIBL ($\Delta V_{th,\mathrm{DIBL}}$) is negative showing SCE due to drain bias V_{ds}.

7.3   GATE-INDUCED DRAIN AND SOURCE LEAKAGE CURRENTS

The gate-induced drain and source leakage currents, I_{gidl} and I_{gisl}, respectively, are caused when a FinFET device is operated at high drain voltage $|V_{ds}|$ and low gate voltage $|V_{gs}| \leq 0$. Thus, for an n-channel DG-FinFET device, when $V_{gs} \leq 0$ and a high value of V_{ds} is applied to the device shown in Figure 7.2, the resulting high electric field causes a large band bending near the silicon surfaces in the gate-drain overlap region triggering *band-to-band tunneling* (BTBT) of carriers. As a result, a significant amount of drain leakage current is observed in FinFET devices.

Figure 7.2(a) shows a two-dimensional (2D) cross-section of an nFinFET device along with the corresponding energy band diagram (Figure 7.2(b)) in the gate-drain overlap depletion region under high $V_{ds} \gg V_{gs}$ condition causing GIDL current I_{gidl} due to BTBT. This BTBT current is due to the generation of carriers in the gate-drain overlap region as shown in Figure 7.2(a). From the basic device physics (Chapter 3), we know that an applied $V_{gs} > 0$ tends to induce depletion and inversion layer in the p-type body of a metal-oxide-semiconductor (MOS) system, whereas $V_{gs} < 0$ tends to do the same in the n-type body of an MOS system. Therefore, under the biasing condition $V_{ds} \gg 0$ and $V_{gs} \leq 0$, the n-type gate-drain overlap region represents an MOS system and tends to deplete and invert the surface of the region. However, the inversion of the gate-drain overlap region does not easily take place since the drain is very heavily doped compared to the channel. Nevertheless, when $V_{gd} \ll 0$, the applied V_{ds} at least causes the overlap region to be depleted of carriers inducing a large band bending at the surfaces as shown in Figure 7.2(b). If the band bending at the Si/gate oxide interface in the gate-drain overlap region is larger than the energy gap E_g of the drain region, BTBT occurs.

Under the BTBT condition, the electrons in the valence band (VB) of the n^+ drain region tunnel through the narrowed bandgap into the conduction band (CB) leaving behind holes in the VB. In addition, defects or traps created by source-drain

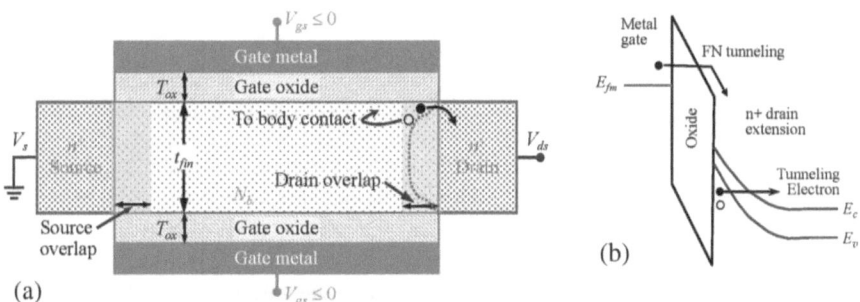

(a)   (b)

FIGURE 7.2   Gate-induced drain leakage current in an nFinFET device: (a) gated diode, at the drain-FinFET only, showing electron-hole pair generation and transport; and (b) FN tunneling due to high lateral electric field by applied drain voltage.

implantation lead to trap-assisted tunneling and enhancing the tunneling probability of carriers across the narrowed energy gap. Thus, as the minority carriers, generated either by BTBT or trap-assisted tunneling, arrive at the surface to attempt to form the inversion layer, they are immediately swept laterally to the substrate. The tunneling electrons are collected at the drain electrode/contact contributing to I_{ds}, whereas the left-behind holes in the VB are collected at the substrate electrode contributing to the substrate current for bulk-FinFETs or at the source contact in the case of SOI-FinFETs contributing to the source leakage current. This phenomenon is defined as the gate-induced drain leakage (GIDL) and is a major contributor to the off-state leakage current in FinFETs.

Thus, from the above description of the physical mechanism of GIDL, we find that under the GIDL phenomena, the semiconductor surface in the gate-drain overlap region is in deep depletion with band bending $\phi_s > 2\phi_B$ due to the lack of sustainable hole inversion layer at the surface. The lack of the formation of the hole inversion layer is due to the fact that any holes generated would drift and diffuse to the body for bulk-FinFETs or source for SOI-FinFETs by the built-in junction potential along with applied substrate-drain (V_{bd}) reverse bias, if any. However, in the case of forward V_{bd}, the holes may remain at the interface and form an inversion layer and cause band bending to be pinned at $2\phi_B < E_g$ by screening V_{ds} from the depletion charge. Thus, in the case of forward bias V_{bd}, the I_{gidl} is suppressed. On the other hand, for SOI-FinFETs, holes buildup in the floating body raise the body potential to forward bias the body-source junction. This enables the holes to be injected into the n^+ source region.

In the framework of the above described physical mechanism, it is to be noted that GIDL is not due to SCE. Figure 7.3 shows a comparison of DIBL due to SCE and GIDL due to BTBT. The ideal $I - V$ plots of an ideal FinFET in Figure 7.3 show the leakage currents due to DIBL and GIDL. The position of minimum in the I_{ds} around V_{gs} due to DIBL depends on V_{dd}, technology parameters of the device, and trap density, whereas the GIDL depends on V_{dd} at $V_{gs} \leq 0$ or $V_{dg} >> 0$.

In reality, the DIBL is measured as the difference between the values of V_{th} measured at $V_{ds} = V_{ds,low}$ (e.g., 50 mV) and $V_{ds} = V_{dd}$ as shown in Figure 7.3. Here, V_{th} is measured at a constant current level in the subthreshold region of the device. Since V_{th} decreases with increasing V_{ds}, $\Delta V_{th,\text{DIBL}}$ is negative as shown in Figure 7.3.

7.3.1 Formulation of Gate-Induced Drain Leakage Current

From the general discussions on the physical mechanism of GIDL, the requirements for the generation of GIDL current in an nFinFET device are:

1. The biasing condition of the device must satisfy $V_{ds} >> V_{gs}$ so that $V_{dg} >> 0$ to cause a large band bending at the surface of the gate-drain overlap region in the drain
2. The band bending at the Si/SiO$_2$ interface of the gate-drain overlap region must be higher than E_g so that VB energy states overlap the CB energy states as shown in Figure 7.2(b). In this case, the semiconductor surface in

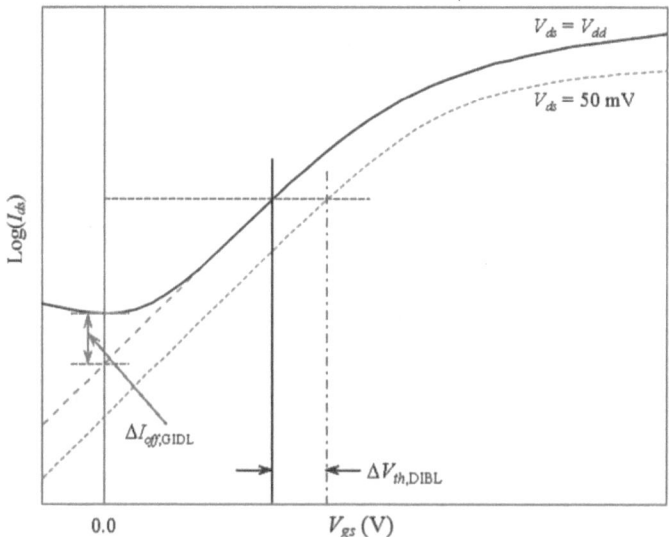

FIGURE 7.3 $I_{ds} - V_{gs}$ characteristics of a typical DG-FinFET device at $V_{ds} = 50$ mV and V_{dd} (supply voltage): $\Delta I_{off,GIDL}$ and $\Delta V_{th,DIBL}$ are the increase in the off-state leakage current due to GIDL and decrease in V_{th} due to DIBL, respectively.

the gate-drain overlap region is in deep depletion with the band bending more than E_g and surface potential $\phi_s > 2\phi_B$; where ϕ_B is the bulk potential defined in Equation 2.37. Thus, the onset of GIDL can be defined at $\phi_s = E_g$, that is, when the band bending at the Si/SiO$_2$ in the gate-drain overlap region is equal to the energy gap

3. The electric field due to the applied V_{ds} must be large (~MV cm^{-1}) so that the carrier tunneling barrier in the CB and VB overlap region is narrow to facilitate BTBT

Assuming that the above described requirements hold, we can derive a mathematical expression for the leakage current I_{gidl}, due to GIDL using Wentzel-Kramers-Brillouin (WKB) approximation for the quantum mechanical tunneling of particles. Thus, applying WKB approximation, the BTBT tunneling current density J_{gidl} can be shown as is [2]

$$J_{gidl} = A \cdot E_s \exp\left(-\frac{B}{E_s}\right) \tag{7.2}$$

where:

A is a constant that depends on the density of states of both the emitting and receiving sides

B is a physical parameter that depends on E_g and effective mass of the carriers in the direction of tunneling

E_s is the surface electric field at the gate-drain overlap region affected by BTBT

Now, from Gauss' law, we know that at the Si/SiO$_2$ interface

$$\varepsilon_{ox} E_{ox} = \varepsilon_{si} E_s$$

$$\text{or,} \quad E_s = \frac{\varepsilon_{ox}}{\varepsilon_{si}} \frac{V_{ox}}{T_{ox}} \tag{7.3}$$

where:

ε_{ox} is the permittivity of SiO$_2$
ε_{si} is the permittivity of silicon
E_{ox} is the electric field in oxide = V_{ox}/T_{ox}
V_{ox} is the voltage drop in oxide
T_{ox} is the gate oxide thickness

Then, considering an ideal *n*FinFET structure with applied gate bias V_{gs} and drain bias V_{ds} such that $V_{dg} \gg 0$ (requirement 1), we get at the Si/SiO$_2$ interface of the n^+ gate-drain overlap MOS capacitor region

$$V_{ds} = \left(V_{gs} - V_{fbsd} \right) + \phi_s + V_{ox} \tag{7.4}$$

where:

V_{fbsd} is the flat band voltage that represents the part of V_{gs} used to make the band flat at the n^+ surface of the n^+ gate-drain overlap MOS system
ϕ_s is the surface potential at the n^+ gate-drain overlap region

Now, using requirement (2), $\phi_s = E_g$ at the onset of BTBT, we get from Equation 7.4

$$V_{ox} = V_{ds} - V_{gs} + V_{fbsd} - E_g \tag{7.5}$$

Then substituting the expression for V_{ox} from Equation 7.5 into Equation 7.3, we get the expression for surface electric field as

$$E_s = \frac{V_{ds} - V_{gs} + V_{fbsd} - E_g}{\left(\varepsilon_{si} / \varepsilon_{ox} \right) \cdot T_{ox}} \tag{7.6}$$

If W_{eff} is the effective width of the device, then the expression for $I_{gidl} = (J_{gidl} \times W_{eff})$ due to BTBT can be obtained by substituting the expression for E_s from Equation 7.6 into Equation 7.2 and is given by

$$I_{gidl} = A \cdot W_{eff} \cdot \left(\frac{V_{ds} - V_{gs} + V_{fbsd} - E_g}{\left(\varepsilon_{si} / \varepsilon_{ox} \right) \cdot T_{ox}} \right) \exp\left(-\frac{B \cdot \left(\varepsilon_{si} / \varepsilon_{ox} \right) \cdot T_{ox}}{V_{ds} - V_{gs} + V_{fbsd} - E_g} \right) \tag{7.7}$$

It is to be noted that for undoped or lightly doped ultrathin-body FinFETs, the potential at both sides of the body is changed by the same V_g and the high electric field (MV cm^{-1}) condition due to V_{ds} being not strictly applicable to FinFETs. Therefore, I_{gidl} is negligibly small in lightly doped thin body FinFETs.

7.3.2 FORMULATION OF GATE-INDUCED SOURCE LEAKAGE CURRENT

Following the procedure used in Section 7.3.1, we can derive an expression for GISL current I_{gisl} due to BTBT for a symmetric nFinFET in the gate-source overlap region as

$$I_{gisl} = A \cdot W_{eff} \cdot \left(\frac{V_{sd} - V_{gs} + V_{fbsd} - E_g}{\left(\varepsilon_{si} / \varepsilon_{ox} \right) \cdot T_{ox}} \right) \exp\left(-\frac{B.\left(\varepsilon_{si} / \varepsilon_{ox} \right) \cdot T_{ox}}{V_{sd} - V_{gs} + V_{fbsd} - E_g} \right) \quad (7.8)$$

where:

V_{sd} is the source-drain bias with $V_{sd} \gg V_{gs}$ so that $V_{sg} \gg 0$

For bulk-FinFET devices, both the drain to substrate and source to substrate biases affect the BTBT leakage currents. Therefore, for accurate characterization of I_{gidl} and I_{gisl} in bulk-FinFETs, drain to substrate (V_{db}) and source to substrate (V_{sb}) biases must be considered.

7.4 IMPACT IONIZATION CURRENT

Impact ionization is a physical phenomenon of ionizing the lattice atoms by highly energetic electrons knocking out VB electrons from the lattice atoms creating electrons and holes. Thus, when nFinFET devices operate in the strong inversion regime, the channel electron traveling through high electric field near the drain end of the channel can become highly energetic. These highly energetic electrons are called *hot electrons*. These hot electrons with sufficient kinetic energy, when they collide with the lattice atoms, can ionize the lattice atoms by knocking out electrons from the VB leaving behind holes [7–10]. The holes go into the substrate creating substrate leakage current I_{sub} as shown in Figure 7.4. Some of the electrons have enough energy to overcome the Si/SiO$_2$ energy barrier to reach the gate oxide and generate gate current I_g as shown in Figure 7.4. And, some are collected to the drain contributing to the drain current I_{ds}. The maximum electric field E_m near the drain has the greatest control of hot-carrier effects.

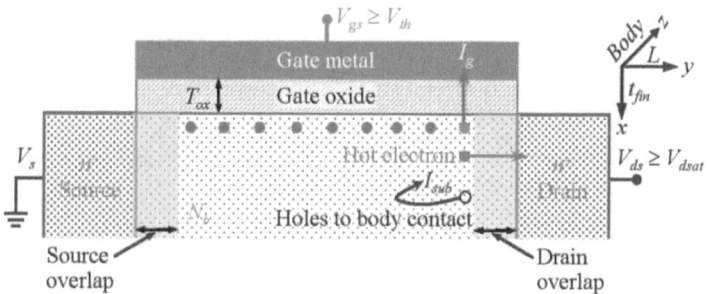

FIGURE 7.4 Hot-carrier effect in FinFETs showing channel hot electrons in an nFinFET device contributing to drain current and generating substrate current I_{sub} and gate current I_g.

The detailed hot-carrier mechanism in MOSFET devices is described in [15] and is summarized below. When a high drain bias $V_{ds} > V_{dsat}$ (drain saturation voltage) is applied to an nFinFET device in strong inversion with $V_{gs} > V_{th}$, the inversion layer electrons traveling under high channel electric field cause the following:

- high energetic electrons traveling along the channel acquire sufficient kinetic energy from the electric field and become hot
- hot electrons cause carrier multiplication by colliding with the silicon atoms in the lattice and breaking covalent bonds, thus creating electrons and holes
- holes are swept into the substrate for bulk-FinFETs due to the favorable electric field producing substrate leakage current I_{sub}
- I_{sub} flowing through the bulk causes a potential drop in the body which forward biases the source-channel pn-junction, thus reducing the source-channel potential barrier $\phi_{bi}(s)$ and enabling more carrier injection from the source to channel
- Additional carrier injection due to reduced $\phi_{bi}(s)$ causes more carrier flow in the drain, thus, increasing I_{ds} referred to as the *substrate current-induced body effect* (SCBE) discussed in Section 6.6

From the above discussions, we find that the impact ionization leakage current in an nFinFET device is due to the holes that are generated by ionization of lattice atoms by channel hot electrons as they travel from the source to the drain. The local impact ionization current $I_{ii}(y)$ along the channel of a device increases with the increase in the channel current I_{ds} and the strength of the electric field since a higher electric field increases the kinetic energy of the channel electrons. Therefore, $I_{ii}(y)$ can be written as

$$I_{ii}(y) = I_{ds}\alpha_n \tag{7.9}$$

where:
  α_n is the electron impact ionization coefficient per unit length and is a strong function of the channel electric field $E(y)$ along the transport direction of the carriers and is given by [9]

$$\alpha_n = A_i \exp\left[-\frac{B_i}{E(y)}\right] \tag{7.10}$$

where:
  A_i is a material constant that represents the occurrence of impact ionization event per unit length in the pinch-off region near the drain end of the channel
  B_i is a material constant that represents the critical field required to initiate the impact ionization event

Most of the reported data on α_n have been measured in bulk silicon and the constants A_i and B_i show a wide range of values [9,10,18]. Slotboom *et al.* [18] measured α_n at the surface and in the bulk silicon and reported the values for the constants as shown in Table 7.1.

TABLE 7.1
Surface and Bulk Impact
Ionization Coefficients in Silicon

α_n	A_i (cm^{-1})	B_i (V cm^{-1})
Surface	2.45×10^6	1.92×10^6
Bulk	7.03×10^5	1.23×10^6

Due to the exponential dependence of α_n on electric field as shown in Equation 7.10, it is easy to see that the impact ionization will dominate at the position of the maximum electric field. In a FinFET, the maximum electric field E_m is present at the drain end of the device [15]. Therefore, we expect the impact ionization current to be dominated by the maximum electric field E_m at the drain end of the channel. Substituting Equation 7.10 in Equation 7.9 and then integrating along the length of the channel in the velocity saturation region l_i, we can write the total impact ionization current in nFinFETs as

$$I_{ii} = I_{ds} A_i \int_0^{l_i} \exp\left(-\frac{B_i}{E(y)}\right) dy. \tag{7.11}$$

where:
  y is the distance along the channel length with $y = 0$ representing the start of the impact ionization region
  l_i is the length of the drain section where impact ionization takes place as shown in Figure 7.5

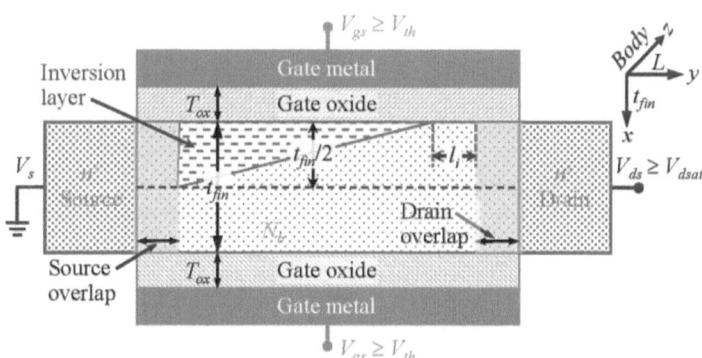

FIGURE 7.5 Hot-carrier current effect in an nFinFET on bulk substrate showing the impact ionization region l_i at the drain end of the device; here, the impact ionization region for top gate is shown for illustration only.

The parameter l_i can be treated as an effective impact ionization length and for DG-FinFETs can be obtained by pseudo-2D analysis of the electric field near the drain end of the channel as described in Section 6.2.2 and is given by

$$l_i^2 = \frac{\varepsilon_{si}}{2\varepsilon_{ox}} T_{ox} t_{fin} \tag{7.12}$$

where:
T_{ox} is the gate oxide thickness
t_{fin} is the thickness of the fin body

In order to solve Equation 7.11, we first calculate the electric field in the channel. Based on a pseudo-2D analysis [16], it can be shown that the channel electric field $E(y)$ is given by

$$E(y) = -\frac{dV}{dy} = \sqrt{\frac{\left(V(y) - V_{dsat}\right)^2}{l_i^2} + E_c^2} \tag{7.13}$$

where E_c represents the channel electric field at which the carriers reach velocity saturation (at $y = 0$, $E(y) = E_c$) and the corresponding voltage at that point is the saturation voltage V_{dsat}. The value of E_c is about 2×10^4 V cm^{-1} for electrons.

The maximum field E_m, which occurs at the drain end, can easily be obtained by replacing $V(y)$ with V_{ds} in Equation 7.13. Since at impact ionization condition, $E_c^2 \ll \left(V_{ds} - V_{dsat}\right)^2 / l_i^2$, therefore, neglecting E_c, Equation 7.13, the approximate expression for E_m can be shown as

$$E_m \cong \frac{\left(V_{ds} - V_{dsat}\right)}{l_i}. \tag{7.14}$$

Now, replacing dy in Equation 7.11 with $\left(\dfrac{dy}{dE}\right) dE = -E^2 \left(\dfrac{dy}{dE}\right) d\left(\dfrac{1}{E}\right)$, we get

$$I_{ii} = -I_{ds} A_i \int_{E_c}^{E_m} \exp\left(-\frac{B_i}{E(y)}\right) E^2 \frac{dy}{dE} d\left(\frac{1}{E}\right) \tag{7.15}$$

Here, $y = 0$, $E(y) = E_c$ and $y = l_i$, $E(y) = E_m$.

Again, from pseudo-2D analysis of the velocity saturation region, we can show

$$E(y) = E_c \cosh\left(\frac{y}{l_i}\right) = E_c \frac{\exp(y/l_i) - \exp(-y/l_i)}{2} \cong E_c \frac{1}{2} \exp\left(\frac{y}{l_i}\right) \tag{7.16}$$

Since l_i is very small resulting in y/l_i a very large number, therefore, $\exp(-y/l_i)$ is negligibly small; then differentiating Equation 7.16, we get

$$\frac{dE}{dy} = E_c \frac{1}{2} \exp\left(\frac{y}{l_i}\right) \cdot \left(\frac{1}{l_i}\right) = E(y)\left(\frac{1}{l_i}\right) \tag{7.17}$$

Then, from Equation 7.17, we can show

$$-E^2\left(\frac{dy}{dE}\right) = -E^2\left(\frac{l_i}{E}\right) = l_iE \tag{7.18}$$

Substituting Equation 7.18 in Equation 7.15, we get

$$I_{ii} = -I_{ds}A_i\int_{E_c}^{E_m} l_iE(y)\exp\left(-\frac{B_i}{E(y)}\right)d\left(\frac{1}{E}\right) \tag{7.19}$$

Since the exponential term in Equation 7.19 has a pronounced peak at $E = E_m$, we evaluate it at $E(y) = E_m$ and let it be constant over the region so that we can remove it from the integral. After this simplification, Equation 7.19 can be solved for I_{sub}

$$I_{ii} = -I_{ds}A_il_iE_m\int_{Ec}^{E_m} \exp\left(-\frac{B_i}{E(y)}\right)d\left(\frac{1}{E}\right) \tag{7.20}$$

After integration and simplification, we can show assuming $E_c \ll E_m$,

$$I_{ii} \cong I_{ds}\frac{A_i}{B_i}l_iE_m\exp\left(-\frac{B_i}{E_m}\right) \tag{7.21}$$

Substituting Equation 7.14 into Equation 7.21, we get

$$I_{ii} \cong I_{ds}\frac{A_i}{B_i}\left(V_{ds} - V_{dsat}\right)\exp\left(-\frac{l_iB_i}{V_{ds} - V_{dsat}}\right) \tag{7.22}$$

Equation 7.22 is used for substrate current calculation in FinFET devices. Note that Equation 7.22 is independent of device geometry. In order to model channel length dependence of I_{ii}, the ratio A_i/B_i is replaced by $(\alpha_0 + \alpha_1/L_{eff})$ to express

$$I_{ii} \cong \left(\alpha_0 + \frac{\alpha_1}{L_{eff}}\right)\left(V_{ds} - V_{dsat}\right)\exp\left(-\frac{\beta}{V_{ds} - V_{dsat}}\right)\cdot I_{dsa} \tag{7.23}$$

where;
α_0 is a geometry-independent parameter
α_1 is a channel-length-dependent parameter
$\beta = l_iB_i$
I_{dsa} is the drain current without the impact ionization

Thus, the geometry-dependent impact ionization leakage current can be characterized by a basic set of parameters $\{\alpha_0, \alpha_1, \beta\}$ obtained from the measurement data.

7.5   SOURCE-DRAIN *pn*-JUNCTION LEAKAGE CURRENT

The source-drain *pn*-junction currents add as the leakage currents to the intrinsic I_{ds} of FinFETs and cause a noticeable change in the device performance. The detailed discussions on *pn*-junction current are presented in Section 2.3.6. As discussed in Section 2.3.6, the current (I_{jn}) through an ideal *pn*-junction is given by Equation 2.119

$$I_{jn} = I_{jn0}\left[\exp\left(\frac{V_{jn}}{v_{kT}}\right) - 1\right]$$ (7.24)

where:

I_{jn0} is the reverse saturation current and depends on the geometry, materials, and temperature of the *pn*-junction

V_{jn} is the applied forward bias

v_{kT} is the thermal voltage

The ideal expression for I_{jn} given by Equation 2.119 becomes inaccurate over a significant range of device operations both in the forward and reverse biased modes due to several reasons including generation recombination in the depletion region and high-level injection as described in Section 2.3.6.2. Thus, considering the physical effects and depending on the magnitude of the applied forward voltage, the current through a *pn*-junction is represented by an empirical expression

$$I_{jn} = I_{jn0}\left[\exp\left(\frac{V_{jn}}{n_{js}v_{kT}}\right) - 1\right]$$ (7.25)

where:

n_{js} is called the *ideality factor* and is a measure of the deviation of the real and ideal $I - V$ plots

Typically, for a *pn*-junction, $n_{js} = 1$ when diffusion current dominates and $n_{js} = 2$ when recombination current dominates or there is high level injection.

7.6   GATE OXIDE TUNNELING LEAKAGE CURRENTS

As the gate oxide becomes progressively thinner in each generation of VLSI technology, the magnitude of the direct tunneling currents through the gate oxide becomes more significant. In direct tunneling, the carriers from the inversion layer of the silicon surface can tunnel directly through the energy gap of the SiO_2 layer instead of tunneling into the conduction band of the SiO_2 layer. In order to control direct tunneling through the gate oxide, a thicker dielectric layer with a higher dielectric constant (κ) is desired. A thick, high-*k* gate oxide retains the control of the gate over the channel with orders of magnitude reduction in dielectric leakage current compared with SiO_2 of the same effective oxide thickness (EOT). Furthermore, the use of a metal gate eliminates the polysilicon gate depletion effect which was effectively increasing the gate dielectric thickness, thus reducing the gate control of the channel.

High-*k* oxides, in general, are also found to form a better interface with metal gates than the traditional polysilicon gate. However, the gate tunneling leakage through the oxide remains a significant and increasing concern as each new technology generation requires thinner EOT.

The direct-tunneling current can be very large for advanced CMOS technologies with oxide thickness of about one nanometer [11]. Reported data show that I_{gate} is extremely high for thinner $T_{ox} < 2$ nm due to direct tunneling gate leakage current. Therefore, it is critical to characterize gate current of advanced MOSFETs for circuit design.

There are five tunneling components of gate current I_g as shown in Figure 7.6:

1. I_{gd} = leakage current through gate-to-drain overlap region
2. I_{gcd} = part of gate-to-channel current collected at the drain
3. I_{gs} = leakage current through gate-to-source overlap region
4. I_{gcs} = part of gate-to-channel current collected at the source
5. I_{gb} = gate-to-substrate leakage current (accumulation and inversion)

The detailed analysis of these tunneling currents unavoidably involves quantum mechanical analysis [11]. The detailed mathematical formulations of gate current are available in the available publications [2].

7.7 SUMMARY

This chapter presented the physical mechanisms and mathematical formulations of different components of leakage currents in FinFET devices during operation in VLSI circuits and systems. First of all, the subthreshold leakage current during device operation in the weak-inversion regime is described. This leakage current is due to the close proximity of the drain region to the source in short-channel devices, thus, lowering the barrier height of the source-channel *pn*-junction resulting in more carrier injection from the source to the channel and increased current flow through the device. And, the overall effects of subthreshold leakage current due to SCEs such as V_{th} degradation and DIBL are discussed. Then the physical mechanism of

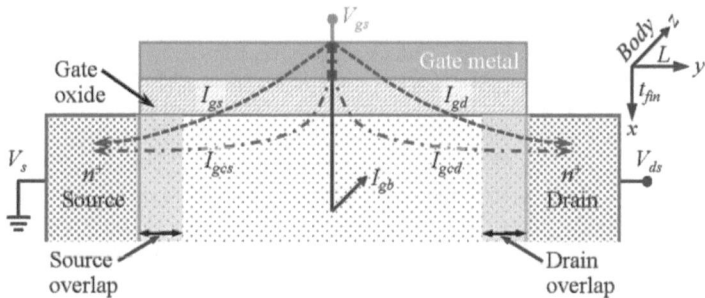

FIGURE 7.6 Gate current in FinFETs: different components of tunneling current in nanometer scale MOSFETs.

the gate-induced drain and source leakage currents due to BTBT is discussed and mathematical expressions for GIDL and GISL currents are presented. After the derivation of I_{gidl} and I_{gisl}, the leakage current due to impact ionization of the substrate atoms by highly energetic hot-carriers in the drain end of the channel is described and a mathematical expression to analyze this leakage current is presented. Then the source-drain pn-junction leakage current is briefly discussed. Finally, an overview of the leakage currents generated by direct tunneling of channel carriers to gate oxide of the transistor is presented.

REFERENCES

1. F.-L. Yang, D.-H. Lee, H.-Y. Chen, *et al.*, "5nm-gate nanowire FinFET." In: *Symposium on VLSI Technology*, pp. 196–197, 2004.
2. Y.S. Chauhan, D.D. Lu, S. Venugopalan, *et al.*, *FinFET Modeling for IC Simulation and Design: Using the BSIM-CMG Standard*, Academic Press, San Diego, CA, 2015.
3. T.Y. Chan, J. Chen, P.K. Ko, and C. Hu, "The impact of gate-induced drain-leakage current on MOSFET scaling." In: *IEEE International Electron Devices Meeting Technical Digest*, pp. 718–721, 1987.
4. S.A. Parke, J.E. Moon, H.C. Wann, *et al.*, "Design for suppression of gate-induced drain leakage in LDD MOSFET's using a quasi-two dimensional analytical model," *IEEE Transactions on Electron Devices*, 39(7), pp. 1694–1703, 1992.
5. H.J. Wann, P.K. Ko, and C. Hu, "Gate-induced band-to-band tunneling leakage current in LDD MOSFET's." In: *IEEE International Electron Devices Meeting Technical Digest*, pp. 147–150, 1992.
6. N. Lindert, M. Yoshida, C. Wann, and C. Hu, "Comparison of GIDL in p+-poly PMOS and n+-poly PMOS devices," *IEEE Electron Device Letters*, 17(6), pp. 285–287, 1996.
7. T.Y. Chan, P.K. Ko, and C. Hu, "A simple method to characterize substrate current in MOSFET's," *IEEE Electron Device Letters*, 5(12), pp. 505–507, 1984.
8. C. Hu, "Hot carrier effects." In: *Advanced MOS Device and Physics*, N.G. Einspruch and G. Gildenblat, (eds.), vol. 18, pp. 119–139, VLSI Electronics Microstructure Science, Academic Press, New York, 1989.
9. S. Saha, C.S. Yeh, and B. Gadepally, "Impact ionization rate of electrons for accurate simulation of substrate current in submicron devices," *Solid-State Electronics*, 36(10), pp. 1429–1432, 1993.
10. S. Saha, "Hot-carrier reliability in sub-0.1 μm nMOSFET devices." *Materials Research Society Symposium Proceedings*, 428, pp. 379–384, 1996.
11. S.-H. Lo, D.A. Buchanan, Y. Taur, and W. Wang, "Quantum-mechanical modeling of electron tunneling current from the inversion layer of ultra-thin-oxide nMOSFET's," *IEEE Electron Device Letters*, 18(5), pp. 209–211, 1997.
12. N. Yang, W.K. Hension, J.R. Hauser, and J.J. Wortman, "Modeling study of ultrathin gate oxides using direct tunneling current and CV measurements in MOS devices," *IEEE Transactions on Electron Devices*, 46(7), pp. 1464–1471, 1999.
13. W.-C. Lee and C. Hu, "Modeling gate and substrate currents due to conduction- and valence-band electron and hole tunneling." In: *Symposium on VLSI Technology*, pp. 198–199, 2000.
14. W.-C. Lee and C. Hu, "Modeling CMOS tunneling currents through ultrathin gate oxide due to conduction- and valence-band electron and hole tunneling," *IEEE Transactions on Electron Devices*, 48(7), pp. 1366–1373, 2001.
15. S.K. Saha, *Compact Models for Integrated Circuit Design: Conventional Transistors and Beyond*, CRC Press, Taylor & Francis Group, Boca Raton, FL, 2015.

16. N.D. Arora, *MOSFET Models for VLSI Circuit Simulation: Theory and Practice*, Springer – Verlag, Wien, 1993.

17. S. Saha, "Scaling considerations for high performance 25 nm metal-oxide-semiconductor field-effect transistors," *Journal of Vacuum Science and. Technology B*, 19(6), pp. 2240–2246, 2001.

18. J.W. Slotboom, G. Streutker, G.J.T. Davids, and P.B. Hartong, "Surface impact ionization in silicon devices." In: *IEEE International Electron Devices Meeting Technical Digest*, pp. 494–497, 1987.

8 Parasitic Elements in FinFETs

8.1 INTRODUCTION

In Chapters 5 and 6, we discussed the intrinsic characteristics of ideal fin field-effect transistor (FinFET) devices. In real FinFET devices, the parasitic elements such as extrinsic resistance and capacitance that arise due to its complex three-dimensional (3D) structure significantly affect the device performance. In real devices, the parasitic resistance becomes comparable to the intrinsic channel resistance and degrades the electrostatic behavior of real devices [1]. Thus, it is crucial to estimate the parasitic resistance and capacitance of FinFET devices for accurate characterization of the performance of real devices in very large scale integrated (VLSI) circuits and systems. However, the parasitic elements are difficult to predict in a FinFET device due to its complex 3D geometry as described in Chapter 4. Therefore, a simplified device structure is used to derive parasitic elements in FinFET devices.

In FinFET devices, the parasitic resistance consists of source-drain series resistance and gate resistance. However, the impact of source-drain series resistance has become more significant on the electrostatic behavior of these devices than the gate resistance due to high-k gate dielectric and metal-gate (MG) electrode process technology [2]. Therefore, it is important to understand the basic theory of source-drain parasitic resistance and develop an appropriate mathematical formulation for accurate estimation of the electrostatic performance of FinFET devices [3]. On the other hand, the gate resistance impacts the transient behavior of FinFETs, such as switching delay in complementary logic circuits when the devices are operating in a time-varying electrical field. Thus, it is also important to consider the effect of gate resistance for accurate characterization of the performance of FinFET devices during transient operation [4,5].

Again, the parasitic capacitance elements of FinFETs affect the transient response of the digital as well as analog VLSI circuits and systems. Therefore, it is important to understand the basic theory of parasitic capacitances of the complex 3D structure of FinFET devices [5–9]. In FinFETs, the parasitic capacitances include different components of fringe capacitance, gate overlap capacitance, and source-drain pn-junction capacitance. Several reports on the mathematical analysis of FinFET parasitic elements are available in the literature [1,3–9]. In this chapter, a comprehensive mathematical formulation of the parasitic elements in FinFET devices is presented to assess the static as well as transient behavior of FinFET devices.

8.2 SOURCE-DRAIN PARASITIC RESISTANCE

In the FinFET fabrication process discussed in Chapter 4, the raised source-drain (RSD) regions are formed using the selective epitaxial growth (SEG) process to reduce the source-drain series resistance. Thus, in order to derive mathematical expressions for FinFET source-drain series resistance, we first define the structural parameters of the RSD region in Section 8.2.1.

8.2.1 RAISED SOURCE-DRAIN FinFET STRUCTURE

Figure 8.1 shows a single-fin FinFET device with RSD regions as described in Chapter 4. In Section 4.3.3, we discussed the fabrication of multiple parallel fins to achieve sufficient current driving capability for the target design specifications of FinFET VLSI circuits and systems [10]. In addition, high layout density can be achieved by merging multiple fins to enlarge the RSD region using the SEG process as shown in Figure 8.2. Thus, it important to consider both the isolated single fin as

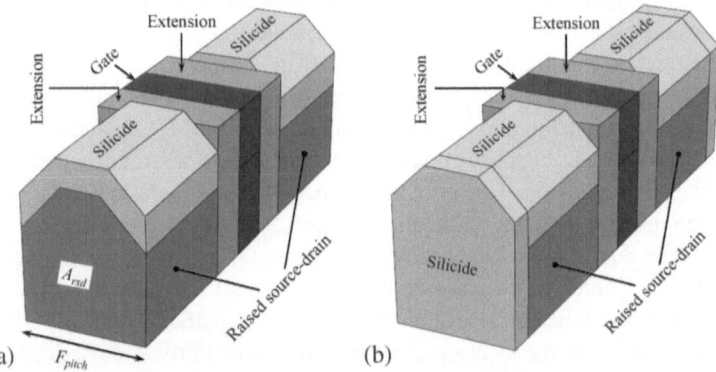

FIGURE 8.1 Ideal three-dimensional cross-section of a selective epitaxially grown ideal silicon raised source-drain structures of a FinFET: (a) non-rectangular epitaxial layer and silicide on the top; (b) non-rectangular epitaxial layer and silicide on the top as well as at two ends of the device.

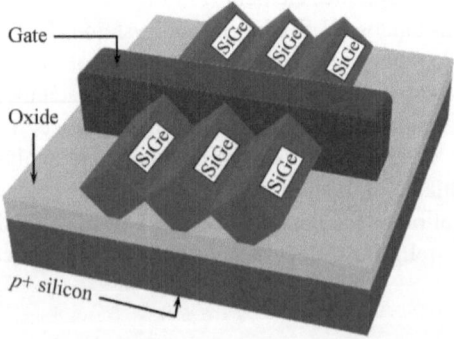

FIGURE 8.2 Merged raised source-drain structure formed using selective epitaxial growth of SiGe layer to reduce the source-drain series resistance.

well as multiple fins merged RSD structures in deriving mathematical expressions to estimate the parasitic resistance of FinFET devices.

The single-fin FinFET structure shown in Figure 8.1 consists of SEG RSD, source-drain extensions (SDEs), and a thin-body fin underneath the gate stack. The channel portion of the fin is wrapped on three sides by the gate stack. The source-drain regions are enlarged by SEG to reduce parasitic resistances. However, as discussed in Chapter 4, the shape of the SDE regions outside the gate are not affected during the SEG process since these SDE regions are protected by the source-drain spacers. The metallic region on the top of the RSD is the silicide. In a typical FinFET fabrication process, the silicide may wrap around the RSD in three sides as shown in Figure 8.1(b).

Figure 8.3 shows the two-dimensional (2D) cross-section of the RSD region through the cutline along the direction of the fin pitch (F_{pitch}) of the FinFET device structure shown in Figure 8.1(a).

In order to derive mathematical expressions for the source-drain series resistance, first of all, let us estimate the area components of different segments shown in Figure 8.3.

The area of the rectangular RSD section bounded by parameters F_{pitch} and fin height H_{fin} is given by

$$A_1 = F_{pitch} \times H_{fin} \tag{8.1}$$

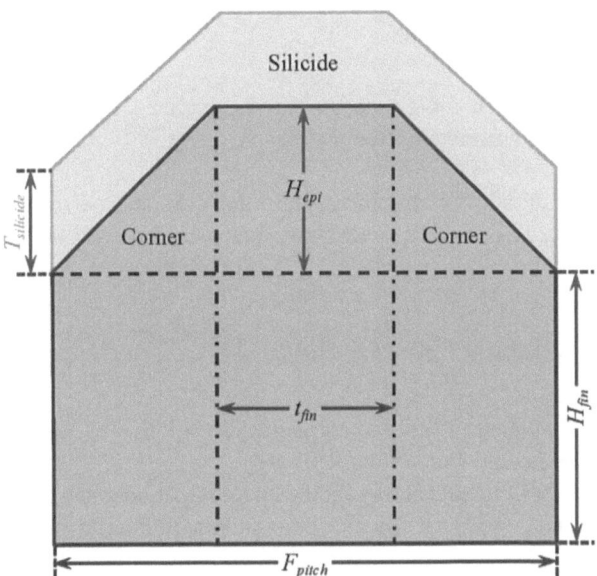

FIGURE 8.3 Two-dimensional front view of the non-rectangular raised source-drain region of the FinFET device structure described in Figure 8.1a; t_{fin}, H_{fin}, H_{epi}, and $T_{silicide}$ are the fin thickness, fin height, epitaxial layer thickness, and thickness of the silicide layer, respectively; corners represent the non-rectangular shape of the structure.

Similarly, the area of the rectangular epitaxial section with dimensions t_{fin} and H_{epi} is given by

$$A_2 = t_{fin} \times H_{epi} \qquad (8.2)$$

where:

t_{fin} = fin thickness
H_{epi} = height of the epitaxial layer over the fin

Now, in order to estimate the area of the SEG corner sections shown in Figure 8.3, let us assume that they are symmetrical triangles with height H_{epi} and base $(F_{pitch} - t_{fin})/2$. Then the total area of the triangular corners is given by

$$A_3 = \frac{1}{2}\left(F_{pitch} - t_{fin)}\right) \times H_{epi} \qquad (8.3)$$

Thus, adding Equations 8.1–8.3, the total area A_{rsd} of the RSD region shown in Figure 8.1(a) is given by [1]

$$A_{rsd,sum} = F_{pitch} \times H_{fin} + \left[t_{fin} + \frac{1}{2}\left(F_{pitch} - t_{fin}\right)\right] \times H_{epi} \qquad (8.4)$$

Now, to account for the non-ideal triangular shape of the corners, we can write a generalized expression for the total area A_{rsd} as

$$A_{rsd} = F_{pitch} \times H_{fin} + \left[t_{fin} + C_r\left(F_{pitch} - t_{fin}\right)\right] \times H_{epi} \qquad (8.5)$$

where:

C_r is a parameter that represents the non-triangular shape of the SEG corner regions and accounts for the real device effect

Then for the multiple-fin structures, the total area and perimeter components $A_{rsd,total}$ and $P_{rsd,total}$, respectively, of RSD resistance can be expressed as [1]

$$A_{rsd,total} = N_{fin} \times A_{rsd} + A_{rsd,end}$$

$$P_{rsd,total} = N_{fin} \times \left(F_{pitch} + \Delta L_{sil-epi}\right) A_{rsd,total} + P_{rsd,end} \qquad (8.6)$$

where:

N_{fin} is the total number of fins and merged RSD regions
$A_{rsd,end}$ is the end component of the RSD area
$\Delta L_{sil-epi}$ is the correction term for silicide/epitaxial-silicon interface length per fin
$P_{rsd,end}$ is the end component silicide/epitaxial-silicon interface length

8.2.2 Components of Source-Drain Series Resistance

For the simplicity of mathematical formulation, we consider an ideal rectangular FinFET device structure as shown in Figure 8.4, identify the components of the source-drain series resistance, and derive appropriate mathematical expressions.

Then we modify these expressions to include the non-rectangular structural effects for accurate estimation of the parasitic resistance of real FinFET devices. Thus, with reference to Figure 8.4, the source-drain series resistance of an ideal FinFET structure can be broadly divided into three basic components:

1. Contact resistance (R_{con}): R_{con} is the total resistance of the bulk RSD region and silicon-silicide interface area
2. Spreading resistance (R_{sp}): R_{sp} describes the resistance due to current crowding or spreading from the thin SDE-fin into the large RSD region
3. Extension resistance (R_{sde}): R_{sde} characterizes the bias-dependent resistance in the thin SDE-fin region under the spacer

In Sections 8.2.2.1–8.2.2.3, the detailed discussions on each component of the FinFET source-drain series resistance are presented.

8.2.2.1 Contact Resistance

The source-drain *contact resistance* R_{con} consists of the bulk resistivity of the RSD region and resistance of the silicon-silicide interface region. In order to estimate R_{con}, let us assume that the entire RSD region is the sum of elemental vertical sections each with length dx as shown in Figure 8.5(a). Each element has a bulk resistance component ΔR_{cb} and a contact resistance component ΔR_{cc}. These resistance elements are connected in a distributed resistance network as shown in Figure 8.5(b) [1]. Then from Ohm's law (Section 2.2.5.1), the bulk component of a rectangular RSD contact resistance $\Delta R_{cb'}$ for a small elemental length dx is given by

$$\Delta R_{cb'} = \rho \frac{dx}{H_{rsd} \cdot W_{rsd}} \tag{8.7}$$

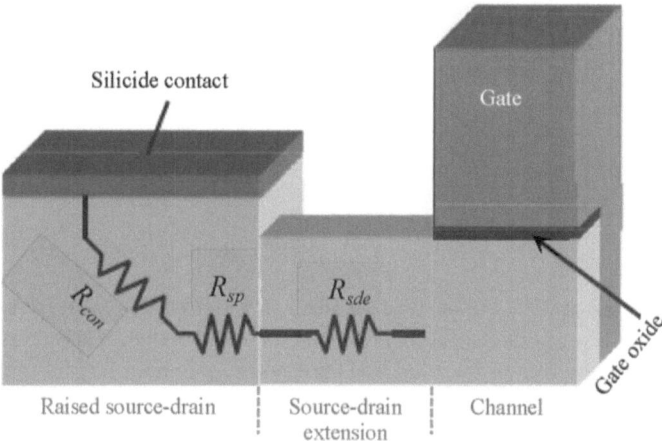

FIGURE 8.4 The components of the source-drain series resistance for an ideal rectangular raised source-drain structure of a FinFET device: here, R_{con}, R_{sp}, and R_{sde} are the contact, spreading, and source-drain extension resistance components of the overall source-drain series resistance.

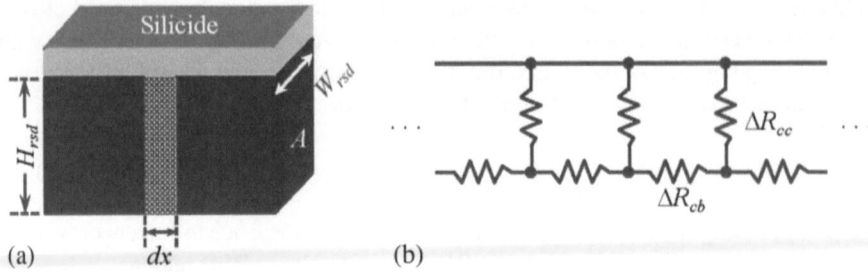

FIGURE 8.5 RSD contact resistance estimation: (a) a typical RSD region divided into infinitesimally small elements of length dx for the evaluation of the distributed resistance; (b) an equivalent resistance network for contact resistance estimation; in the figure, ΔR_{cb} and ΔR_{cc} are the bulk and contact resistance components of the overall contact resistance.

where:

H_{rsd} is the height of the RSD region
W_{rsd} is the width of the RSD region
ρ is the bulk resistivity of the material of the RSD region (Section 2.2.5.1)

From Equation 2.46, we can write the expression for ρ for an n^+ RSD region as

$$\rho = \frac{1}{qN_{rsd} \cdot \mu_{rsd}} \tag{8.8}$$

where:

q is the electronic charge
N_{rsd} is the uniformly doped doping concentration of the RSD region
μ_{rsd} is the effective mobility of electrons in the RSD region

Again, using Ohm's law, the contact resistance component $\Delta R_{cc'}$ of the element dx of the rectangular section of the RSD region shown in Figure 8.5(a) is given by

$$\Delta R_{cc'} = \frac{\rho_c}{dx \times W_{rsd}} \tag{8.9}$$

where:

ρ_c is the specific resistivity of the RSD contact in Ω-cm^2

Equations 8.7 and 8.9 are derived for ideal rectangular contacts. In order to consider the real device effects, the expressions for $\Delta R_{cb'}$ and $\Delta R_{cc'}$ are generalized as ΔR_{cb} and ΔR_{cc} in terms of the total RSD cross-sectional area $A_{rsd,total}$ and perimeter length $P_{rsd,total}$ given by

$$\Delta R_{cb} = \rho \frac{dx}{A_{rsd,total}}$$

$$\tag{8.10}$$

$$\Delta R_{cc} = \frac{\rho_c}{dx \times P_{rsd,total}}$$

The expressions in Equation 8.10 represent the elemental bulk RSD and contact resistance components of the distributed resistance network as shown in Figure 8.b(b). The final expression for the total contact resistance is obtained by using transmission line analysis [11] and can be shown as [1]

$$R_{con} = \rho \cdot \frac{L_T}{A_{rsd,total}} \frac{\eta \cosh \alpha + \sinh \alpha}{\eta \sinh \alpha + \cosh \alpha} \tag{8.11}$$

where:

$$L_T = \sqrt{\frac{\rho_c \times A_{rsd,total}}{\rho \times P_{rsd,total}}}$$

$$\alpha = \frac{L_{rsd}}{L_T} \tag{8.12}$$

$$\eta = \frac{\rho_c \times A_{rsd,total}}{\rho \times L_T \times A_{term}}$$

In Equation 8.12, L_{rsd} is the length of the RSD region; and A_{term} is the area of silicon-silicide at the end of the device as shown in Figure 8.1(b). For devices without the end contact, $A_{term} = 0$, then noting that $\eta \cosh \alpha \gg \sinh \alpha$ and $\eta \sinh \alpha \gg \cosh \alpha$, Equation 8.11 can be simplified to

$$R_{con} = \rho \cdot \frac{L_T}{A_{rsd,total}} \coth \alpha \tag{8.13}$$

8.2.2.2 Spreading Resistance

The *spreading resistance* arises due to the current spreading in the bulk RSD region and can be described using the cross-sectional view of the source-drain to the SDE region along the *xy*-plane as shown in Figure 8.6. When drain current flows from the RSD region into the SDE-fin, it spreads out gradually crowding within the ultrathin SDE-fin. Such current spreading is characterized by the *spreading resistance* or *current crowding* component of the source-drain series resistance.

Figure 8.6 shows a typical top view of the RSD, SDE, and the current spreading regions. For the simplicity of mathematical formulation, let us assume all geometries of the structure under consideration as well as current flow cross-section are squares. Then the area of the SDE-fin is given by

$$A_{fin} = H_{fin} \times t_{fin} \tag{8.14}$$

Since the SDE region is assumed to be square, then the length l_{sde} and width w_{sde} of the SDE-fin is given by

$$l_{sde} = w_{sde} = \sqrt{A_{fin}} = \sqrt{H_{fin} \times t_{fin}} \tag{8.15}$$

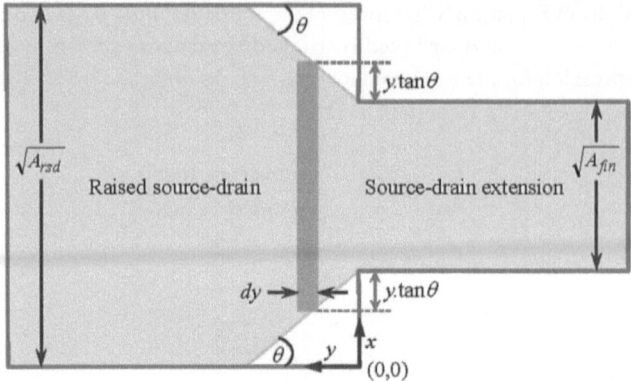

FIGURE 8.6 Spreading resistance estimation: the shaded region represents current spreading from the source-drain extension region into the raised source-drain region; θ is the angle of the current spreading with the boundary of raised source-drain region along the fin direction; and dy is an elemental length of the current spreading region in the RSD region.

Let θ be the angle of current spreading boundary to the fin direction with reference to the top or bottom surface as shown in Figure 8.6. Then the current spreading length along x-direction inside the RSD region at any point y is $2y.\tan\theta$; the factor "2" is due to two triangular segments above and below the SDE-fin as shown in Figure 8.6. Thus, the spreading length at any point y is given by

$$l_{sp} = w_{sp} = \sqrt{A_{fin}} + 2y \cdot \tan\theta \qquad (8.16)$$

Now, if dy is an elemental length between two pints y and $(y + dy)$ of the gradually increasing current spreading length inside the RSD region, then the resistance of each element is given by

$$\Delta R_{sp} = \rho \frac{dy}{\left(\sqrt{A_{fin}} + 2y \cdot \tan\theta\right)^2} \qquad (8.17)$$

where:
   ρ is the bulk resistivity of the material of the RSD region and is given by Equation 8.8

Again, considering RSD region as a square, the length l_{rsd} and width w_{rsd} of the RSD region is given by

$$l_{rsd} = w_{rsd} = \sqrt{A_{rsd}} \qquad (8.18)$$

Then from Equations 8.16 and 8.18, the total length L_1 of the crowded region is given by

$$\sqrt{A_{rsd}} - \sqrt{A_{fin}} = 2L_1 \cdot \tan\theta$$

$$\text{or,} \quad L_1 = \frac{\cot\theta}{2}\left(\sqrt{A_{rsd}} - \sqrt{A_{fin}}\right) \tag{8.19}$$

Therefore, integrating Equation 8.17, the total spreading resistance is given by

$$
\begin{aligned}
R'_{sp} &= \rho \int_0^{L_1} \frac{dy}{\left(\sqrt{A_{fin}} + 2y \cdot \tan\theta\right)^2} \\
&= \rho \frac{-1}{\left(\sqrt{A_{fin}} + 2y \cdot \tan\theta\right) \cdot 2\tan\theta} \bigg|_0^{L_1} \\
&= \frac{\rho\cot\theta}{2}\left(\frac{1}{\sqrt{A_{fin}}} - \frac{1}{\sqrt{A_{fin}} + 2L_1\tan\theta}\right)
\end{aligned} \tag{8.20}
$$

Substituting the expression for $2L_1\tan\theta$ from Equation 8.19 into Equation 8.20, we get

$$R'_{sp} = \frac{\rho\cot\theta}{2}\left(\frac{1}{\sqrt{A_{fin}}} - \frac{1}{\sqrt{A_{rsd}}}\right) \tag{8.21}$$

Equation 8.21 is obtained assuming square SDE-fin and square RSD region. In order to consider real device effect, we can write a general expression for spreading resistance as [1]

$$R'_{sp,real} = \frac{\rho\cot\theta}{s}\left(\frac{1}{\sqrt{A_{fin}}} - \frac{1}{\sqrt{A_{rsd}}}\right) \tag{8.22}$$

where:
 s is the shape parameter of the SDE and RSD regions

Equation 8.22 is the general expression to analyze the total source-drain series resistance due to current crowding.

In an ideal RSD region without current crowding, the total resistance of the same region is given by

$$R'_{sp,ideal} = \rho\frac{L_1}{A_{rsd}} \tag{8.23}$$

Now, substituting for L_1 from Equation 8.19 into Equation 8.23 and using shape parameter s, we can get a general expression for spreading resistance in the ideal case as

$$R'_{sp,ideal} = \frac{\rho \cot \theta}{s} \frac{1}{A_{rsd}} \left(\sqrt{A_{rsd}} - \sqrt{A_{fin}} \right) \qquad (8.24)$$

We know that the spreading resistance R_{sp} is defined as the increase in resistance due to current spreading in the RSD region. Therefore, subtracting Equation 8.24 from Equation 8.22, we get the spreading resistance of FinFETs as

$$R_{sp} = \frac{\rho \cot \theta}{s} \left(\frac{1}{\sqrt{A_{fin}}} - \frac{1}{\sqrt{A_{rsd}}} \right) - \frac{\rho \cot \theta}{s} \frac{1}{A_{rsd}} \left(\sqrt{A_{rsd}} - \sqrt{A_{fin}} \right)$$

$$(8.25)$$

$$\text{or,} \quad R_{sp} = \frac{\rho \cot \theta}{s} \left(\frac{1}{\sqrt{A_{fin}}} - \frac{2}{\sqrt{A_{rsd}}} + \frac{\sqrt{A_{fin}}}{A_{rsd}} \right)$$

Again, for the simplicity of the characterization of spreading resistance in FinFETs, let us define

$$R_{sp0} = \rho \left(\frac{1}{\sqrt{A_{fin}}} - \frac{2}{\sqrt{A_{rsd}}} + \frac{\sqrt{A_{fin}}}{A_{rsd}} \right) \qquad (8.26)$$

Then, Equation 8.25 can be expressed as

$$R_{sp} = K \times R_{sp0} \qquad (8.27)$$

where:
　K = slope factor = $\cot \theta / s$

　From the layout data of the target FinFET device technology, the slope factor K can be obtained from R_{sp} versus R_{sp0} plots using Equations 8.25 and 8.26 and therefore, the shape factor s of the target device technology can be calculated from the relation, $s = \cot \theta / K$ given in Equation 8.27.

8.2.2.3　Source-Drain Extension Resistance

The source-drain extension or SDE resistance R_{sde} component of the source-drain series resistance is due to the resistance to the flow of drain current I_{ds} through the extension region of the fin under the source-drain spacer. The value of R_{sde} depends on the distribution of impurity doping in the SDE region which in turn depends on the processing conditions and surface accumulation due to fringe-field originating from the gate. Thus, R_{sde} is a bias- and technology-dependent component of the source-drain series resistance.

　Figure 8.7 shows only the source-drain spacer configurations on both ends of the gate of length L_g. It is observed from Figure 8.7 that the total spacer of length

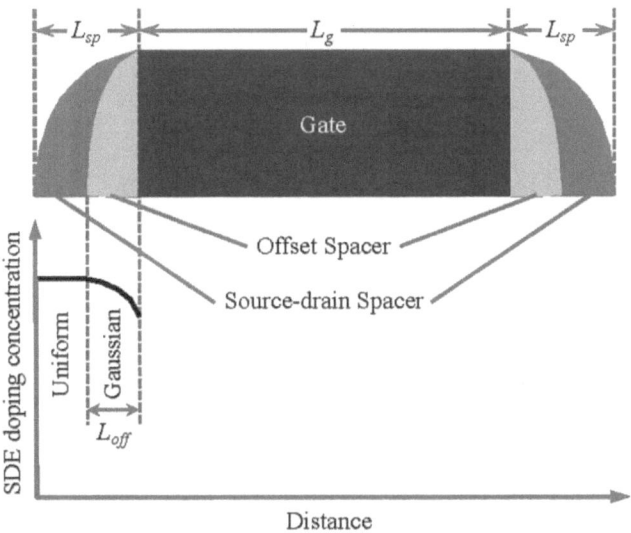

FIGURE 8.7 Estimation of the source-drain extension resistance R_{sde}: spacer configurations and doping profile within the SDE regions under the spacers; L_{off} is the width of the SDE offset spacer, L_{sp} is the total width of source-drain spacer, and L_g is the gate length of FinFET device.

L_{sp} consists of an offset spacer of length L_{off} and the source-drain spacer of length $L_{sp} - L_{off}$. As described in Section 4.3.6, the offset spacer is used to offset the SDE implant and minimize the dopant encroachment in the channel under the gate. Thus, the composite spacer is formed sequentially as (i) offset spacer formation; (ii) SDE implantation; and (iii) source-drain spacer formation followed by RSD growth as described in Section 4.3.7. Since the offset spacer L_{off} is formed over the undoped fin-channel, therefore, we can safely assume a Gaussian doping profile for the SDE dopants in the SDE-fins under the offset spacer, whereas the RSD region is a uniformly heavily doped region as illustrated in Figure 8.7.

Thus, assuming the above described doping distributions within the SDE-fin, R_{sde} can be considered to consist of (1) a bias-dependent accumulation resistance R_{acc} at the surface of the SDE-fin; and (2) two bias-independent bulk resistances, R_{sde1} under the surface accumulation region of the SDE-fin and R_{sde2} under the uniformly doped bulk region as shown in the circuit schematic in Figure 8.8.

Thus, from the schematic representation of the SDE region in Figure 8.8, the total R_{sde} is given by

$$R_{sde} = R_{sde1} \| R_{acc} + R_{sde2} \qquad (8.28)$$

It is to be noted that R_{acc} is due to the conductive path at the surface of the SDE-fin caused by the accumulation of the induced charge by gate fringing fields and is important for the undoped or lightly doped SDE gate-drain underlap FinFETs at SDE/gate edge junction as shown in Figure 8.7 [3].

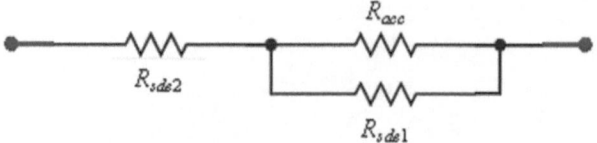

FIGURE 8.8 Circuit schematic for the estimation of source-drain extension resistance: sub-circuit for resistance modeling in the accumulation region; R_{acc} is the bias-dependent accumulation resistance at the surface region of the SDE-fins, R_{sde1} is a bias-independent bulk resistance under the accumulation region, and R_{sde2} is a bias-independent bulk resistance in the uniformly doped bulk region.

Formulation of bias-dependent accumulation resistance, R_{acc}: As described earlier, R_{acc} represents the bias-dependent resistance to the current flow at the surface of the SDE-fin due to the accumulation of the induced charge by gate fringing fields. In order to derive a mathematical expression for R_{acc}, let us consider only the drain end of the channel of a FinFET device with SDE overlap under the gate. Then the voltage at the SDE surface of the accumulation region under the gate is given by

$$V_{acc} = V_{gd} - V_{fbsd} \tag{8.29}$$

where:

V_{gd} is the gate-to-drain voltage
V_{fbsd} is the flat band voltage of the metal-gate/SiO_2/Si MOS system

Therefore, the accumulation charge Q_{acc} at the surface within the accumulation length l_{acc} is given by

$$Q_{acc} = C_{acc}\left(V_{gd} - V_{fbsd}\right) \tag{8.30}$$

where:

C_{acc} is the accumulation capacitance at the surface

Now, from Chapter 5, Equation 5.22, we can show that the drain current I_{ds} flowing through the accumulation region is given by

$$I_{ds} = \mu_{eff}\frac{H_{fin}}{l_{acc}}C_{acc}\left(V_{gd} - V_{fbsd}\right)\times V_{ds} \tag{8.31}$$

where:

μ_{eff} is the effective mobility of current carriers
V_{ds} is the drain voltage

Then from Equation 8.31, the resistance of the current flow through the accumulation region is given by

$$R_{acc} = \frac{V_{ds}}{I_{ds}} = \left(\frac{l_{acc}}{\mu_{eff}\times C_{acc}}\right)\frac{1}{H_{fin}\times\left(V_{gd} - V_{fbsd}\right)} \tag{8.32}$$

Now, if we define R_{acc0} as a technology-dependent parameter given by

$$R_{acc0} = \left(\frac{l_{acc}}{\mu_{eff} \times C_{acc}} \right) \tag{8.33}$$

Then substituting Equation 8.33 into Equation 8.32, the generalized expression for the accumulation resistance can be written as

$$R_{acc} = \frac{R_{acc0}}{\left(V_{gs(d)} - V_{fbsd} \right) \times H_{fin}} \tag{8.34}$$

where:

$V_{gs(d)}$ represents the gate-to-source voltage V_{gs} for the calculation of source extension resistance and gate-to-drain voltage V_{gd} for the calculation of drain extension resistance

Bias-independent bulk resistance components: The bias-independent component R_{sde1} is the bulk resistance underneath the SDE accumulation region (partially under the offset spacer as well as the source-drain spacer). Then, from Section 2.2.5.1, the current density J flowing through this region is given by

$$J = -n(y)qv = -n(y) \times q \times \left(\mu(y)E(y) \right) = qn(y)\mu(y)\frac{dV}{dy} \tag{8.35}$$

where:

$n(y)$ is the space-dependent carrier concentration of the Gaussian doping profile
q is the electronic charge
$v = \mu(y) \times E(y)$ is the drift velocity of carriers
$\mu(y)$ is the concentration-dependent carrier mobility
V is the voltage along the y-direction in the SDE
$E(y) = -dV/dy$ is the potential field gradient (space-dependent variation of V within SDE along the y-direction)

Therefore, the drain current I_{ds} flowing through R_{sde1} is given by

$$I_{ds} = (H_{fin} \times t_{fin}) \times J = (H_{fin} \times t_{fin}) \times qn(y)\mu(y)\frac{dV}{dy} \tag{8.36}$$

Equation 8.36 can be expressed as

$$dV = \frac{I_{ds}}{q(H_{fin} \times t_{fin})} \frac{dy}{n(y)\mu(y)} \tag{8.37}$$

Integrating Equation 8.37, from $V = 0$ to V_{ds} with corresponding $y = 0$ to $y = \Delta L_{sde}$ (effective length of the surface accumulation region), we get

$$R_{sde1} = \frac{V_{ds}}{I_{ds}} = \frac{1}{(H_{fin} \times t_{fin})} \frac{1}{q} \int_0^{\Delta L_{sde}} \frac{dy}{n(y)\mu(y)} \tag{8.38}$$

Thus, the bias-dependent component of R_{sde} can be expressed as

$$R_{sde1} = \frac{R_{sde10}}{H_{fin} \times t_{fin}} \tag{8.39}$$

where:

$$R_{sde10} = \frac{1}{q} \int_0^{\Delta L_{sde}} \frac{dy}{n(y)\mu(y)} \tag{8.40}$$

Here, R_{sde10} is a parameter that depends on the distribution of doping concentration in the SDE region and is determined from the measurement data.

In order to estimate R_{sde2}, we assume that the RSD region is uniformly doped and completely under the source-drain spacer so that the effective length of the uniformly RSD doped region is $L_{sp} - \Delta L_{sde}$. Then following the same mathematical steps used to formulate R_{sde1} [Equation 8.39], we can show

$$R_{sde2} = \frac{V_{ds}}{I_{ds}} = \frac{1}{H_{fin} \times t_{fin}} \frac{L_{sp} - \Delta L_{sde}}{qn\mu} \tag{8.41}$$

In Equation 8.41, the basic parameters q, n, and μ are constants due to the uniformly doped RSD region. Therefore, we can express Equation 8.41 as

$$R_{sde2} = \frac{R_{sde20}\left(L_{sp} - \Delta L_{sde}\right)}{H_{fin} \times t_{fin}} \tag{8.42}$$

where:
   R_{sde20} is a technology-dependent constant and is given by

$$R_{sde20} = \frac{1}{qn\mu} \tag{8.43}$$

Note that in Equation 8.43, n and μ are independent of space due to uniformly doped SDE region of effective length $L_{sp} - \Delta L_{sde}$. Thus, the components of R_{sde} are given by Equations 8.34, 8.39, and 8.42 as shown below

$$R_{acc} = \frac{R_{acc0}}{\left(V_{gs(d)} - V_{fbsd}\right) \times H_{fin}} \tag{8.44}$$

$$R_{sde1} = \frac{R_{sde10}}{H_{fin} \times t_{fin}} \equiv a \tag{8.45}$$

$$R_{sde2} = \frac{R_{sde20}\left(L_{sp} - \Delta L_{sde}\right)}{H_{fin} \times t_{fin}} \tag{8.46}$$

Finally, from Equation 8.28, we can obtain the total source-drain extension resistance R_{sde} using Equations 8.44–8.46.

Now, in order to derive an expression for total R_{sde}, let us derive the component of R_{sde} from Equations 8.44 and 8.45 to get

$$
\begin{aligned}
\frac{1}{R_{sde1} \| R_{acc}} &= \frac{1}{a} + \frac{\left(V_{gs(d)} - V_{fbsd}\right) \times H_{fin}}{R_{acc0}} \\
&= \frac{R_{acc0} + a \times \left(V_{gs(d)} - V_{fbsd}\right) \times H_{fin}}{a \times R_{acc0}} \\
&= \frac{R_{acc0}\left[1 + \left(a / R_{acc0}\right) \times \left(V_{gs(d)} - V_{fbsd}\right) \times H_{fin}\right]}{a \times R_{acc0}} \\
&= \frac{\left[1 + \left(a / R_{acc0}\right) \times \left(V_{gs(d)} - V_{fbsd}\right) \times H_{fin}\right]}{a}
\end{aligned}
\tag{8.47}
$$

Then substituting the expression for a from Equation 8.45 into Equation 8.47, we get

$$
\begin{aligned}
\frac{1}{R_{sde1} \| R_{acc}} &= \frac{\left[1 + \left(\dfrac{R_{sde10}}{H_{fin} \times t_{fin}} / R_{acc0}\right) \times \left(V_{gs(d)} - V_{fbsd}\right) \times H_{fin}\right]}{\dfrac{R_{sde10}}{H_{fin} \times t_{fin}}} \\
&= \frac{\left[1 + \left(\dfrac{R_{sde10}}{R_{acc0} \times t_{fin}}\right) \times \left(V_{gs(d)} - V_{fbsd}\right)\right]}{\dfrac{R_{sde10}}{H_{fin} \times t_{fin}}}
\end{aligned}
\tag{8.48}
$$

Therefore

$$
R_{sde1} \| R_{acc} = \frac{\dfrac{R_{sde10}}{H_{fin} \times t_{fin}}}{1 + \dfrac{R_{sde10}}{R_{acc0} \times t_{fin}} \times \left(V_{gs(d)} - V_{fbsd}\right)}
\tag{8.49}
$$

Then adding the expression for R_{sde2} from Equation 8.42 into Equation 8.49, we get the total SDE resistance at the drain side of a FinFET device from Equation 8.28 as

$$
R_{sde}(D) = \frac{\dfrac{R_{sde10}}{H_{fin} \times t_{fin}}}{1 + \dfrac{R_{sde10}}{R_{acc0} \times t_{fin}} \times \left(V_{gd} - V_{fbsd}\right)} + \frac{R_{sde20}\left(L_{sp} - \Delta L_{sde}\right)}{H_{fin} \times t_{fin}}
\tag{8.50}
$$

Similarly, the total SDE resistance at the source side of the device is given by

$$R_{sde}(S) = \frac{\dfrac{R_{sde10}}{H_{fin} \times t_{fin}}}{1 + \dfrac{R_{sde10}}{R_{acc0} \times t_{fin}} \times \left(V_{gs} - V_{fbsd}\right)} + \frac{R_{sde20}\left(L_{sp} - \Delta L_{sde}\right)}{H_{fin} \times t_{fin}} \qquad (8.51)$$

For circuit analysis, Equations 8.50 and 8.51 can be expressed in terms of technology-dependent parameters given by

$$RS_1 = RD_1 = \frac{R_{sde10}}{H_{fin} \times t_{fin}}$$

$$RS_2 = RD_2 = \frac{R_{sde10}}{R_{acc0} \times t_{fin}} \qquad (8.52)$$

$$RS_3 = RD_{3=} \frac{R_{sde20}\left(L_{sp} - \Delta L_{sde}\right)}{H_{fin} \times t_{fin}}$$

Then using the technology-dependent parameters defined in Equation 8.52, the total SDE resistances can be expressed as

$$R_{sde}(S) = \frac{R_{S1}}{1 + R_{S2} \times \left(V_{gs} - V_{fbsd}\right)} + R_{S3}$$

$$R_{sde}(D) = \frac{R_{D1}}{1 + R_{D2} \times \left(V_{gd} - V_{fbsd}\right)} + R_{D3} \qquad (8.53)$$

The parameter set $\{RS_1, RS_2, RS_3, RD_1, RD_2, RD_3\}$ are obtained from the measurement data of a target technology.

8.3 GATE RESISTANCE

The gate resistance becomes important during transient analysis of FinFET devices and we can use the mathematical formulations used for conventional MOSFET devices [12]. At any low frequency, the gate resistance of a device can be calculated from the sheet resistance of the gate material and from Equation 2.52 can be shown as

$$R_g = \rho_{sh,gate} \frac{W_{eff}}{L_{eff}} \qquad (8.54)$$

where:
L_{eff} is the effective channel length of FinFETs
W_{eff} is the effective width of FinFETs
$\rho_{sh,gate}$ is the sheet resistance per square of the metal-gate

At high frequencies, the accurate formulation of the gate resistance is very complex due to the distributed transmission-line effect. Therefore, a lumped equivalent gate resistance with an empirical parameter α_g can be used to account for the distributed RC effects and is given by [13,14]

$$R_g = \rho_{sh,gate} \frac{W_{eff}}{L_{eff}} \alpha_g \tag{8.55}$$

where:

$$\alpha_g = \begin{cases} 1/3; & \text{for single-sided gate electrode contact} \\ 1/12; & \text{for double-sided gate electrode contact} \end{cases} \tag{8.56}$$

It is found that the distributed RC effect of the gate as well as the *non-quasi-static* effect, that is, the distributed RC effect of the channel affects the high frequency characteristics of FinFET devices. Therefore, an additional component of gate resistance must be considered to account for the distributed RC effect in the channel. Thus, at high frequency operation of a FinFET device, the distributed channel resistance *seen* by the signal applied to the gate also contributes to the effective gate resistance in addition to the resistance of the gate electrode. Then the effective gate resistance $R_{g,eff}$ consists of two parts: the distributed gate electrode resistance $R_{g,eltd}$ and the distributed channel resistance R_{gch} seen from the gate [15] such that

$$R_{g,eff} = R_{g,eltd} + R_{gch} \tag{8.57}$$

Typically, $R_{g,eltd}$ is insensitive to bias and frequency and can be expressed [1]

$$R_{g,eltd} = \rho_{sh,geltd} \frac{W_{eff}}{L_{eff}} \alpha_g + \beta_g \tag{8.58}$$

where:
$\rho_{sh,geltd}$ is the gate electrode sheet resistance
β_g is the external gate resistance

$$\beta_g = \begin{cases} 1; & \text{for single-sided gate electrode contact} \\ 2; & \text{for double-sided gate electrode contact} \end{cases} \tag{8.59}$$

Again, R_{gch} consists of the resistance (R_{st}) to account for the static channel resistance and the excess-diffusion channel resistance (R_{ed}) due to the change of channel charge distribution by the transient excitation of the gate voltage [16]. The overall channel resistance seen from the gate is given by

$$\frac{1}{R_{gch}} = \gamma \left(\frac{1}{R_{st}} + \frac{1}{R_{ed}} \right) \tag{8.60}$$

where:
γ is a parameter that accounts for the distributed nature of the channel resistance

8.4 PARASITIC CAPACITANCE ELEMENTS

The mathematical formulation of parasitic capacitance is rigorous due to the complex 3D structure of FinFET devices. In this section, a brief overview of the parasitic capacitances in FinFETs is presented.

The components of FinFET parasitic capacitances are

1. Bias-dependent overlap capacitance C_{ov} due to the gate/SDE overlap region
2. Bias-independent (or, weak bias-dependent) fringing capacitance C_{fr} due to the tall vertical gate geometry

Thus, the total parasitic capacitance is given by

$$C_p = C_{ov} + C_{fr} \tag{8.61}$$

In reality, C_{ov} is used to describe the bias-dependent capacitance–voltage (C–V) characteristics of FinFET devices over the target bias range, whereas C_{fr} is used to account for the scaling of FinFET capacitance across geometries [1]. Figure 8.9 shows an ideal schematic representation of a FinFET device with different parasitic elements.

8.4.1 GATE OVERLAP CAPACITANCE

The source-drain overlap capacitances are parasitic elements that originate due to the encroachment of source-drain implant profile under the gate region during the IC fabrication process. The post-implant thermal processing steps cause lateral

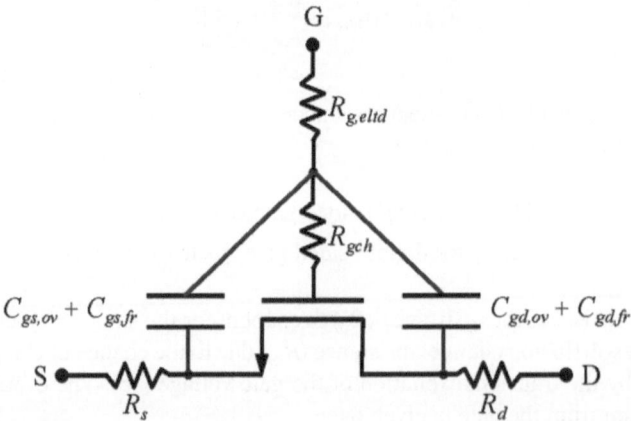

FIGURE 8.9 A typical representation of RC network of a FinFET device: S and D are the source and drain terminals, respectively; $C_{gs,ov}$ and $C_{gd,ov}$ are the gate-to-source and gate-to-drain overlap capacitances, respectively; $C_{gs,fr}$ and $C_{gd,fr}$ are the gate-to-source and gate-to-drain fringing capacitances, respectively; and R_s and R_d are the external source and drain resistances, respectively.

diffusion of dopants under the gate and overlap of the source-drain regions in the final device structure. For the simplicity of quantitative discussions, we consider the symmetrical source-drain regions so that the source overlap distance l_{ov} is the same as that of the drain. Assuming parallel plate formulation, the overlap capacitance C_{GSO} and C_{GDO} for the source and drain regions, respectively, of a single-fin double-gate FinFET can be approximated as

$$C_{GSO} = C_{GDO} = \frac{\varepsilon_{ox}}{T_{ox}} 2H_{fin}l_{ov} = 2C_{ox}H_{fin}l_{ov} \tag{8.62}$$

where:
ε_{ox} is the permittivity of gate oxide
T_{ox} is gate oxide thickness
C_{ox} is gate capacitance $= \varepsilon_{ox}/T_{ox}$
"2" is due to double-gate configuration

From Equation 8.62, the source and drain overlap capacitance per unit width C_{gso} and C_{gdo}, respectively, are given by

$$C_{gs,ov} = C_{gd,ov} = C_{ox}l_{ov} \tag{8.63}$$

The detailed discussions on gate overlap capacitance on multiple-fin FinFET devices are available in published report [5].

8.4.2 Fringe Capacitance

The fringe capacitance C_{fr} arises due to the close proximity of the FinFET structural parameters such as gate, fin region under the source-drain spacer, and RSD as well as RSD contacts. Thus, C_{fr} represents the parasitic capacitance relating the scaling parameters of FinFETs. Thus, it is used to provide the scaling of C_p across the device geometries of a FinFET device technology.

The mathematical formulation of C_{fr} of a FinFET is complex due to the complexities of its tall vertical 3D structure. Therefore, the basic understanding of C_{fr} and its origin are described in this section. The detailed mathematical derivation of C_{fr} is available in published reports [1,5].

Now, let us consider the 2D cross-section of an ideal FinFET structure along the yz-plane as shown in Figure 8.10. Then the major components of C_{fr} are

1. Fin-to-gate capacitance (C_{fg}): this is due to the electric field lines originating from the SDE-fin under the source-drain spacer to the gate edge
2. Source-drain contact to gate capacitance (C_{cg}): this is the capacitance between the gate and the epitaxially grown source-drain contact. C_{cg} includes three major components: C_{cg1}, C_{cg2}, and C_{cg3}, each depending on the electric flux originating from the different surfaces of the gate and contact regions

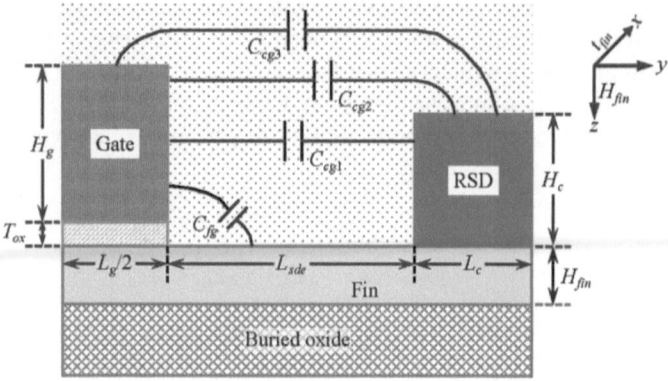

FIGURE 8.10 Components of fringe capacitance in FinFETs: C_{fg} is the fin-to-gate capacitance; C_{cg1}, C_{cg2}, and C_{cg3} are the components of gate-to-contact capacitance; L_g, T_{ox}, are the gate length and gate oxide thickness, respectively; H_g is the height of gate electrode; L_{sde} and L_c are the lengths of source-drain extension and contact regions, respectively; H_{fin} and H_c are the fin height and contact height, respectively.

8.4.2.1 Fin-to-Gate Fringe Capacitance

In order to derive an expression for gate fringe capacitance C_{fg}, we consider the length of each electric field lines originating from the SDE-fin under the source-drain spacer to the gate edge to be the perimeter of a quarter of ellipse. After several approximations, an expression for C_{fg} can be shown as [1]

$$C_{fg} = H_{fin}\left[C_{fg,sat} - \frac{1}{2}\left\{ \left(C_{fg,sat} - C_{fg,\log} - \delta\right) + \sqrt{\left(C_{fg,sat} - C_{fg,\log} - \delta\right)^2 + 4\delta C_{fg,sat}} \right\} \right]$$

$$(8.64)$$

where:

δ is a parameter that defines the transition from $C_{fg,sat}$ to $C_{fg,\log}$

$C_{fg,sat}$ and $C_{fg,\log}$ are geometry-dependent parameters that depend on H_g, T_{ox}, L_{sde}, H_c, and L_c as shown in Figure 8.10 and dielectric constant of source-drain spacer

8.4.2.2 Gate-to-Contact Fringe Capacitance

From Figure 8.10, the expression for total C_{cg} can be written as

$$C_{cg} = H_{fin} \times \left(C_{cg1} + C_{cg2} + C_{cg3} \right)$$

$$(8.65)$$

where:

C_{cg1} is the parallel plate capacitance between the gate and contact as shown in Figure 8.10

C_{cg2} is the capacitance due to the electric field originating from the gate sidewall, which travels a distance L_{sde} horizontally, and then follows a quarter circle until it terminates on top of the contact as shown in Figure 8.10

C_{cg3} describes the capacitance with electric field lines originating from the top of the gate and terminating on the top of the contact as shown in Figure 8.10

Again, C_{cg3} can be represented by two parallel capacitor components: (a) C_{cg3a} considering semicircular electric fields with diameters ranging from L_{sde} to L_{sde} + $L_c + L_g/2$; and (b) C_{cg3b}, a parallel plate capacitance characterized by the area of the contact surface and the horizontal distance. The total C_{cg3} is given by

$$C_{cg3} = \frac{1}{\dfrac{1}{C_{cg3a}} + \dfrac{1}{C_{cg3b}}} \tag{8.66}$$

For the simplicity of analysis, the overall capacitance can be considered as three components: C_{top}, two side components C_{side}, and two corner components C_{corner} [1]. The total capacitance for a multifin FinFET with N_{fin} number of fins is given by

$$C_f = N_{fin} \times \left(2C_{corner} + 2C_{side} + C_{top}\right) \tag{8.67}$$

8.5   SOURCE-DRAIN pn-JUNCTION CAPACITANCE

We know that the source-drain and channel regions are doped with opposite types of impurities (e.g., p-type channel with n^+-type source-drain or n-type channel with p^+-type source-drain), thus forming source-channel and drain-channel pn-junctions. We discussed in Chapter 2 that a small change in the applied voltage to a pn-junction results in a junction capacitance. Thus, the source-drain pn-junction capacitances that result during small-signal operation of a FinFET must be accounted for accurate characterization of FinFET device performance. In a FinFET device, the source-drain to body pn-junction capacitances consists of three components: the bottom junction capacitance, sidewall junction capacitance along the isolation edge, and sidewall junction capacitance along the gate edge. An analogous set of equations can be used for both sides, but each side has a separate set of parameters to estimate the value of pn-junction capacitance.

From Equation 2.139, the expression for the capacitance (C_j) of an isolated pn-junction can be written as

$$C_j = \frac{C_{j0}}{\left(1 - \dfrac{V_{bs(d)}}{\phi_{bi}}\right)^{m_j}} \tag{8.68}$$

where:
$V_{bs(d)}$ is the applied voltage to pn-junction
C_{j0} is the value of capacitance at $V_{bs(d)} = 0$
ϕ_{bi} is the built-in potential of the pn-junction
m_j is co-efficient of doping grading of the junction and typically, $0.2 < m_j < 0.6$

In Equation 8.68, $V_{bs(d)} = V_{bs}$ for the applied source bias with reference to body bias; and $V_{bs(d)} = V_{bd}$ for the applied drain bias with reference to body bias.

8.5.1 REVERSE-BIAS CAPACITANCE

As the channel length scales down, direct coupling between source-drain leads to an increased amount of leakage current. As described in Chapter 4, for bulk-FinFET devices, an anti-punchthrough (APT) implant is used to prevent this coupling [17]. This implant placed just below the lightly doped fin region laterally diffuses under the source-drain pn-junction region. A heavily doped APT implant increases the doping concentration near the junction which leads to an increase in the component of the junction tunneling leakage current. In order to minimize the leakage current as well as the capacitance of the source-drain pn-junctions of bulk-FinFETs, the APT implant conditions are optimized to achieve a graded junction [17]. The APT doping establishes a built-in electric field across APT/well doping boundary such as n^+n (for pFinFETs) and p^+p (for nMOSFETs), thus acting as a junction as described in Section 2.2.5.2. This n^+n (for pFinFETs) or p^+p (for nMOSFETs) barrier due to APT implant along with source-drain pn-junction operate as a double junction as shown in Figure 8.11.

Figure 8.12 shows the cross-section of a pFinFET drain region showing p^+n^+-junction between the p^+ drain and n^+ APT region and n^+n-well high-low doping boundary with APT implant region. When the applied reverse bias ($V_{ds} > 0$) at the drain is increased, the depth of the depletion region into the APT layer will increase; at the same time the n^+n-well (APT/well) high-low boundary is depleted due to electron diffusion from the high concentration region to the low concentration n-well region due to favorable applied electric field. Thus, the net result is that at a certain $V_{bs(d)} = V_x$, the entire APT region is depleted, and the depletion region extends through the APT region and enters the n-well region. This leads to a deviation in the behavior of junction capacitance from ideal uniformly doped p^+n step junction diode observed in Figure 2.28.

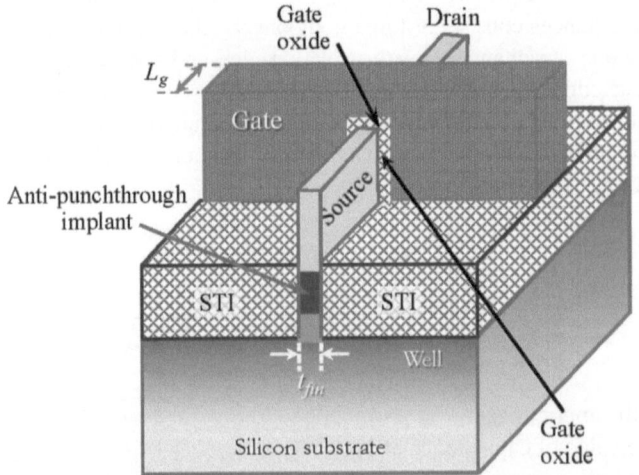

FIGURE 8.11 Three-dimensional cross-section of a bulk-FinFET structure: anti-punchthrough implant below the fin laterally diffused under the source-drain region forming channel/APT pn-junction and APT/well boundary (n^+n for nFinFETs or p^+p for pFinFETs) below the APT region. STI is the shallow trench isolation oxide and Well is well implant in the bulk substrate.

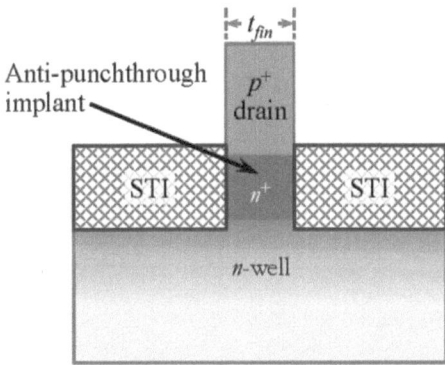

FIGURE 8.12 Two-dimensional cross-section of the source-drain region of a bulk pFinFET structure with laterally diffused n^+ anti-punchthrough implant under the p^+ source-drain region forming p^+n^+-junction above the APT region and n^+n high-low region below the APT region.

The reported data show the deviation of $1/C_j^2$ vs. $V_{bs(d)}$ plot for the source-drain pn-junctions of FinFETs with APT implant from the ideal diode behavior given by Equation 8.68 [18]; here, $V_{bs(d)}$ is the bulk (b) to source (s) (or drain (d)) applied reverse bias at FinFET source-drain pn-junctions. The bulk-FinFET source-drain pn-junction capacitance tends to show two different slopes when an APT implant is used. In order to account for this deviation in the bulk-FinFET source-drain pn-junction capacitance due to APT implant, a new capacitance model has been reported [18] and is given by

$$C_j = \begin{cases} \dfrac{C_{j01}}{\left(1 - \dfrac{V_x}{\phi_{bi}}\right)^{m_j}}; & 0 < V_{bs(d)} < V_x \\[4ex] \dfrac{C_{j02}}{\left(1 - \dfrac{V_{bd} - V_x}{\phi_{bi2}}\right)^{m_{j2}}}; & V_{bd} > V_x. \end{cases} \qquad (8.69)$$

where:

V_x is the transition voltage of the source (drain) reverse bias at which the slope of $1/C_j^2 - V_{bs}$ plots changes as shown in Figure 8.13

C_{j01} is the capacitance at V_{bs} (V_{bd}) = 0

ϕ_{bi} is the built-in potential of the pn-junction given by Equation 2.84 in terms of doping concentration of junctions

m_j is the gradient of dopant distribution under the source-drain pn-junction region

C_{j02} is the capacitance at V_{bs} (V_{bd}) = V_x

ϕ_{bi2} is the effective built-in potential of the pn-junction that depends on the distribution of the punchthrough dopants

m_{j2} is the gradient of dopant distribution under the source-drain pn-junction region due to APT implant

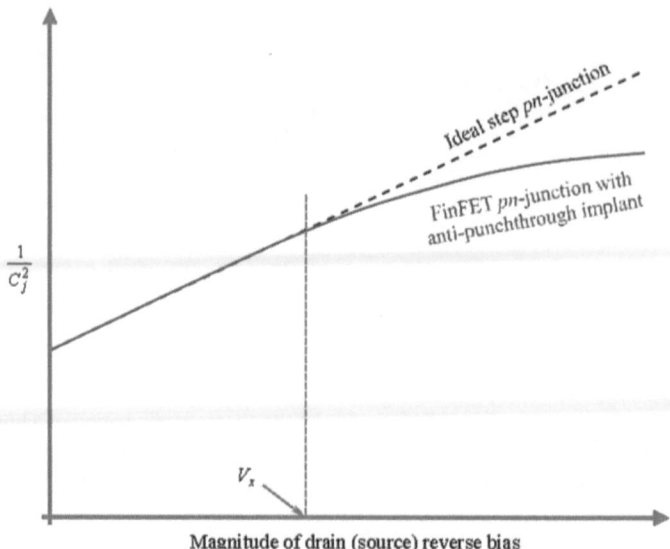

FIGURE 8.13 Illustration of the effect of APT implant on $1/C_j^2$ vs $V_{bs(d)}$ plots of source-drain *pn*-junctions of FinFETs: the broken line represents the single slope of an ideal step *pn*-junction; whereas the solid line represents source-drain *pn*-junction capacitance of a FinFET with APT implant showing two slopes; V_x is the transition voltage representing deviation from ideal *pn*-junction behavior.

All the reverse bias diode parameters can be characterized from the measured $C_j - V_{bs}$ and $C_j - V_{bd}$ characteristics of FinFETs.

8.5.2 Forward Bias Capacitance

Plots of Equation 8.68 show that the capacitance C_j decreases as the reverse biased $|V_d|$ increases ($V_{bs(d)} < 0$) as shown in Figure 2.28. However, Equation 8.68 shows that when the *pn*-junction is forward biased ($V_{bs(d)} > 0$), the capacitance C_j increases and becomes infinite at $V_d = \phi_{bi}$ as observed in Figure 2.28 (continuous line, Curve 2). This is because Equation 8.68 no longer applies due to depletion approximation becoming invalid. For the simplicity of modeling forward biased *pn*-junction capacitances, we can make a series expansion of Equation 8.68. Thus, for modeling the forward biased *pn*-junction, Equation 8.68 is simplified by a series expansion of the denominator and by neglecting the higher order terms, we can show

$$\left(1 - \frac{V_{bs(d)}}{\phi_{bi}}\right)^{-m_j} = 1 + m_j \frac{V_{bs(d)}}{\phi_{bi}} + \cdots \tag{8.70}$$

Thus, the forward bias *pn*-junction capacitance can be calculated by the equation

$$C_j = C_{j0}\left(1 + m_j \frac{V_{bs(d)}}{\phi_{bi}}\right) \tag{8.71}$$

To a first order, Equation 8.71 is valid for FinFETs since the depletion region in the forward bias condition is confined closure to the junction. However, more accurate expressions can be derived from *pn*-junction charge calculations [1].

For device modeling, the total source-drain *pn*-junction capacitance can be calculated from source (or drain) body bottom area junction capacitance along the isolation edge, and source (drain) body sidewall junction capacitance along the gate edge as described in [1,12].

8.6 SUMMARY

This chapter discussed the parasitic resistance and capacitance elements of FinFET devices. First of all, mathematical expressions are derived for contact resistance, spreading resistance, and source-drain extension resistance components of the source-drain series resistance. Then a brief overview of the gate resistance is presented for the analysis of FinFETs operating at non-quasi-static mode. Then the parasitic overlap and fringe capacitances are briefly discussed. Finally, source-drain *pn*-junction capacitance and the impact of APT implant on the source-drain *pn*-junction capacitance are briefly overviewed.

REFERENCES

1. Y.S. Chauhan, D.D. Lu, S. Venugopalan, *et al.*, *FinFET Modeling for IC Simulation and Design: Using the BSIM-CMG Standard*, Academic Press, San Diego, CA, 2015.
2. K. Mistry, W.C. Lee, C. Kuo, *et al.*, "A 45nm logic technology with high-k+metal gate transistors, strained silicon, 9 Cu interconnect layers, 193nm dry patterning, and 100% Pb-free packaging." In: *IEEE International Electron Devices Meeting Technical Digest*, pp. 247–250, 2007.
3. A. Dixit, A. Kottantharayil, N. Collaert, *et al.*, "Analysis of the parasitic S/D resistance in multiple-gate FETs," *IEEE Transactions on Electron Devices*, 52(6), pp. 1132–1140, 2005.
4. W. Wu and M. Chan, "Gate resistance modeling of multifin MOS devices," *IEEE Electron Device Letters*, 27(1), pp. 68–70, 2006.
5. W. Wu and M. Chan, "Analysis of geometry-dependent parasitics in multifin double gate FinFETs," *IEEE Transactions on Electron Devices*, 54(4), pp. 692–698, 2007.
6. A.S. Roy, C.C. Enz, and J.M. Sallese, "Compact modeling of gate sidewall capacitance of DG-MOSFET," *IEEE Transactions on Electron Devices*, 53(10), pp. 2655–2657, 2006.
7. M.J. Kumar, S.K. Gupta, and V. Venkataraman, "Compact modeling of the effects of parasitic internal fringe capacitance on the threshold voltage of high-k gate-dielectric nanoscale SOI MOSFETs," *IEEE Transactions on Electron Devices*, 53(4), pp. 706–711, 2006.
8. J. Lacord, G. Ghibaudo, and F. Boeuf, "Comprehensive and accurate parasitic capacitance models for two- and three-dimensional CMOS device structures," *IEEE Transactions on Electron Devices*, 59(5), pp. 1332–1344, 2012.
9. K. Lee, T. An, S. Joo, *et al.*, "Modeling of parasitic fringing capacitance in multifin trigate FinFETs," *IEEE Transactions on Electron Devices*, 60(5), pp. 1786–1789, 2013.
10. J. Kedzierski, M. Ieong, E. Nowak, *et al.*, "Extension and source/drain design for high-performance FinFET devices," *IEEE Transactions on Electron Devices*, 50(4), pp. 952–958, 2003.

11. H.H. Berger, "Models for contacts to planar devices," *Solid-State Electronics*, 15(2), pp. 145–158, 1972.
12. S.K. Saha, *Compact Models for Integrated Circuit Design: Conventional Transistors and Beyond*, CRC Press, Taylor & Francis Group, Boca Raton, FL, 2015.
13. R. Goyal, *High-Frequency Analog Integrated Circuit Design*, John Willy & Sons, Inc., New York, 1994.
14. W. Liu and M.-C. Chang, "Transistor transient studies including transcapacitive current and distributive gate resistance for inverter circuits," *IEEE Transactions on Circuit and Systems I: Fundamental Theory and Applications*, 45(4), pp. 416–422, 1998.
15. X. Jin, J.-J. Ou, C.-H. Chen, *et al.*, "An effective gate resistance model for CMOS RF and noise modeling." In: *IEEE International Electron Devices Meeting Technical Digest*, pp. 961–964, 1998.
16. C. Enz and Y. Cheng, "MOS transistor modeling issues for RF circuit design." In: *Workshop of Advances in Analog Circuit Design, France*, 1999.
17. K. Okano, T. Izumida, H. Kawasaki, *et al.*, "Process integration technology and device characteristics of CMOS FinFET on bulk silicon substrate with sub-10 nm fin width and 20 nm gate length." In: *IEEE Electron Devices Meeting Technical Digest*, pp. 721–724, 2005.
18. S. Venugopalan, M.A. Karim, A.M. Niknejad, *et al.*, "Compact models for real device effects in FinFETs: Quantum-mechanical confinement and double junctions in FinFETs." In: *Proceedings of the International Conference on Simulation of Semiconductor Devices and Processes*, pp. 292–295, 2012.

9 Challenges to FinFET Process and Device Technology

9.1 INTRODUCTION

In Chapter 1 we discussed Moore's law as the driving force in the microelectronics industry to continue to provide integrated circuits (ICs) with more and more powerful transistors with higher integration density and lower power consumption [1]. However, the continued scaling of transistors in the decananometer regime is constrained by several physical phenomena leading to short-channel effects (SCEs) [2–6]. In order to control SCEs, not only has the device substrate been changed from the bulk silicon to silicon-on-insulator (SOI) [7], but the device structure has also been continuously engineered and transformed from the two-dimensional (2D) planar transistors to three-dimensional (3D) vertical devices [2,3]. Through this technological evolution of transistors, the *fin field-effect transistor* (FinFET) has been adopted to high volume manufacturing as the alternative to 2D planar complementary metal-oxide-semiconductor (CMOS) technology due to its excellent short-channel immunity [2–4,8]. As presented in Chapter 4, the FinFET is a complex 3D device with complex fabrication technology. Therefore, the implementation of such 3D devices in the manufacturing of very large scale integrated (VLSI) circuits requires innovative efforts in process architecture as well as the integration of new materials required for the fabrication processes. Thus, many new challenges have emerged during the transition from the 2D to 3D device manufacturing technology. There are several published reports on the challenges and difficulties of FinFET process technology for VLSI manufacturing [9,10]. In this chapter, a brief overview of the challenges of the FinFET process and device technology is presented.

9.2 PROCESS TECHNOLOGY CHALLENGES

9.2.1 LITHOGRAPHY CHALLENGES

The patterning of fins, as described in Section 4.3.3, poses an enormous challenge in the fabrication of FinFET devices. A state-of-the-art lithography is required to create sharp fin patterns. Thus, for the fabrication of 20 nm and 14 nm node devices, the most developed technique used is the 193 nm ArF immersion lithography with multiple-patterning [11]. And, in 7 nm technology node, the 193 nm ArF immersion lithography with *self-aligned double patterning* (SADP) and *self-aligned quadruple patterning* (SAQP) techniques are required [12]. SADP, as discussed in Section 4.3.3, is a technique that applies a spacer transfer process for small pitch, whereas SAQP is a technique where the SADP is used twice to create the extremely narrow shapes and lines. There are different challenges with multiple-patterning such as edge placement error, pitch walking, and high cost [13]. Therefore, for 7 nm node, *extreme ultraviolet* (EUV) lithography and 193 nm immersion lithography along with multiple-patterning are very promising. The benefit of implementing such an expensive EUV lithography technique is to replace some of the most complex multiple-patterning layers. However, it is expected that ArF immersion lithography will continue to be used for some of the other critical layers in the 10 nm and 7 nm nodes [9].

9.2.1.1   ArF Lithography with Multi-Patterning

Overlay: The overlay accuracy in multiple-patterning lithography for 10 nm and 7 nm technology nodes is challenging. In order to drive the allowed overlay error down to extreme low values, high order overlay correction schemes are required to control the process variability [9]. Furthermore, the increase in the number of split layers increases the complexity of metrology in the total overlay and alignment tree exponentially while increasing the hard-mask steps in the process stack. As a result, the set-up and the verification of overlay metrology recipe becomes more critical and a holistic approach that addresses the total overlay optimization from process design to process setup is required for volume manufacturing [9]. Therefore, overlay accuracy must be improved in metrology, wafer processing, and masking steps for precision wafer patterning [9].

Mask with reticle enhancement techniques (RETs): For tall and narrow FinFET devices, the mask making is challenging. In order to deal with the diffraction issues, various RETs with optical proximity corrections (OPCs) are used to modify the mask patterns and improve the printability on the wafer. Advanced OPCs such as inverse lithography techniques (ILTs) or shapes approaching ILT are required to resolve the target process window in nanometer node technology with increasing structural complexity. Thus, the mask shapes are complex as they require finer geometries and spacing [9].

9.2.1.2   Extreme Ultraviolet Lithography

The extreme ultraviolet or EUV lithography offers the usage of only a single mask exposure instead of multiple exposures. However, there are three major challenges to implement EUV lithography for volume manufacturing of FinFET device technology: *power source*, *resists*, and *mask infrastructure* [9].

The major challenge in implementing EUV lithography for mass production is the source of light for 13.5 nm wavelength that enables cost-effective production capacity of the exposure tool. Though the source issue is considerably matured and several tools are available for the required power at the wafer-level, it is still inadequate for large-scale manufacturing of VLSI circuits.

Another critical technical challenge of EUV lithography is the development of resist material with high resolution and high sensitivity as well as low line-edge roughness (LER) and low outgassing simultaneously [14,15]. In addition, to improve throughput in high volume manufacturing, the resist sensitivity to the 13.5 nm wavelength radiation of EUV needs to be improved, while the line-width roughness (LWR) specification must be held at a low single-digit (nm) [16]. Though resist LER is generally controlled by chemical processes, the replication of the mask pattern roughness and replicated surface roughness or photon noise will still play a significant role as the pattern dimensions continuously shrink [17].

The availability of defect-free reflective mask has been another most critical challenge in EUV lithography for high volume manufacturing [18]. The EUV patterned masks introduce new materials and surfaces that may cause particle adhesion and cleaning [9]. Therefore, a pellicle is needed to protect the mask during the use of EUV scanner to mitigate the risk of particle adhesion. The remaining challenge for EUV mask with pellicle is that the stress from pellicle mounting may cost overlay error.

9.2.2 Process Integration Challenges

As described in Section 4.3.3, the dense fins in FinFETs are patterned by using the SADP technique, followed by oxide filling, planarization, and recessing to pattern the fin active region and form shallow trench isolation (STI). This fin patterning in the overall process sequence (Figure 4.3) causes many challenges in the fabrication of FinFETs [18–22]. These challenges are related to the issues of: (1) *precise and uniform fin patterning*; (2) *3D gate and spacer patterning*; (3) *uniform junction formation in fin*; and (4) *stress engineering*.

9.2.2.1 Precise and Uniform Fin Patterning

We discussed in Chapters 5–8 that the electrical characteristics of FinFETs depend on the fin geometry (thickness, height, and verticality) [22]. On one hand, taller fins are required to achieve higher current that impose severe challenges to FinFET manufacturing. On the other hand, thinner fins are favorable to channel electrostatic control that causes mobility degradation, random discrete doping (RDD) from the source-drain doping gradients, and variation in the off-state leakage current.

For precision fin patterning, the fin etching in the bulk silicon must be controlled by a timing process. In most cases, the fins at the edges of a cluster suffer a higher variability than those in the middle. To achieve uniform fin thickness and height in a cluster, dummy fins are required [23] which are removed at the pitch. As fin pitch shrinks and approaches the overlay limit, fin removal becomes challenging. The fin isolation by STI and anti-punchthrough implant steps are also challenging due to the tighter pitch, difficulty to control STI depth, and doping variation.

Another challenge in fin patterning is the preservation of the structural integrity of fins with a high aspect ratio. The silicon surface of narrow fins appears different than the bulk silicon [24], and excessive silicon loss may occur after the usual wet cleaning step. Also, the oxidation is faster at the corner and at the tip of the fins. Furthermore, the dry etching of fins is more stringent due to the 3D topography. Therefore, a plasma pulsing scheme may be a viable alternative to minimize the silicon loss in fin patterning [22].

9.2.2.2 Gate and Spacer Patterning

The patterning of tall fins increases the process integration complexity of dummy polysilicon gate, spacer, and replacement metal gate. It is difficult to etch the polysilicon gate with high aspect ratio and precise control of dimensions [22]. The charging and micro-loading in etching lead to variable gate length (L_g). Due to the tall vertical fin dimension, a significant over-etch is required to remove the residual polysilicon on the fin sidewalls as well as to remove the offset spacers on the fin sidewalls for selective epitaxial growth (SEG) of the source-drain regions of FinFETs [24,25]. These over-etchings cause damage to the silicon fins. Therefore, careful optimization of the dry and wet etch process is required to fabricate 3D gates with minimum L_g variation and fin damage.

The replacement metal gate module also poses severe challenges as it requires new steps to enable interaction during the chemical-mechanical planarization (CMP) process for STI as discussed in Section 4.3.8. The control of the gate height is essential in a replacement gate process. If the gate is over-polished, the raised source-drain region is exposed to the polish resulting in external resistance and mobility variation. On the other hand, if the gate is under-polished, the contact taper causes variation of the external resistance and may cause an open-contact yield issue. Therefore, a more precise and controllable CMP process is required for FinFET fabrication [26].

9.2.2.3   Uniform Junction Formation in Fin

Impurity doping is one of the most critical integration challenges for the fabrication of FinFETs [27,28]. The issues of doping challenges include: conformal doping in the source-drain contact and extension regions to ensure uniform carrier conduction in the fin channel; the shadowing from the neighboring fins due to the tight pitch of the fins that limits the beam incidence angles; and damage, accumulation, and annealing in the high aspect ratio fins.

It is challenging to dope impurities uniformly in tall and narrow fins with shrunk pitch using the conventional ion implantation processes [29]. The post-implant amorphization of the silicon fins causes poor re-crystallization during the junction anneal process thereby causing poor dopant activation and defected fins [27]. The implant condition for fin doping can also impact the quality and the selective epitaxial growth rate of the source-drain which in turn may affect the source-drain and contact resistance of the FinFET devices. Thus, innovative doping schemes are required to overcome the doping challenge in FinFETs and achieve uniform doping profiles.

9.2.2.4   Stress Engineering

Stress engineering is also a challenging issue for FinFET fabrication technology. The most effective process to induce stress in source-drain is embedded SiGe (compressive stress for p-channel FinFETs), SiC (tensile stress for n-channel FinFETs) or stress in trench contact, and in metal gate [30]. The effectiveness of the gate and source-drain stressors depends on the trade-off between reducing the stressor volume and enhancing the stressor proximity to the channel [31]. In order to further increase the stress and enhance the channel mobility, the Ge content in SiGe source-drain can be increased, similar to the fabrication process used in planar CMOS technology [32–35].

The source-drain SEG layers may suffer from several issues including facet formation [36,37], defects, micro-loading, non-uniform strain distribution, surface roughness, and pattern dependency [38–42]. The pattern dependency occurs due to the variation of packing density and size of the transistor in a chip. The main reason for the pattern dependency of the SEG process is the non-uniform consumption of reactant gas molecules when the exposed silicon area varies in a chip. This problem can be minimized by optimizing the growth parameters and by designing chip layouts such that the exposed silicon is uniformly distributed over the area of the chip to create uniform gas consumption [42]. The uniformity of strain and the control of the defect density in the fin channel region is a huge challenge for FinFET process technology.

9.2.2.5 High-k Dielectric and Metal Gate

The high-k dielectric along with metal gate process is used for advanced CMOS technology due to the high dielectric constant and a relatively large bandgap of the high-k dielectric [43]. Typically, HfO_2 high-k dielectric with high permittivity (a dielectric constant of about 25) and a relatively large bandgap (5.7 eV) is used as the gate dielectric for both the nFinFET and pFinFET devices (Table 9.1) [9]. In addition, the HfO_2 has high heat of formation, good thermal and chemical stability on silicon, and large barrier height at interfaces with silicon. And, at an operation voltage of 1–1.5 V, the leakage current through HfO_2 dielectric films is several orders of magnitude lower than that through SiO_2 films with the same equivalent oxide thickness (EOT) [35,43]. However, one of the main challenges of HfO_2 integration for sub-22 nm FinFETs is the thermal instability of HfO_2/silicon interface. There is an inevitable $SiOx$ interlayer between HfO_2 and silicon substrate [43–45], even though the HfO_2/silicon is theoretically found to be thermodynamically stable [46]. Table 9.1 shows the relevant high-k HfO_2 dielectric and metal gate technology parameters for 22 nm and 14 nm technology nodes [9]. The table also shows the use of TiAlN and TiN metal gates for nFinFETs and pFinFETs, respectively.

Table 9.2 summarizes the typical technology parameters of the high-k dielectric and metal gate for 22 nm and 14 nm FinFET technology nodes [9]. It is to be noted from Table 9.2 that the thicknesses of the $SiOx$ interlayer thermal oxide have been reduced significantly from about 1.1 nm in 22 nm node to about 0.6 nm in 14 nm node. On the other hand, the thickness of the high-k dielectric increased from about 1.0 nm in 22 nm node to about 1.2 nm in 14 nm node. However, the overall EOT of the gate dielectric is decreased.

In FinFET architecture, the aspect ratio of the replacement gate structure is larger which makes it a huge challenge to fill the trench. Thus, alternative metal gate

TABLE 9.1

Typical Materials for High-k Dielectric and Metal Gate for 22 nm and 14 nm FinFET Technology Nodes

Technology nodes	Device structure	High-k dielectric		Metal gate	
		nFinFET	pFinFET	nFinFET	pFinFET
22 nm	FinFET	HfO_2	HfO_2	TiAlN	TiN
14 nm	FinFET	HfO_2	HfO_2	TiAl	TiN

TABLE 9.2

Typical Technology Parameters of the High-k Dielectric and Metal Gate for 22 nm and 14 nm Technology Nodes

Technology nodes	Film thickness (nm)			
	Thermal oxide	High-k	TiAl(N)	TiN
22 nm	~1.1	~1.0	~1.2	~1.4
14 nm	~0.6	~1.2	~3.7	*

processing such as atomic layer deposition (ALD) is considered as a solution for the metal gate deposition because of its excellent capability to achieve conformal step coverage [47,48]. However, for the implementation of ALD, the appropriate metal gate materials are required for workfunction engineering of nFinFET and pFinFET devices as well as good capability of step coverage.

9.2.2.6 Variability Control

Process variability control is more critical and has become increasingly challenging for FinFET devices [49,50]. The electrical variation in FinFET devices is very sensitive to the variations in fin thickness t_{fin} and fin height H_{fin}. The variation in H_{fin} occurs from fin etching, STI deposition, STI CMP, and STI recessing processing steps as described in Chapter 4.

In general, gate etching profile and L_g variation over fin topography are difficult to control. Source-drain epitaxy is a sensitive process over fin topography [51] and the resistance and stress fluctuations occur due to the change in the shape of fins. Furthermore, in the ion implantation process, the defected layers are another source of variability [50]. In a fin with channel doping concentration of 6×10^{17} cm^{-3}, about a third of the variability is due to random discrete doping (RDD) as observed in planar MOSFETs [52]. Although the channel doping can be avoided in FinFETs, RDD from the source-drain doping gradient causes variability for devices with $L_g <$ 10 nm [29,53]. The channel interface and gate stack workfunction variations have a remarkable negative effect on the transistor performance. Double patterning raises concerns about the way in which individual polygons are split across two masks. Since overlay is not scaled as fast as the minimum feature size, the mask-alignment issues introduce a new source of variability in the spacing between the polygons [23].

9.2.2.7 Spatial Challenges

The reduction in contacted gate pitch for sub-22 nm FinFET devices requires a trade-off between gate length L_g, source-drain spacer thickness, and source-drain contact area. A minimum spacer thickness is defined by reliability requirements as well as on the target specification of capacitance between the gate and source-drain electrodes. Narrow source-drain contacts increase the access resistance of the device.

It is well known that the downscaling of device technology degrades the performance of interconnect wiring. At and below 22 nm technology nodes, the interconnect resistance is expected to increase significantly due to the reduction in the wire cross-section. In addition to a reduction in the wire cross-section, carrier scattering from the boundaries of the individual copper crystal grains and interfaces with barrier layers rapidly increase the interconnect resistivity as well as the resistance of the individual wires. Also, the double patterning causes routing challenges and difficulties in the access to standard cell pins due to the constraints on the interconnect pitch. In sub-22 nm FinFET technology nodes, the interconnect resistance and capacitance (RC) start to dominate the delay. It is challenging to reduce the line via resistance and capacitance and maintain reliability (electromigration (EM)), time-dependent dielectric breakdown (TDDB), bias temperature instability (BTI), and hot carrier injection (HCI)) to acceptable values. In order to achieve low interconnect resistance while maintaining reliability, the metal filling process must be defect-free. However, reducing line resistance requires enough space for actual wiring material, leaving less room for barrier, thus causing degradation in reliability. Furthermore, to achieve lower via resistance, a thinner barrier at the bottom of the via is required which causes the EM blocking boundary to be insufficient [54]. Thus, there are significant challenges for FinFET process integration and innovative engineering solutions are needed to overcome them.

9.2.3 DOPANT IMPLANTATION CHALLENGES

The challenges in the FinFET device fabrication process include conformal doping of source-drain areas and lowering fin damage due to ion implantation [55].

9.2.3.1 Conformal Doping

For FinFETs, the major challenge is the conformal distribution of dopants within the fins irrespective of the source-drain extension (SDE) dopant activation, diffusion, and profile abruptness [55–57]. The non-conformal doping profile causes degradation in the drive current of FinFETs. The conformal distribution of dopants can be achieved by a large tilt beamline implantation. However, the shadowing by the neighboring fins limits the conformal doping. An alternative technique to achieve conformal distribution of dopants within the fin is plasma doping with optimized process parameters.

9.2.3.2 Damage Control

In FinFETs, the damage control during doping is another challenge to achieve the target performance of the devices. In FinFET structure, a narrow fin on the pedestal of the well is isolated from large crystal volume. Thus, the surface proximity and 3D structure pose a severe constraint on the recrystallization of post-implant amorphous silicon fin. In the case of a complete amorphization of silicon fins, only a very small seed for recrystallization can cause a defective growth, causing degradation in the resistivity and drive current of the devices [58,59]. Therefore, in FinFET fabrication it is critical to reduce the amorphization depth created by ion implantation and annealing to minimize the fin damage. In order to achieve damage-free SDE fin doping, hot implantation can be used. Using high temperature implantation, damage accumulation can be significantly reduced, thus decreasing the self-amorphization of the implant. The hot implant technique also improves the fin line conductance and junction leakage [60].

9.2.4   THE ETCHING CHALLENGES

9.2.4.1   Depth Loading Control of Fin Etching

During the fin etch (Section 4.3.3), only a small part of the process gases is *ionized* into plasma under the radio frequency power. Most of the gas molecules exist in the chamber as *neutrals* which induce deposition during the etching process. The neutral molecules are easy to stick onto the surfaces before entering at the bottom of the trench, inducing taper profile (Chapter 4) that blocks ions from reaching the bottom [9]. On the other side, new stored ions are formed at trench bottom to react upon the incoming ions. As a result, as the etching reaction continues and etching depth increases, the flux ratio of ion and neutral molecules decreases and etching bombardment is weakened [9]. Within the critical dimension (CD) sizes, the smaller CD induces weaker bottom reaction. Therefore, the etching depth depends on the opening CD size: bigger size induces deeper depth. The CD dependence of etching depth can be improved using the bias-pulsing technique [9]. The bias-pulsing etching process can be used to achieve a smaller depth loading effect that is caused by different opening CD size [61,62].

9.2.4.2   Gate Etch Control

The selective and residue-free etch processes are challenging in FinFET device fabrication. There are a few new process and materials challenges in FinFET fabrication as summarized below. First of all, an excessive Si loss is observed after the usual pre-gate-oxide clean as discussed in Chapter 4. Therefore, wet cleaning must be optimized with dilute concentration and lower temperatures. Secondly, the oxidation of fins is also faster at the corner and tip of the fins. And, the dry etching on fins is more stringent due to the 3D structures, and a plasma bias-pulsing scheme may be a viable alternative for minimizing silicon loss [63,64]. The gate etch control is critical to maintaining uniform fin height H_{fin}. The variation in H_{fin} impacts the electrical properties of FinFETs such as the threshold voltage V_{th}. This indicates that the (dry or wet) etching step is crucial for 3D transistors compared to the planar ones.

In addition, controlling the selectivity of self-aligned etching (Section 4.3.3) is challenging for FinFETs of gate length below 14 nm [65] to enable appropriate contact slit opening for local transistor contacts.

9.2.4.3   STI Process for Gate

As discussed in Section 4.3.3, the fin height is defined by etching STI oxide (TEOS). It is one of the critical processes to control the fin height H_{fin}. The oxide is etched back using a highly selective etch process and it is challenging to have full control to form a silicon fin with defined dimensions. Thus, it is especially critical to define H_{fin} by etching STI oxide. The variation in H_{fin} severely impacts the electrical properties of transistors such as the threshold voltage V_{th}. This indicates that an appropriate etching step must be used for STI recess to define H_{fin}.

9.2.4.4   Gate Process

In Section 4.3.8, we described the replacement metal gate (RMG) process by removing the polysilicon dummy gate and SiO_2 dummy gate oxide with the wet process [66–69].

The RMG process is complicated and challenging as described in Chapter 4. Typically, due to the chemical nature of the HF-based wet etchant, the dummy gate oxide etching process cannot remove the polysilicon without any oxidizer [70–72]. However, it is critical to completely remove the polysilicon dummy gate without leaving any residues in the narrow and steep trenches [67–69]. The residues occupy the space of the intended location of the high-k and metal gates and may cause device failure.

Typically, the wet processes use aqueous solutions using water as the solvent and the final rinses using deionized water or ultra-pure water to clean away the chemicals from the surface of the wafers. However, it is well known that some defects can be generated due to the surface tension of water. During the drying process, the high capillary force of water could pull nearby structures to form permanent defects, so-called pattern collapse [73] or stiction [74]. In order to eliminate pattern collapse, gas phase etching such as the HF vapor process can be used where the intermolecular force is not too strong compared to the liquid phase.

In FinFET-like vertical structures, "new materials" including multilayers of SiGe/Si are used. Since it is needed to selectively etch either silicon or SiGe layers, a lot of effort is made to use $HF:H_2O_2:CH_3COOH$ mixtures to etch SiGe selectively to Si [75–77] or to use tetra-methyl-ammonium hydroxide (TMAH) based to remove silicon from SiGe [78].

9.3   DEVICE TECHNOLOGY CHALLENGES

9.3.1   Multiple Threshold Voltage Devices

Threshold voltage V_{th} control and multiple-V_{th} device options of a manufacturing CMOS technology are important for analog applications. A typical planar CMOS technology at a node offers different options for V_{th} including low-V_{th} for high performance VLSI circuits, standard-V_{th} for logic design, and high-V_{th} for analog and radio frequency (RF) applications. This V_{th} control of a planar-MOSFET device can be defined from the expression [3] given by

$$V_{th} = V_{fb} + 2\phi_B + \frac{1}{C_{ox}}\sqrt{2q\varepsilon_{si}\left(2\phi_B + V_{bs}\right)N_b} \qquad (9.1)$$

where:
   V_{fb} is the flat band voltage
   ϕ_B is the bulk potential
   C_{ox} is the gate capacitance $= \varepsilon_{ox}/T_{ox}$; ε_{ox} being the permittivity of gate dielectric
   V_{bs} is the body bias
   N_b is the channel doping concentration

Equation 9.1 shows that V_{th} depends on T_{ox}, N_b, and V_{bs}. Thus, multiple-V_{th} devices of a planar CMOS technology node can be achieved by changing the substrate doping concentration N_b and/or by gate dielectric thickness T_{ox}. In addition, V_{bs} can be used to modulate V_{th} of a MOSFET device.

For multigate FinFET devices, the body is generally undoped or lightly doped. It is also difficult to implement multiple dielectric thicknesses in 3D structures due to lithography and etching challenges. Therefore, it is difficult to achieve multiple-V_{th} FinFET devices in a non-planar CMOS technology using the conventional fabrication process. In order to analyze the options for multiple-V_{th} FinFET devices in a technology node, let us derive a first order expression for V_{th} of FinFET devices from Equation 3.30 given by

$$V_{gs} = V_{fb} + \phi_s - V_{ch}(y) - \frac{Q_s}{C_{ox}}$$
(9.2)

where:
V_{gs} is the gate voltage
ϕ_s is the surface potential
$V_{ch}(y)$ is the channel potential due to applied drain bias V_{ds}
Q_s is the total charge density in the silicon fin channel

Now, for metal gate FinFET devices, V_{fb} is given by Equation 3.21 as

$$V_{fb} = \Phi_{ms} - \frac{Q_0}{C_{ox}}$$
(9.3)

where:
Φ_{ms} $(= \Phi_m - \Phi_s)$ is the difference between the metal gate electrode workfunction (Φ_m) and bulk-silicon workfunction (Φ_s)
Q_0 is the oxide charge at the Si/gate oxide interface and ≈ 0 for ideal defect-free oxide

Again, if Q_i and Q_b are the inversion and bulk charges, respectively, then $Q_s = Q_i + Q_b$. Since at the subthreshold region $Q_i << Q_b$, we have $Q_s \cong Q_b$. Also, since at threshold condition, the drain voltage V_{ds} and hence $V_{ch}(y)$ is negligibly small, therefore from Equations 9.2 and 9.3 we can express V_{th} of FinFET devices as

$$V_{th} \approx \Phi_m - \Phi_s + 2\phi_B + \left(\frac{qN_b t_{fin}}{C_{ox}} \right) H_{fin}$$
(9.4)

where:
t_{fin} is the fin thickness
H_{fin} is the fin height

Thus, from Equation 9.4, we observe that there are several possible *options* for multiple-V_{th} FinFET devices in a non-planar CMOS technology. First of all, by changing C_{ox} using multiple oxide thickness and patterning multiple fin height H_{fin} (that is, taller fins to achieve higher-V_{th} devices). However, as described in Chapter 4, achieving multiple T_{ox} is challenging in complex FinFET fabrication technology. Also, since fins are patterned by spacer defined technology, the various spacer techniques required to patterning multiple height fins on the same substrate is difficult.

Thus, to have full control to form a silicon fin with defined dimensions is a difficult task. It is especially critical to define the fin height by etching the shallow trench isolation oxide. Therefore, it is challenging to achieve multiple-V_{th} devices in FinFET technology using conventional methods. Alternatively, as shown in Equation 9.4, an innovative technique such as metal workfunction Φ_m engineering can be used to achieve multiple-V_{th} FinFETs in non-planar CMOS technology.

In a FinFET technology, with HfO_2 high-k gate dielectric and TiN metal gate, V_{th} control and multiple-V_{th} device technology are achieved by workfunction engineering using aluminum (Al) implantation as described in Chapter 4. In the high-k metal gate process, aluminum ($1 \times 10^{15} - 1 \times 10^{16}$ cm^{-2}) is implanted into TiN metal (not into the high-k) using ultralow energy implanter. The effective workfunction (EWF) of MG is modulated by Al implantation via Al-induced dipole at the HfO_2/SiO_x interface layer. Al diffuses differently in/through TiN depending on its growth method. Since Al-rich TiN has a more n-type effective workfunction, stacks with higher amount of Al diffused into TiN translate into lower EWF values (that is, more n-type EWF). And, TiN (least in Al-rich) is selected for p-type workfunction as shown in Tables 9.1 and 9.2.

9.3.2 WIDTH QUANTIZATION

One of the major differences between a FinFET and a planar MOSFET device is the fact that the FinFET device consists of multiple small unit fins of height H_{fin} and fin thickness t_{fin}. As discussed in Chapter 5, the width of a FinFET device is given by

$$W = n \times \left(2H_{fin} + t_{fin}\right) \qquad (9.5)$$

where:
n is the number of fins of a device and $n = 1, 2, 3, ..., n$
The factor "2" in H_{fin} is due to two sidewall gates

Equation 9.5 shows that the width W of a FinFET depends on the integer n, the number of fins used to build a device and thus quantized. This is referred to as the "width quantization." This unique *width quantization* property of FinFETs is due to the constraint in patterning multiple height FinFETs at a technology node [79]. Due to the technology complexies, only fins of constant height are patterned by lithography techniques as discussed in Chapter 4. Thus, a large device is designed using multiple unit fins. As a result, the larger devices with multiple unit fins are susceptible to random V_{th} variation among fins due to RDD within each fin channel and therefore, the effect of dopant fluctuations must be considered on device performance [80].

The width quantization severely impacts the device leakage current distribution due to random V_{th} variation [79]. The width quantization-induced FinFET leakage current can potentially lead to the chip failure due to insufficient noise margin, inaccurate full chip power estimates, and improper guidelines for leakage-sensitive circuits [81]. Also, width quantization is critical for analog application. In analog IC design, W is a circuit design parameter and is a continuous variable. However, due to width quantization, the design parameter W is a set of small positive integers instead of a continuous variable. Thus, the width quantization poses a severe challenge in FinFET circuit design and therefore, requires width quantization aware FinFET circuit design with advanced statistical modeling techniques for accurate prediction of the leakage current distribution due to width quantized FinFET devices in VLSI circuits [81]. Furthermore, the width quantization increases the complexities and number of layout design rules including spacing rules to reduce coupling, SADP rules for fin patterning, and dummy gate rules for replacement metal gate.

9.3.3 CRYSTAL ORIENTATION

Traditionally, CMOS ICs are fabricated on silicon substrates with a ⟨100⟩ crystalline orientation due to their high electron mobility compared to hole mobility [82–85] as shown in Figure 9.1 and reduced interface trap density. Similar mobility trends are observed for high-k gate dielectric as gate oxide as shown in Figure 9.2 [86].

Figure 9.3(a) shows the layout of IC devices at different crystal orientations to optimize electron and hole mobilities. As shown in Figure 9.3(a), the device orientation, (1) *perpendicular* and (4) *parallel* to the wafer flat or notch, the channel surface lies in the ⟨110⟩ plane where the hole mobility is the highest, whereas the electron mobility is the lowest (Figures 9.1 and 9.2). On the other hand, the device orientation, (2) at a 45-degree angle, the channel surface lies in the ⟨100⟩ plane, the hole mobility is the lowest and electron mobility is the highest (Figures 9.1 and 9.2). For any intermediate orientations, such as position (3), the electron and hole mobilities are at intermediate values [86,87].

Figure 9.3(b) shows a FinFET structure standing vertically on the wafer pedestal, hence the fin surface is orientated in the multiple crystal planes. Thus, the orientation of the fins depends on layout. This poses a serious challenge in layout design and fin patterning. In a FinFET, the conducting channel lies on the sidewall of a silicon

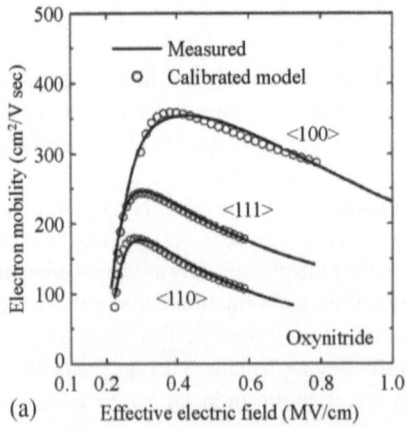
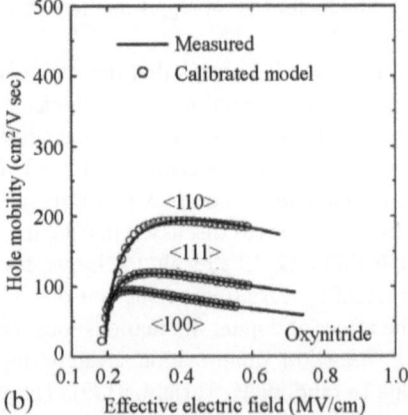

(a)   (b)

FIGURE 9.1 Carrier mobility for ⟨100⟩, ⟨110⟩, and ⟨111⟩ bulk-silicon crystal surfaces as a function of effective electric field for conventional nitrided oxide: (a) electron mobility; (b) hole mobility [86].

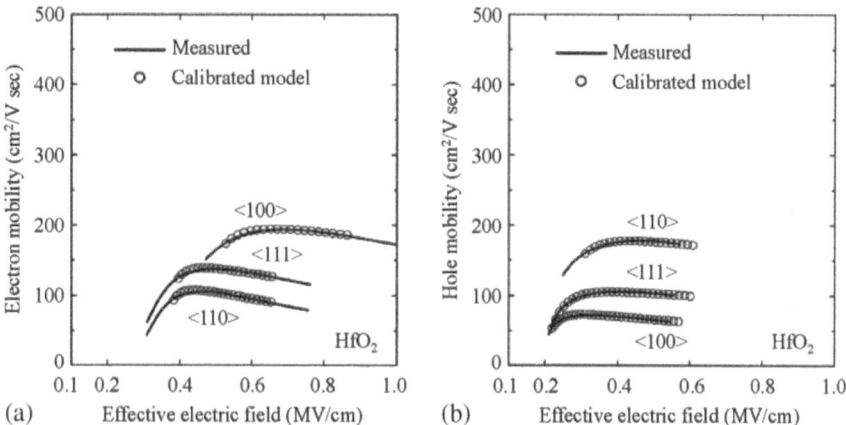

FIGURE 9.2 Carrier mobility for ⟨100⟩, ⟨110⟩, and ⟨111⟩ bulk-silicon crystal surfaces as a function of effective electric field for high-*k* gate dielectric: (a) electron mobility; (b) hole mobility [86].

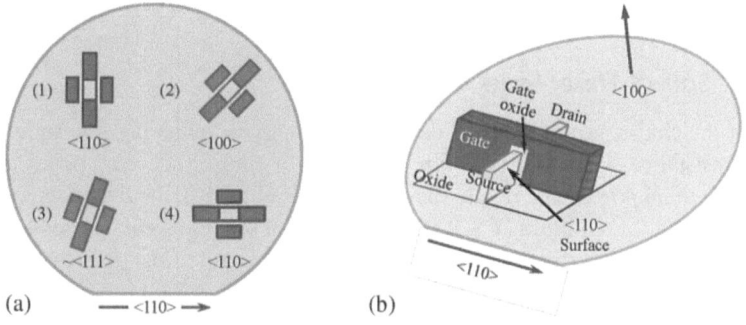

FIGURE 9.3 Device orientation on wafer for mobility optimization: (a) device orientation, (1) perpendicular or (4) parallel to the wafer flat offers highest hole mobility, (2) at a 45-degree angle to the wafer flat offers highest electron mobility, and at any intermediate position, intermediate values of electron and hole mobilities can approximate the ⟨111⟩ plane; (b) FinFET ⟨110⟩ orientation parallel to wafer flat, the electron and hole mobilities are at intermediate values.

pillar (fin channel). Thus, in a standard ⟨100⟩ wafer, where the gate and active fin area are aligned either perpendicular or parallel to the wafer flat, the device channel lies in the ⟨110⟩ plane. However, if the transistor layout is rotated by a 45-degree in the plane of the wafer, then the resulting orientation of the device channel is ⟨100⟩. With an intermediate rotation, the electron and hole mobilities are at intermediate values between ⟨100⟩ and ⟨110⟩ orientations, which can be approximated to ⟨111⟩ surface.

In FinFET layout design, all devices can be drawn at different angles relative to the wafer flat or notch to achieve ⟨100⟩, ⟨110⟩, and ⟨111⟩ orientations and maximize mobility values for both *n*FinFET and *p*FinFET devices. The optimum mobility scheme to align the channels of *n*FinFET and *p*FinFET to lie in the ⟨100⟩ and ⟨110⟩ planes, respectively, can be achieved by rotating the layout of one of the two device types by a 45-degree angle. Although either the *n*FinFET or *p*FinFET device could be rotated depending on the wafer flat direction, however, the rotation of the *n*Fin-FET is preferable due to its smaller size. In either case, it is a lithography challenge to define minimum line-widths in both 0- and 45-degree rotations. In addition, under this orientation scheme of ⟨100⟩ for *n*FinFET and ⟨110⟩ for *p*FinFET devices, a 40% area penalty is incurred for small devices due to the area overhead required for implementation of a 45-degree device rotation [86].

Again, compared to the traditional ⟨100⟩ orientation, the wafers with ⟨110⟩ and ⟨111⟩ orientations improve the area efficiency due to reduced requirement for *p*Fin-FET width and hence the stronger *p*FinFET current drive per unit width. For a ⟨110⟩ wafer substrate, the layout area penalty could be minimized since the FinFET side-walls of ⟨100⟩ and ⟨110⟩ orientations are at right angles to each other. Then *n*Fin-FET and *p*FinFET devices can be drawn parallel and perpendicular to ⟨110⟩ wafer flat direction. This may still incur a minor area penalty, but it would be significantly reduced from the orientation scheme in which one device type is rotated by 45 degrees [86].

9.3.4   SOURCE-DRAIN SERIES RESISTANCE

One of the challenges of FinFET device technology is achieving low source-drain series resistance. High source-drain resistance of FinFET devices poses a severe challenge to analog design due to the degradation of device conductance. Typically, merged raised source-drain (RSD) regions as shown in Figure 9.4 are used to achieve lower source-drain series resistance. However, the control of the growth of the RSD epitaxial layer can be challenging and may cause an increase in the defect density. Furthermore, stress provided by merged fins for strained-silicon fin channel is more difficult to control than unmerged fins.

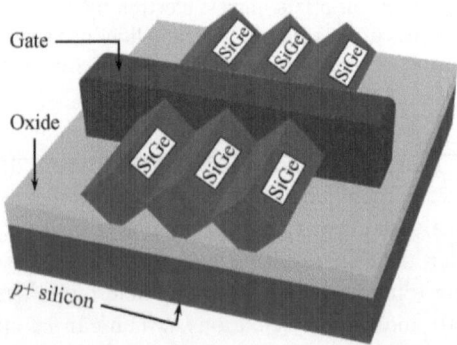

FIGURE 9.4   Merged raised source-drain structure using selective epitaxial growth SiGe layer to reduce the source-drain series resistance.

9.4 CHALLENGES IN FinFET CIRCUIT DESIGN

As the critical dimension of VLSI circuits approaches atomic size near the 3 nm regime, the VLSI circuit designs are becoming increasingly challenging. In designing VLSI circuits using near atomic scale FinFET devices, many issues including high leakage current, low gain, width quantization, and the sensitivity and tolerance of the manufacturing process become extremely critical and must be pre-estimated. With the increasing technical challenges in the circuit design to achieve the CD limit near the 3 nm regime, the circuit design methodologies are continuously evolving to provide novel solutions to address many of these challenges.

9.5 SUMMARY

This chapter presented an overview of the major challenges of the FinFET process, device, and circuit design in manufacturing VLSI circuits and systems. First of all, the lithography challenges such as overlay, pitch walking, and edge placement error, as well as high cost to the currently used 193 nm ArF immersion lithography with SADP and SAQP techniques for 22 nm technology node and beyond are highlighted. Though these issues of the current lithography can be minimized using the EUV lithography technique, it is challenging to implement the EUV lithography technique in high volume manufacturing of FinFETs due to its high cost.

After discussions on lithography challenges, the FinFET process integration challenges related to gate and spacer patterning, patterning of uniform fins, and the SEG SiGe stressor material on raised source-drain regions are overviewed. The SEG layers on FinFETs may cause serial problems including strain relaxation on fins-faceted shape leading to higher defect density and pattern dependency effect due to different transistor architecture and their density in the chip. Then the challenges in the high-k and metal gate processing and thermal instability of HfO_2/silicon interface for the integration of HfO_2 as the gate dielectric for FinFETs are discussed. In addition, the challenge in controlling gate workfunction using Al implant in TiN metal gate is highlighted.

Next, the challenges to fin doping and controlling fin dimensions of FinFETs are overviewed. It is discussed that the conformal doping challenge is due to the shadowing caused by the neighboring fins during ion implant of the fins in an array. The conformal distribution of dopants is required to mitigate the risk of drive current degradation. To overcome the challenges of conformal doping, plasma doping with optimized process parameters can be used. On controlling fin dimensions, it is challenging to define the fin height by etching the STI oxide. The variation in fin height influences electrical properties of transistors such as V_{th}. This indicates that (dry or wet) etching must be controlled in the FinFET fabrication process.

After discussing the process technology challenges, the major device technology challenges such as width quantization, multiple-V_{th} transistors at a technology node, and optimal crystal orientation in FinFET layout design to achieve high mobility for both nFinFETs and pFinFETs are overviewed. The width quantization imposes some challenges on VLSI circuit design, especially for analog applications. Limitation of

V_{th}-tuning or multiple-V_{th} transistors limits the application of FinFETs in analog circuit and mixed-signal applications. The channel orientation, such as ⟨110⟩ sidewall planes for higher hole mobility and ⟨100⟩ sidewall planes for higher electron mobility, poses challenges in layout design. A hybrid orientation scheme might be difficult to implement in practice. Finally, a brief overview of the challenges in circuit design is presented.

Most of the challenges discussed in this chapter are in development or partially solved. However, the most critical issue of achieving atomic scale FinFETs near 3 nm is the significant increase in the manufacturing cost due to the necessity for the implementation of new materials and innovative techniques.

REFERENCES

1. G.E. Moore, "Cramming more components onto integrated circuits," *Electronics*, 38(8), pp. 114–117, 1965.
2. Y.S. Chauhan, D.D. Lu, S. Venugopalan, *et al.*, *FinFET Modeling for IC Simulation and Design: Using the BSIM-CMG Standard*, Academic Press, San Diego, CA, 2015.
3. S.K. Saha, *Compact Models for Integrated Circuit Design: Conventional Transistors and Beyond*, CRC Press, Taylor & Francis Group, Boca Raton, FL, 2015.
4. K.J. Kuhn, "Considerations for ultimate CMOS scaling," *IEEE Transactions on Electron Devices*, 59(7), pp. 1813–1828, 2012.
5. S. Saha, "Scaling considerations for high performance 25 nm metal-oxide-semiconductor field-effect transistors," *Journal of Vacuum Science and. Technology B*, 19(6), pp. 2240–2246, 2001.
6. Y. Taur, D.A. Buchanan, W. Chen, *et al.*, "CMOS scaling into the nanometer regime," *Proceedings of the IEEE*, 85(4), pp. 486–504, 1997.
7. J.-P. Colinge (ed.), "The SOI MOSFET: From single gate to multigate." In: *FinFETs and Other Multi-Gate Transistors*, pp. 2–4, Tyndall National Institute, Cork, Ireland, 2008.
8. A. Veloso, L.A. Ragnarsson, M.J. Cho, *et al.*, "Gate-last vs. gate-first technology for aggressively scaled EOT logic/RF CMOS." In: *Symposium on VLSI Technology*, pp. 34–35, 2011.
9. H.H. Radamson, Y. Zhang, X. He, *et al.*, "The challenges of advanced CMOS process from 2D to 3D," *Applied Sciences*, 7(10), p. 1047, 2017.
10. N. Collaert (ed.), *CMOS Nanoelectronics*, Pan Stanford Publishing, Singapore, 2013.
11. F.G. Pikus and A. Torres, "Advanced multi-patterning and hybrid lithography techniques." In: *Proceedings of the Asia and South Pacific Design Automation Conference*, pp. 611–616, 2016.
12. H. Yaegashi, "Pattern fidelity control in multi-patterning towards 7 nm node." In: *Proceedings of the IEEE International Conference on Nanotechnology*, pp. 452–455, 2016.
13. J. Jiang, S. Chakrabarty, M. Yu, and C.K. Ober, "Metal oxide nanoparticle photoresists for EUV patterning," *Journal of Photopolymer Science and Technology*, 27(5), pp. 663–666, 2014.
14. D.D. Simone, M. Mao, F. Lazzarino, and G. Vandenberghe, "Metal containing resist readiness for HVM EUV lithography," *Journal of Photopolymer Science and Technology*, 29(3), pp. 501–507, 2016.
15. D. Mamezaki, M. Watanabe, T. Harada, and T. Watanabe, "Development of the transmittance measurement for EUV resist by direct resist coating on a photodiode," *Journal of Photopolymer Science and Technology*, 29(5), pp. 749–752, 2016.
16. Y. Yoda, A. Hayakawa, S. Ishiyama, *et al.*, "Next-generation immersion scanner optimizing on-product performance for 7 nm node." In: *Proceedings of the SPIE Conference on Optical Microlithography XXIX*, vol. 9780, 2016.

17. A. Wojdyla, A. Donoghue, M.P. Benk, *et al.*, "Aerial imaging study of the mask-induced line-width roughness of EUV lithography masks." In: *Proceedings of the SPIE Conference on Extreme Ultraviolet (EUV) Lithography VII*, vol. 9776, 2016.

18. A.O. Antohe, D. Balachandran, L. He, *et al.*, "SEMATECH produces defect-free EUV mask blanks: Defect yield and immediate challenges." In: *Proceedings of the SPIE Conference on Extreme Ultraviolet (EUV) Lithography VI*, vol. 9422, 2015.

19. C.-H. Jan, U. Bhattacharya, R. Brain, *et al.*, "A 22 nm SoC platform technology featuring 3-D tri-gate and high-k/metal gate, optimized for ultra low power, high performance and high density SoC applications." In: *IEEE International Electron Devices Meeting Technical Digest*, vol. 21, pp. 44–47, 2012.

20. C.-H. Jan, M. Agostinelli, M. Buehler, *et al.*, "A 32 nm SoC platform technology with 2nd generation high-k/metal gate transistors optimized for ultra low power, high performance, and high density product applications." In: *IEEE International Electron Devices Meeting Technical Digest*, pp. 647–650, 2009.

21. K.J. Kuhn, A. Murthy, R. Kotlyar, and M. Kuhn, "Past, present and future: SiGe and CMOS transistor scaling," *ECS Transactions*, 33(6), pp. 3–17, 2010.

22. N. Kise, H. Kinoshita, A. Yukimachi, *et al.*, "Fin width dependence on gate controllability of InGaAs channel FinFETs with regrown source/drain," *Solid-State Electronics*, 126, pp. 92–95, 2016.

23. A.P. Jacob, R. Xie, M.G. Sung, *et al.*, "Scaling challenges for advanced CMOS devices," *International Journal of High Speed Electronics and Systems*, 26(1/2), pp. 2–76, 2017.

24. T. Matsukawa, Y. Liu, K. Endo, *et al.*, "Variability origins of FinFETs and perspective beyond 20 nm node." In: *Proceedings of the IEEE International SOI Conference*, pp. 1–28, 2011.

25. J. Kavalieros, B. Doyle, S. Datta, *et al.*, "Tri-gate transistor architecture with high-k gate dielectrics, metal gates and strain engineering." In: *Symposium on VLSI Technology*, pp. 50–51, 2006.

26. K.J. Kuhn, "CMOS transistor scaling past 32 nm and implications on variation." In: *Proceedings of the IEEE/SEMI Advanced Semiconductor Manufacturing Conference*, pp. 241–246, 2010.

27. A. Veloso, A. De Keersgieter, P. Matagne, *et al.*, "Advances on doping strategies for triple-gate finFETs and lateral gate-all-around nanowire FETs and their impact on device performance," *Material Science in Semiconductor Processing*, 62, pp. 2–12, 2017.

28. M.I. Current, "Ion implantation of advanced silicon devices: Past, present and future," *Material Science and in Semiconductor Processing*, 62, pp. 13–22, 2017.

29. K.-W. Ang, J. Barnett, W.-Y. Loh, *et al.*, "300 mm FinFET results utilizing conformal, damage free, ultra shallow junctions (Xj_5 nm) formed with molecular monolayer doping technique." In: *IEEE International Electron Devices Meeting Technical Digest*, pp. 837–840, 2011.

30. S. Pidin, T. Mori, K. Inoue, *et al.*, "A novel strain enhanced CMOS architecture using selectively deposited high tensile and high compressive silicon nitride films." In: *IEEE International Electron Devices Meeting Technical Digest*, pp. 213–216, 2004.

31. N. Xu, B. Ho, M. Choi, *et al.*, "Effectiveness of stressors in aggressively scaled FinFETs," *IEEE Transactions on Electron Devices*, 59(6), pp. 1592–1598, 2012.

32. T. Ghani, M. Armstrong, C. Auth, *et al.*, "A 90 nm high volume manufacturing logic technology featuring novel 45 nm gate length strained silicon CMOS transistors." In: *IEEE International Electron Devices Meeting Technical Digest*, pp. 978–980, 2003.

33. C. Auth, C. Allen, A. Blattner, *et al.*, "A 22 nm high performance and low-power CMOS technology featuring fully-depleted tri-gate transistors, self-aligned contacts and high density MIM capacitors." In: *IEEE International Electron Devices Meeting Technical Digest*, pp. 131–132, 2012.

34. S. Thompson, G. Sun, K. Wu, *et al.*, "Key differences for process-induced uniaxial vs. substrate-induced biaxial stressed Si and Ge channel MOSFETs." In: *IEEE International Electron Devices Meeting Technical Digest*, pp. 221–224, 2004.

35. K. Mistry, C. Allen, C. Auth, *et al.*, "A 45 nm logic technology with high-k + metal gate transistors, strained silicon, 9 Cu interconnect layers, 193 nm dry patterning, and 100% Pb-free packaging." In: *IEEE International Electron Devices Meeting Technical Digest*, pp. 247–250, 2007.

36. H. Xiao, *3D IC Devices Technologies, and Manufacturing*, SPIE Press, Bellingham, WA, 2016.

37. D. Dutartre, A. Talbot, and N. Loubet, "Facet propagation in Si and SiGe epitaxy or etching," *ECS Transaction*, 3(7), pp. 473–487, 2006.

38. S. Mujumdar, K. Maitra, and S. Datta, "Layout-dependent strain optimization for p-channel trigate transistors," *IEEE Transactions on Electron Devices*, 59(1), pp. 72–78, 2012.

39. J. Hållstedt, M. Kolahdouz, R. Ghandi, *et al.*, "Pattern dependency in selective epitaxy of B-doped SiGe layers for advanced metal oxide semiconductor field effect transistors," *Journal of Applied Physics*, 103(5), p. 4907, 2008.

40. M. Kolahdouz, J. Hållstedt, A. Khatibi, *et al.*, "Comprehensive evaluation and study of pattern dependency behavior in selective epitaxial growth of B-doped SiGe layers," *IEEE Transactions on Nanotechnology*, 8(3), pp. 291–297, 2009.

41. G. Wang, M. Moeen, A. Abedin, *et al.*, "Impact of pattern dependency of SiGe layers grown selectively in source/drain on the performance of 22 nm node pMOSFETs," *Solid-State Electronics*, 114(12), pp. 43–48, 2015.

42. C. Qin, G. Wang, M. Kolahdouz, *et al.*, "Impact of pattern dependency of SiGe layers grown selectively in source/drain on the performance of 14 nm node FinFETs," *Solid-State Electronics*, 124(10), pp. 10–15, 2016.

43. M. Johansson, M.Y.A. Yousif, P. Lundgren, *et al.*, "HfO$_2$ gate dielectrics on strained-Si and strained-SiGe layers," *Semiconductor Science and Technology*, 18(9), pp. 820–826, 2003.

44. J.H. Choi, Y. Mao, and J.P. Chang, "Development of hafnium based high-k materials - A review," *Material Science and Engineering: R: Report*, 72(6), pp. 97–136, 2011.

45. Y.B. Zheng, S.J. Wang, and C.H.A. Huan, "Microstructure-dependent band structure of HfO$_2$ thin films," *Thin Solid Films*, 504(1–2), pp. 197–200, 2006.

46. W.S. Hwang, C. Shen, X. Wang, *et al.*, "A novel hafnium carbide HfCx metal gate electrode for NMOS device application." In: *Symposium on VLSI Technology*, pp. 156–157, 2007.

47. S.M. George, "Atomic layer deposition: An overview," *Chemical Reviews*, 110(1), pp. 111–131, 2010.

48. R.L. Puurunen, "Surface chemistry of atomic layer deposition: A case study for the trimethylaluminum/water process," *Journal of Applied Physics*, 97(12), p. 1301, 2005.

49. W.P. Maszara and M.R. Lin, "FinFETs-Technology and circuit design challenges." In: *Proceedings of the ESSCIRC*, pp. 3–8, 2013.

50. T. Mérelle, G. Curatola, A. Nackaerts, *et al.*, "First observation of FinFET specific mismatch behavior and optimization guidelines for SRAM scaling." In: *IEEE International Electron Devices Meeting Technical Digest*, pp. 241–244, 2008.

51. S. Maeda, Y. Ko, J. Jeong, *et al.*, "3 dimensional scaling extensibility on epitaxial source drain strain technology toward Fin FET and Beyond." In: *Symposium on VLSI Technology*, pp. T88–T89, 2013.

52. S.K. Saha, "Modeling process variability in scaled CMOS technology," *IEEE Design and Test of Computers*, 27(2), pp. 8–16, 2010.

53. R. Huang, R. Wang, J. Zhuge, *et al.*, "Characterization and analysis of gate-all-around Si nanowire transistors for extreme scaling." In: *Proceedings of the IEEE Custom Integrated Circuits Conference*, pp. 1–8, 2011.

54. C. Basaran and M. Lin, "Damage mechanics of electromigration in microelectronics copper interconnects," *International Journal of Materials and Structural Integrity*, 1(1/2/3), pp. 16–39, 2007.

55. B.J. Pawlak, R. Duffy, and A. De Keersgieter, "Doping strategies for FinFETs," *Materials Science Forum*, 573/574, pp. 333–338, 2008.

56. J.O. Borland, "Smartphones: Driving technology to more than Moore 3-D stacked devices/chips and more Moore FinFET 3-D Doping with High Mobility Channel Materials from 20/22 nm production to 5/7 nm exploratory Research," *ECS Transactions*, 69(10), pp. 11–20, 2015.

57. W. Vandervorst, P. Eyben, M. Jurzack, *et al.*, "Conformal doping of FINFETs: A fabrication and metrology challenge," *AIP Conference Proceedings of the International Symposium on VLSI Technology, Systems and Applications*, 1066(1), pp. 449–456, 2008.

58. R. Duffy and M. Shayesteh, "FinFET doping; material science, metrology, and process modeling studies for optimized device performance," *AIP Conference Proceedings of the International Conference on Ion Implantation Technology*, 1321(1), pp. 17–22, 2011.

59. B. Colombeau, B. Guo, H.J. Gossmann, *et al.*, "Advanced CMOS devices: Challenges and implant solutions," *Physica Status Solidi (a)*, 211(1), pp. 101–108, 2014.

60. B.S. Wood, F.A. Khaja, B.P. Colombeau, *et al.*, "Fin doping by hot implant for 14 nm FinFET technology and beyond," *ECS Transactions*, 58(9), pp. 249–256, 2013.

61. S. Barraud, V. Lapras, M.P. Samson, *et al.*, "Vertically stacked-nanowires MOSFETs in a replacement metal gate process with inner spacer and SiGe source/drain." In: *IEEE International Electron Devices Meeting Technical Digest*, pp. 464–467, 2016.

62. S. Banna and A. Agarwal, "Pulsed high-density plasmas for advanced dry etching processes," *Journal of Vacuum Science & Technology A*, 30(4), pp. 801–829, 2012.

63. K.J. Kanarik, G. Kamarthy, and R.A. Gottscho, "Plasma etch challenges for FinFET transistors," *Solid State Technology*, 55(3), pp. 15–20, 2012.

64. K. Endo, S. Noda, M. Masahara, *et al.*, "Fabrication of FinFETs by damage-free neutral-beam etching technology," *IEEE Transactions on Electron Devices*, 53(8), pp. 1826–1832, 2006.

65. M. Honda and T. Katsunuma, "Etch challenges and evolutions for atomic-order control." In: *Proceedings of the IEEE 16th International Conference on Nanotechnology*, pp. 448–451, 2016.

66. C. Auth, A. Cappellani, J.-S. Chun, *et al.*, "45 nm High-k + metal gate strain-enhanced transistors." In: *Symposium on VLSI Technology*, pp. 128–129, 2008.

67. F. Sebaai, J.I. Del Agua Borniquel, R. Vos, *et al.*, "Poly-silicon etch with diluted ammonia: Application to replacement gate integration scheme," *Solid State Phenomena*, 145/146, pp. 207–210, 2009.

68. F. Sebaai, A. Veloso, M. Claes, *et al.*, "Poly-silicon wet removal for replacement gate integration scheme: Impact of process parameters on the removal rate," *Solid State Phenomena*, 187, pp. 53–56, 2012.

69. H. Takahashi, M. Otsuji, J. Snow, *et al.*, "Wet etching behavior of poly-Si in TMAH solution," *Solid State Phenomena*, 195, pp. 42–45, 2012.

70. D.M. Knotter, "Etching mechanism of vitreous silicon dioxide in HF-based solutions," *Journal of American Chemical Society*, 122(18), pp. 4345–4351, 2000.

71. H. Kikyuama, N. Miki, K. Saka, *et al.*, "Principles of wet chemical processing in ULSI microfabrication," *IEEE Transactions on Semiconductor Manufacturing*, 4(1), pp. 26–35, 1991.

72. H. Robbins and B. Schwartz, "Chemical etching of silicon: I. The system HF, HNO_3, and H_2O," *Journal of Electrochemical Society*, 106(6), pp. 505–508, 1959.

73. K. Yoshimoto, M.P. Stoykovich, H.B. Cao, *et al.*, "A two-dimensional model of the deformation of photoresist structures using elastoplastic polymer properties," *Journal of Applied Physics*, 96(4), pp. 1857–1865, 2004.

74. N. Tas, T. Sonnenberg, H. Jansen, *et al.*, "Stiction in surface micromachining," *Journal of Micromechanics and Microengineering*, 6(4), pp. 385–397, 1996.

75. G.K. Chang, T.K. Carns, S.S. Rhee, *et al.*, "Selective etching of SiGe on SiGe/Si heterostructures," *Journal of Electrochemical Society*, 138(1), pp. 202–204, 1991.

76. T.K. Cams, M.O. Tanner, and K.L. Wang, "Chemical etching of $Si_{1-x}Ge_x$ in $HF:H_2O_2:CH_3COOH$," *Journal of Electrochemical Society*, 142(4), pp. 1260–1266, 1995.

77. B. Holländer, D. Buca, S. Mantl, and J.M. Hartmann, "Wet chemical etching of Si, $Si_{1-x}Ge_x$, and Ge in $HF:H_2O_2:CH_3COOH$," *Journal of Electrochemical Society*, 157(6), pp. 643–646, 2010.

78. K.W. Ostyn, F. Sebaai, J. Rip, *et al.*, "Selective etch of Si and SiGe for gate all-around device architecture," *ECS Transactions*, 69(8), pp. 147–152, 2015.

79. E.J. Nowak, I. Aller, T. Ludwig, *et al.*, "Turning silicon on its edge [double gate CMOS/FinFET technology]," *IEEE Circuits and Devices Magazine*, 20(1), pp. 20–31, 2004.

80. S.K. Saha, "Modeling statistical dopant fluctuations effect on threshold voltage of scaled JFET devices," *IEEE Access*, 4, pp. 507–513, 2016.

81. J. Gu, J. Keane, S. Sapatnekar, and C. Kim, "Width quantization aware FinFET circuit design." In: *IEEE Custom Integrated Circuits Conference*, pp. 337–340, 2006.

82. T. Sato, Y. Takeishi, and H. Hara, "Mobility anisotropy of electrons in inversion layers on oxidized silicon surfaces," *Physical Review Part B, Condensed Matter*, 4(6), pp. 1950–1960, 1971.

83. M. Yang, E.P. Gusev, M. Ieong, *et al.*, "Performance dependence of CMOS on silicon substrate orientation for ultrathin oxynitride and HfO_2 gate dielectrics," *IEEE Electron Device Letters*, 24(5), pp. 339–341, 2003.

84. M. Kinugawa, M. Kakumu, T. Usami, and J. Matsunaga, "Effects of silicon surface orientation on submicron CMOS devices." In: *IEEE Electron Devices Meeting Technical Digest*, pp. 581–584, 1985.

85. S. Takagi, A. Toriumi, M. Iwase, and H. Tango, "On the universality of inversion layer mobility in Si MOSFETs: Part II – Effects of surface orientation," *IEEE Transactions on Electron Devices*, 41(12), pp. 2363–2368, 1994.

86. L. Chang, M. Ieong, and M. Yang, "CMOS circuit performance enhancement by surface orientation optimization," *IEEE Transactions on Electron Devices*, 51(10), pp. 1621–1627, 2004.

87. L. Chang, Y.-K. Choi, D. Ha, *et al.*, "Extremely scaled silicon nano-CMOS Devices," *Proceeding of the IEEE*, 91(11), pp. 1860–1873, 2003.

10 FinFET Compact Models for Circuit Simulation

10.1 INTRODUCTION

This chapter presents an overview of compact model for fin field-effect transistor (FinFET) devices. Compact model of an integrated circuit (IC) element is the mathematical description of its characteristics that are used for computer-aided design (CAD) and analysis of ICs. In essence, *compact model* of an IC manufacturing technology describes the characteristics of its active and passive devices by a set of physics-based analytical expressions with technology-dependent model parameters that are solved by a circuit simulator like SPICE (*Simulation Program with Integrated Circuit Emphasis*) [1] for the design and analysis of ICs. Whereas, *compact modeling* is an art of generating compact model of an IC manufacturing technology by extracting the technology-dependent device model parameters of the circuit elements for the design and analysis of very large scale integrated (VLSI) circuits [2]. In reality, a composite compact model of a VLSI manufacturing technology includes a set of models for transistors along with their parasitic components that runs robustly for realistic assessment of the technology in circuit CAD [2–4]. Today, compact model is the most important part of the process design kit (PDK) [2,5,6] which is the interface between the circuit designers and process technologists. Thus, the compact model for circuit CAD is the bridge between the circuit design and process technology groups [2,5–7]. An effective compact model must accurately capture all real-device effects and simultaneously produce them in a form suitable for maintaining high computational efficiency [5,8]. Thus, the compact model of a manufacturing technology includes models for both transistor devices as well as interconnection layers [2,9,10]. However, the objective of this chapter is to provide a brief overview of compact FinFET device model only. Again, in VLSI circuits and systems, FinFET devices can be used in two different configurations: common multiple-gate, referred to as the *multigate*, where a common gate terminal is used to bias the devices and the gate dielectric thicknesses is the same for all gates; or as an independent multigate where all gates are independently biased and the gate dielectric thickness may also be different for each gate [2,5]. In this chapter, we derive compact FinFET model for common multigate devices only.

10.2 COMPACT DEVICE MODEL

Before discussing the formulation of compact FinFET model for circuit CAD, let us overview the general composition of compact model. Compact device model of an IC device describes its terminal behavior in terms of the current-voltage ($I - V$), capacitance-voltage ($C - V$), and the carrier transport processes within the device.

Figure 10.1 shows the basic features of a typical compact device model of a representative IC technology. As shown in Figure 10.1, a compact model is made of a *core model* along with the various models to account for the effects of geometry and physical phenomena in the device. For a FinFET, the core model describes $I - V$ and $C - V$ behavior of an ideal large geometry isolated device [5,8,10] of a target IC fabrication technology. The core model represents about 20% of the overall model code both in terms of execution time as well as the number of lines in the code [10]. The rest of the model code comprises multiple models that describe the numerous real-device effects that are responsible for the accuracy of the compact model.

For FinFET devices, the real device effects accompanying the core model include short-channel effects (SCEs), quantum mechanical (QM) effects, output conductance, leakage currents, band-to-band tunneling, noise, process variability, non-quasi-static (NQS) effect, and strain effect as illustrated in Figure 10.1 [2,5,8,10]. Thus, in deriving the compact FinFET model, first of all, we formulate the core FinFET device model and then discuss the mathematical formulation of real device effects to present a complete compact FinFET model for circuit CAD. In Section 10.3.1, we present the core compact model for common multigate FinFET devices. In deriving the compact FinFET device model, all the theory developed in Chapters 3 and 5 is applicable, with appropriate modifications for the channel doping concentration and applied terminal voltages.

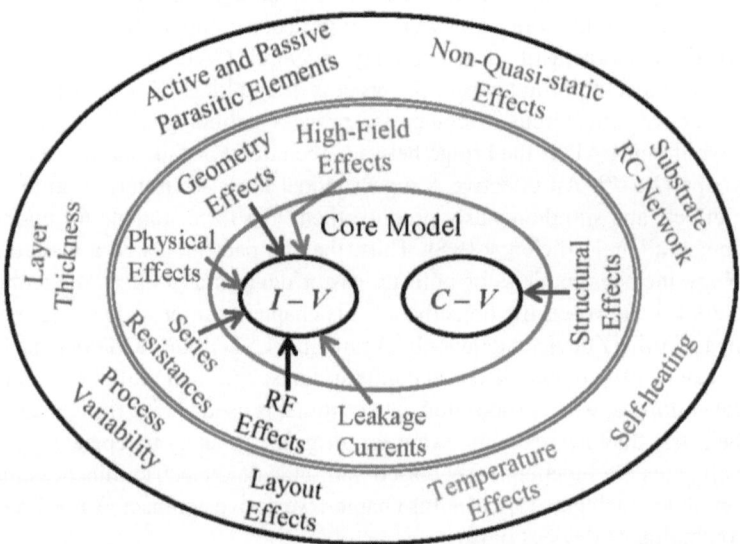

FIGURE 10.1 A typical composition of compact models of an IC technology: the core model includes the basic $I - V$ and $C - V$ characteristics of a large geometry device shown in the inner circle; the core model is accompanied by the models for physical phenomena affecting the device and models for geometry and structural effects as shown in the middle circle; the models for external phenomena including the ambient temperature, layout effects, process variability, and non-quasi-static effects of the complete compact device model are shown in the outer circle.

10.3   COMMON MULTIPLE-GATE COMPACT FinFET MODEL

In common multigate configuration, all gates of a FinFET device are electrically interconnected and biased at the same electrical terminal voltage. It is also assumed that the gate workfunctions and the gate dielectric thicknesses are the same. However, the carrier mobilities at the inversion condition are dependent on the crystal orientations and/or strain level [11].

10.3.1   CORE MODEL

The core model is derived using gradual-channel approximation (GCA) [12] described in Chapter 5. It is also assumed that the physical effects such as mobility degradation on device performance can, safely, be neglected. In compact FinFET core model formulation, both the charge-based [13] and surface potential-based [14,15] modeling approaches have been used. The core model described in the following section is based on the solution of Poisson and drift-diffusion equations for a long-channel double-gate (DG) FinFET device with moderately doped channel [16]. The model is shown to predict accurate device performance of FinFET devices at the nanometer nodes [15,17].

10.3.1.1   Electrostatics

For the simplicity of model formulation, let us consider the two-dimensional (2D) cross-section of an ideal double-gate (DG) n-type FinFET, hereafter referred to as the nFinFET device structure as a common multigate transistor as shown in Figure 10.2. First of all, we obtain the electrostatic potential, $\phi(x, y)$ at any point (x, y) by solving 2D Poisson's equation given by [Equation 3.37]

$$\frac{d^2\phi(x,y)}{dx^2} = -\frac{q}{\varepsilon_{si}}\left[p(x,y) - n(x,y) + N_d^+(x,y) - N_a^-(x,y) \right] \qquad (10.1)$$

where:

    q is the magnitude of the electronic charge

    ε_{si} is the permittivity of silicon fin channel

    $p(x,y)$, $n(x,y)$, $N_d^+(x,y)$, and $N_a^-(x,y)$ are the hole, electron, ionized donor, and ionized acceptor concentrations at any point (x,y) of the fin channel, respectively

Now, from Equation 3.50, the minority carrier concentration, $n(x, y)$ at any point (x, y) of a p-type substrate is given by

$$n(x,y) = n_i \exp\left(\frac{\phi(x,y) - \phi_B}{v_{kT}} \right) \qquad (10.2)$$

where:

    n_i is the intrinsic carrier concentration

    ϕ_B is the bulk potential

    v_{kT} is the thermal voltage given by kT/q with k and T being the Boltzmann constant and ambient temperature, respectively

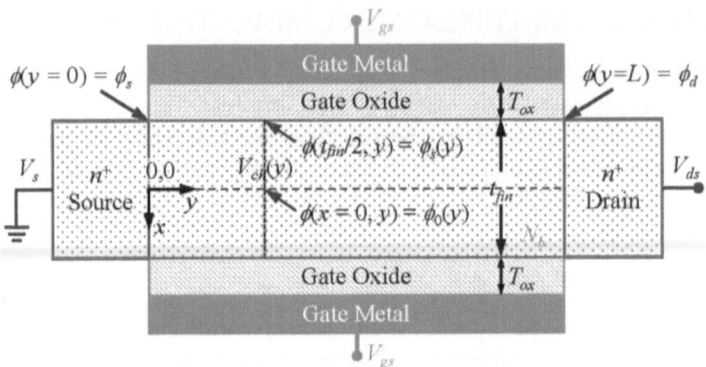

FIGURE 10.2 Schematic of an ideal symmetric common DG-nFinFET device used to formulate device model: T_{ox}, t_{fin}, and N_b are the gate oxide thickness, fin or body thickness, and body doping concentration, respectively; the origin of the coordinate system $(0,0)$ is at the center at $(L = 0, t_{fin}/2)$; ϕ_s and ϕ_d are the surface potentials at the source and drain ends of the channel, respectively.

Typically, FinFETs are undoped (not intrinsic) or lightly doped channel devices; therefore, we consider only the inversion carrier electron concentration $n(x, y)$ at any point (x, y) given by Equation 10.2 and uniformly doped p-type body doping concentration, $N_a(x, y) \equiv N_b$ (as the negatively charged ionized acceptors). Now, let us assume that $V_{ch}(y)$ is the channel potential at any point y as described in Section 5.3 and assume GCA is valid [12]. Then for a DG-nFinFET device shown in Figure 10.2, we can express Equation 10.1 as

$$\frac{d^2\phi(x, y)}{dx^2} = \frac{q}{\varepsilon_{si}}\left[n_i \exp\left(\frac{\phi(x, y) - \phi_B - V_{ch}(y)}{v_{kT}}\right) + N_b\right] \tag{10.3}$$

where:

$$V_{ch}(y) = \begin{cases} V_s; & \text{at } y = 0 \quad \text{(source-end)} \\ V_d; & \text{at } y = L \quad \text{(drain-end)} \end{cases} \tag{10.4}$$

In Chapter 5, we solved for $\phi(x, y)$ by integrating Poisson's equation twice for an undoped channel, $N_b \approx 0$. However, due to the finite doping concentration N_b, Equation 10.3 cannot be integrated. Therefore, for the simplicity of an analytical solution, we assume that the contribution of N_b is a small perturbation to the overall electrostatic potential $\phi(x, y)$ at any point (x, y) so that we can write

$$\phi(x, y) \cong \phi_1(x, y) + \phi_2(x, y) \tag{10.5}$$

where:
  $\phi_1(x, y)$ is the contribution to $\phi(x, y)$ due to the inversion carriers without the effect of the ionized body dopants
  $\phi_2(x, y)$ is a small contribution to $\phi(x, y)$ due to body dopants N_b

With the above described assumption, we can write Poisson's equation for potential $\phi_1(x, y)$ as

$$\frac{d^2\phi_1(x,y)}{\partial x^2} = \frac{qn_i}{\varepsilon_{si}}\exp\left(\frac{\phi_1(x,y)-\phi_B-V_{ch}(y)}{v_{kT}}\right) \tag{10.6}$$

And Poisson's equation for potential $\phi_2(x, y)$ as

$$\frac{d^2\phi_2(x,y)}{\partial x^2} = \frac{qN_b}{\varepsilon_{si}} \tag{10.7}$$

Now, if the thickness of the fin t_{fin} is less than the width of the depletion region by the gate, then for a certain gate bias V_{gs}, the silicon fin is fully depleted and consequently, the inversion carriers will spread through the entire body. Under this condition, the inversion charge, $Q_i \gg Q_b$ (bulk charge) and therefore, we can safely neglect the term containing N_b in Equation 10.3 and the channel potential is obtained by solving Equation 10.6 following the procedure described in Section 5.4.1.

Next, we integrate Equation 10.6 from the center point $(x = 0, y)$ to a point (x, y) to find the potential $\phi_1(x, y)$ along the thickness of the structure shown in Figure 10.2. Since for a symmetric DG-FinFET structure the vertical component of the electric field E_x is zero at the center, that is, at $x = 0$, $d\phi_1/dx = 0$ and $\phi_1(x = 0, y) = \phi_0(y)$, we use the following boundary conditions to integrate Equation 10.6

$$\begin{cases} \phi_1(x,y) = \phi_0(y), \dfrac{d\phi_1(x,y)}{dx} = 0; & \text{at } x = 0 \\[2ex] \phi_1(x,y) = \phi_1(x), \dfrac{d\phi_1(x,y)}{dx} = \dfrac{d\phi_1(x,y)}{dx}; & \text{at } x = x \end{cases} \tag{10.8}$$

Then using the boundary conditions in Equation 10.8 and following the procedure described in Section 5.4.1, we get $\phi_1(x, y)$ by integrating Equation 10.6 twice as

$$\phi_1(x,y) = \phi_0(y) - 2v_{kT}\ln\left[\cos\left(\sqrt{\frac{q}{2\varepsilon_{si}v_{kT}}\frac{n_i^2}{N_b}\exp\left(\frac{\phi_0(y)-V_{ch}(y)}{v_{kT}}\right)}\cdot x\right)\right] \tag{10.9}$$

where:

$\phi_0(y)$ is the potential at the center of the body as shown in Figure 10.2

ϕ_B is expressed in terms of n_i and N_b using Equation 3.49, $n_i\exp\left(-\phi_B/v_{kT}\right) = n_i^2/N_b$

In order to solve for $\phi_2(x, y)$, we use Equation 3.55 to express Equation 10.7 as

$$\frac{d}{dx}\left(\frac{d\phi_2(x,y)}{\partial x}\right)^2 = 2\frac{qN_b}{\varepsilon_{si}}\frac{d\phi_2(x,y)}{dx} \tag{10.10}$$

Similarly, we can solve Equation 10.10 for $\phi_2(x, y)$ by integrating twice and applying the boundary conditions given by

$$\begin{cases} \phi_2(x,y) = 0, \dfrac{d\phi_2(x,y)}{dx} = 0; & \text{at } x = 0 \\[2mm] \phi_2(x,y) = \phi_2(x,y), \dfrac{d\phi_2(x,y)}{dx} = \dfrac{d\phi_2(x,y)}{dx}; & \text{at } x = x \end{cases} \qquad (10.11)$$

Now, integrating Equation 10.10 using the boundary conditions in Equation 10.11 we get

$$\int_0^{\frac{d\phi_2}{dx}} d\left(\frac{\partial\phi_2(x,y)}{\partial x}\right)^2 = 2\frac{qN_b}{\varepsilon_{si}}\int_0^{\phi_2(x,y)} d\phi_2(x,y) \qquad (10.12)$$

After integration and simplification, we get from Equation 10.12

$$\frac{d\phi_2(x,y)}{\sqrt{\phi_2(x,y)}} = \pm\sqrt{\frac{2qN_b}{\varepsilon_{si}}}\cdot dx \qquad (10.13)$$

Then integrating Equation 10.13, from the center point, $x = 0$, $\phi_2(x = 0, y) = 0$ to any point x, $\phi_2(x, y)$ in the fin, we can show after simplification

$$\phi_2(x,y) = \frac{qN_b x^2}{2\varepsilon_{si}} \qquad (10.14)$$

Thus, from Equation 10.5, the surface potential $\phi_s(y)$ at any point y along the surface of the channel is obtained by evaluating the sum of $\phi_1(x, y)$ [Equation 10.9] and $\phi_2(x, y)$ [Equation 10.14] at the surface ($x = -t_{fin}/2$) such that

$$\phi_s(x = -t_{fin}/2, y) \cong \phi_1\left(-\frac{t_{fin}}{2}, y\right) + \phi_2\left(-\frac{t_{fin}}{2}, y\right) \qquad (10.15)$$

In order to derive an expression relating $\phi_s(y)$ and the gate voltage V_{gs}, we use the boundary condition given by Equation 5.53 as

$$\varepsilon_{ox}\frac{V_{gs} - V_{fb} - V_{ch}(y) - \phi(x = \pm t_{fin}/2)}{T_{ox}} = \pm\varepsilon_{si}\frac{d\phi}{dx}\bigg|_{x=\pm t_{fin}/2} \qquad (10.16)$$

where:
   ε_{ox} is the permittivity of gate oxide
   V_{fb} = flat band voltage of the gate oxide fin system
   T_{ox} = gate oxide thickness

Now, we can find $d\phi/dx$ by solving Equation 10.3, following the procedure used in Section 5.4.1 and using Equation 3.55, we can show for a DG-FinFET device

$$\frac{d}{dx}\left(\frac{d\phi}{dx}\right)^2 = \frac{2q}{\varepsilon_{si}}\left[n_i\exp\left(\frac{\phi(x,y) - \phi_B - V_{ch}(y)}{V_{kT}}\right) + N_b\right]\frac{d\phi}{dx} \qquad (10.17)$$

We integrate Equation 10.17 from center potential $\phi(x = 0, y)$ to any point $\phi(x, y)$ using the limits of integration

$$
\begin{cases}
\phi(x, y) = 0, \dfrac{d\phi(x, y)}{dx} = 0; & \text{at } x = 0 \\[2ex]
\phi(x, y) = \phi(x, y), \dfrac{d\phi(x, y)}{dx} = \dfrac{d\phi(x, y)}{dx}; & \text{at } x = x
\end{cases}
\tag{10.18}
$$

Therefore, using Equation 10.18 we get from Equation 10.17

$$
\int_0^{\frac{d\phi}{dx}} d\left(\frac{d\phi}{dx}\right)^2 = \frac{2qn_i}{\varepsilon_{si}} \int_{\phi_0(y)}^{\phi(x,y)} \left[\exp\left(\frac{\phi(x, y) - \phi_B - V_{ch}(y)}{v_{kT}} \right) + \frac{N_b}{n_i} \right] d\phi
\tag{10.19}
$$

Now, we get from Equation 3.43

$$
\frac{N_b}{n_i} = \exp\left(\frac{\phi_B}{v_{kT}} \right)
\tag{10.20}
$$

Then substituting for N_b/n_i from Equation 10.20 into Equation 10.19, we get after integration of Equation 10.19 and simplification

$$
\left(\frac{d\phi(x, y)}{dx} \right)^2 = \frac{2qn_i}{\varepsilon_{si}}
\left[
\begin{array}{l}
v_{kT} \exp\left(\left(\dfrac{\phi_s(y)}{v_{kT}} \right) - \exp\dfrac{\phi_0(y)}{v_{kT}} \right) \exp\left(\dfrac{-\phi_B - V_{ch}(y)}{v_{kT}} \right) \\[2ex]
+ \exp\left(\dfrac{\phi_B}{v_{kT}} \right)(\phi_s(y) - \phi_0(y))
\end{array}
\right]
\tag{10.21}
$$

Thus, the vertical electric field at any point y along the surface of the channel is given by

$$
\left. \frac{d\phi}{dx} \right|_{x = \pm \frac{t_{fin}}{2}} = \sqrt{\frac{2qn_i}{\varepsilon_{si}} \left[v_{kT} \left(\exp\left(\frac{\phi_s(y)}{v_{kT}} \right) - \exp\left(\frac{\phi_0(y)}{v_{kT}} \right) \right) \exp\left(\frac{-\phi_B - V_{ch}(y)}{v_{kT}} \right) + \exp\left(\frac{\phi_B}{v_{kT}} \right)(\phi_s(y) - \phi_0(y)) \right]}
\tag{10.22}
$$

Combining Equations 10.16 and 10.22, we get

$$
V_{gs} = V_{fb} + \phi_s(y) + \frac{\varepsilon_{si}}{C_{ox}} \sqrt{\frac{2qn_i}{\varepsilon_{si}} \left[v_{kT} \left(\exp\left(\frac{\phi_s(y)}{v_{kT}} \right) - \exp\left(\frac{\phi_0(y)}{v_{kT}} \right) \right) \cdot \exp\left(\frac{-\phi_B - V_{ch}(y)}{v_{kT}} \right) + \exp\left(\frac{\phi_B}{v_{kT}} \right) \cdot (\phi_s(y) - \phi_0(y)) \right]}
\tag{10.23}
$$

Equations 10.15 and 10.23 represent a self-consistent system of equations that can be solved to obtain $\phi_0(y)$ and $\phi_s(y)$ for a fully depleted DG-FinFET structure under a set of external biases.

In a partially depleted DG-FinFET, the depletion width X_d is bias-dependent. At the edge of the depletion region, $\phi_1(x = X_d, y) = 0$. With these changes, the surface potential can be derived for the partially depleted devices similar to the fully depleted devices. It can be shown that for the partially depleted body

$$\phi_1\left(x = \frac{t_{si}}{2}, y\right) = -2v_{kT} \cdot \ln\left[\cos\left(\sqrt{\frac{qn_i}{2\varepsilon_{si}v_{kT}}} \exp\left(\frac{-\phi_B - V_{ch}(y)}{v_{kT}}\right) \cdot \frac{X_d}{2}\right)\right] \tag{10.24}$$

And

$$V_{gs} = V_{fb} + \phi_s(y) + \frac{\varepsilon_{si}}{C_{ox}} \sqrt{\frac{2qn_i}{\varepsilon_{si}} \left[\begin{array}{c} v_{kT}\left(\exp\left(\dfrac{\phi_s(y)}{v_{kT}}\right) - 1\right) \cdot \exp\left(\dfrac{-\phi_B - V_{ch}(y)}{v_{kT}}\right) \\[2ex] + \exp\left(\dfrac{\phi_B}{v_{kT}}\right) \cdot \phi_s(y) \end{array} \right]} \tag{10.25}$$

Equation 10.15 along with $\phi_1(x)$ from Equation 10.24 and Equation 10.25 represents a self-consistent system of equations for partially depleted fin that can be solved to obtain $\phi_s(y)$ for a fully depleted DG-FinFET structure under a set of external biases. However, for the convenience of mathematical formulation, let us derive a single surface potential function following the procedure described in Section 3.4.2.4.

Surface potential function: In order to obtain continuous expressions for terminal currents and charges, it is necessary to capture the transition between the fully depleted and partially depleted regimes in a smooth manner. Also, the solution of Equations 10.24 and 10.25 is computationally intensive due to complex $\phi_2(x, y)$ term. To overcome these issues, a simplified expression for $\phi_2(x, y) = \phi_{pert}$ which is continuous between the partially depleted and fully depleted regimes is used as a small perturbation term. Thus, using ϕ_{pert}, a surface potential in both the regimes is calculated through a single continuous equation. As described in Chapter 5, the transformation variable β is the argument of the cosine function in $\phi_1(t_{fin}/2, y)$ in Equation 10.9

$$\beta \equiv \frac{t_{fin}}{2} \sqrt{\frac{q}{2\varepsilon_{si}v_{kT}} \frac{n_i^2}{N_b} \exp\left(\frac{\phi_0(y) - V_{ch}(y)}{v_{kT}}\right)}. \tag{10.26}$$

and, from Equation 10.14, $\phi_{pert} \equiv \phi_2(t_{fin}/2, y)$ is given by

$$\phi_{pert} \equiv \phi_2\left(\frac{t_{fin}}{2}, y\right) = \frac{qN_b}{2\varepsilon_{si}} \frac{t_{fin}^2}{4} \tag{10.27}$$

Thus, through a change of variable and following the procedure described in Section 5.4.1, the unified surface potential ϕ_s equation can be written as

$$f(\beta) \equiv \ln(\beta) - \ln\big(\cos(\beta)\big) - \frac{V_{gs} - V_{fb} - V_{ch}(y)}{2v_{kT}} + \ln\left(\frac{2}{t_{fin}}\sqrt{\frac{2\varepsilon_{si}v_{kT}}{qn_i}}\right)$$

$$+ \frac{2\varepsilon_{si}}{t_{fin}C_{ox}}\sqrt{\beta^2\left[\frac{\exp\left(\dfrac{\phi_{pert}}{v_{kT}}\right)}{\cos^2(\beta)} - 1\right] + \frac{\phi_{pert}}{v_{kT}^2}\Big[\phi_{pert} - 2v_{kT}\ln\big(\cos\beta\big)\Big]} = 0 \qquad (10.28)$$

Equation 10.28 (implicit in β) is the basic surface potential equation for modeling common multigate FinFETs [5,18]. It is solved by first using an analytical approximation for the initial guess [17], followed by two Householder's cubic iterations (3rd order Newton-Raphson iterations); together these make the model numerically robust and accurate. The surface potentials at the source end ϕ_{s0} and drain end ϕ_{sL} are calculated by setting $V_{ch}(y = 0) = V_s$ and $V_{ch}(y = L) = V_d$, respectively. For a lightly doped body, Equation 10.28 can be further simplified [19] to speed up the simulation as shown below.

If ϕ_{pert} (given by Equation 10.27) ≈ 0, we get $\exp\left(\dfrac{\phi_{pert}}{v_{kT}}\right) = 1$. Then in Equation 10.28

$$\left[\frac{\exp\left(\dfrac{\phi_{pert}}{v_{kT}}\right)}{\cos^2(\beta)} - 1\right] = \frac{1}{\cos^2(\beta)} - 1 = \tan^2\beta$$

$$(10.29)$$

And, $$\frac{\phi_{pert}}{v_{kT}^2}\Big[\phi_{pert} - 2v_{kT}\ln\big(\cos\beta\big)\Big] \approx 0$$

Therefore, for $\phi_{pert} = 0$, Equation 10.29 shows that the surface potential function given by Equation 10.28 becomes the same as the undoped body FinFETs given by Equation 5.58, given as

$$\ln(\beta) - \ln\big(\cos(\beta)\big) - \frac{V_{gs} - V_{fb} - V_{ch}(y)}{2v_{kT}} + \ln\left(\frac{2}{t_{fin}}\sqrt{\frac{2\varepsilon_{si}v_{kT}}{qn_i}}\right) + 2r\beta\tan\beta = 0 \quad (10.30)$$

where:
r is structural parameter of a FinFET device defined in Equation 5.57 and is given by

$$r = \frac{\varepsilon_{si}}{C_{ox}t_{fin}} = \frac{\varepsilon_{si}T_{ox}}{\varepsilon_{ox}t_{fin}} \qquad (10.31)$$

A separate surface potential expression is used for the cylindrical gate geometry [20].

10.3.1.2 Drain Current Model

In Chapter 5, the expression for drain to source current I_{ds} for long-channel DG-FinFETs is derived from the solution of drift-diffusion equation using Fermi potential and is given by Equation 5.22. Thus, from Equation 5.22, we can write

$$I_{ds} = \left(\frac{W}{L}\right)\mu(T)\int_{Q_{is}}^{Q_{id}} Q_i(y)\left(\frac{dV_{ch}}{dQ_i}\right)dQ_i \qquad (10.32)$$

where:
  $\mu(T)$ is the low-field and temperature-dependent mobility
  W is the total effective width
  L is the effective channel length
  Q_i is the inversion charge per unit area in the upper half part of the body
  Q_{is} is the inversion charge density at the source end of the channel at $y = 0$
  Q_{id} is the inversion charge density at the drain end of the channel at $y = L$

Now, if Q_b is the bulk charge density in the body, then the total induced charge density in the semiconductor, $Q_s = (Q_i + Q_b)$. Therefore, using Equation 3.30 we get

$$Q_{is} = -C_{ox}\left(V_{gs} - V_{th} - \phi_{s0}\right) - Q_b$$

$$Q_{id} = -C_{ox}\left(V_{gs} - V_{th} - \phi_{sL}\right) - Q_b \qquad (10.33)$$

Again, from Gauss' law and Equation 10.22, we can show the total charge in the fin channel as

$$Q_s(y) = \varepsilon_{si}\frac{d\phi}{dx}\Big|_{x=t_{fin}/2} = \sqrt{2qn_i\varepsilon_{si}\left[v_{kT}\left(e^{\frac{\phi_s(y)}{v_{kT}}} - e^{\frac{\phi_0(y)}{v_{kT}}}\right)\cdot e^{\frac{-\phi_B - V_{ch}(y)}{v_{kT}}} + e^{\frac{\phi_B}{v_{kT}}}\cdot\left(\phi_s(y) - \phi_0(y)\right)\right]} \qquad (10.34)$$

Note that the second term within the square brackets is the bulk charge due to doping concentration N_b [Equation 10.17]. For a lightly doped body at inversion condition $Q_b \ll Q_i$; therefore, neglecting the bulk charge term in Equation 10.34, we can express the inversion charge as

$$Q_i(y) \approx \sqrt{2qn_i\varepsilon_{si}\left[v_{kT}\left(e^{\frac{\phi_s(y)}{v_{kT}}} - e^{\frac{\phi_0(y)}{v_{kT}}}\right)\cdot e^{\frac{-\phi_B - V_{ch}(y)}{v_{kT}}}\right]} \qquad (10.35)$$

Equation 10.35 can be further simplified as

$$Q_i(y) = \sqrt{2qn_i\varepsilon_{si}v_{kT}}\, e^{\frac{\left[\phi_s(y) - \phi_B - V_{ch}(y)\right]}{2v_{kT}}}\sqrt{1 - e^{\frac{\phi_0(y) - \phi_s(y)}{v_{kT}}}} \qquad (10.36)$$

In *strong inversion*, $\phi_s(y) \gg \phi_0(y)$, therefore, $\sqrt{1 - e^{\frac{\phi_0(y) - \phi_s(y)}{v_{kT}}}}$ approaches 1.

In *weak inversion*, we can simplify, $\sqrt{1 - e^{\frac{\phi_0(y) - \phi_s(y)}{v_{kT}}}}$ assuming liner potential profile from $x = 0$ to $x = -t_{fin}/2$. Thus, if E_{avg} is the average electric field in the region between $x = -t_{fin}/2$ to the center potential at $x = 0$, then using Gauss' law, we can write:

$$E_{avg} = -\frac{d\phi(y)}{dx} = \frac{Q_i}{\varepsilon_{si}} \tag{10.37}$$

Now, assuming that surface potential varies linearly from center potential $\phi_0(y)$ to the surface potential $\phi_s(y)$, then Equation 10.37 can be expressed as

$$-\frac{d\phi(y)}{dx} = \frac{\phi_s(y) - \phi_0(y)}{t_{fin}/2} = \frac{Q_i}{\varepsilon_{si}} \tag{10.38}$$

Then from Equation 10.38, the inversion charge is given by

$$\phi_s(y) - \phi_0(y) = \frac{Q_i}{(2\varepsilon_{si})/t_{fin}} = \frac{Q_i}{2C_{si}} \tag{10.39}$$

where:
$C_{si} = \varepsilon_{si}/t_{fin}$ is the capacitance of the silicon fin of a FinFET device

Now, substituting Equation 10.39 into Equation 10.36 and using Taylor's series expansion described in Section 3.4.2.5, the inversion charge for lightly doped DG-FinFETs is given by

$$Q_{i,LD}(y) \cong \sqrt{2qn_i\varepsilon_{si}v_{kT}} \cdot \exp\left(\frac{(\phi_s(y) - \phi_B - V_{ch}(y))}{2v_{kT}}\right) \cdot \sqrt{\frac{Q_{i,LD}(y)}{Q_{i,LD}(y) + 2C_{si}v_{kT}}} \tag{10.40}$$

Equation 10.40 is an implicit equation in $Q_{i,LD}(y)$ and must be solved iteratively to obtain drain current from Equation 10.32. Using $Q_s \approx Q_{i,LD}(y)$ into Equation 10.30, we can compute V_{gs} versus inversion charge density for a lightly doped body, $Q_{i,LD} = -C_{ox}(V_{gs} - V_{fb} - \phi_s)$.

Similarly, following the procedure described in Section 3.4.2.5, the inversion charge density for heavily doped DG-FinFETs can be shown as

$$Q_{i,HD}(y) \approx \sqrt{2qn_i\varepsilon_{si}v_{kT}} \cdot \exp\left(\frac{(\phi_s(y) - \phi_B - V_{ch}(y))}{2v_{kT}}\right) \cdot \sqrt{\frac{Q_{i,HD}(y)}{Q_{i,HD}(y) + 2Q_b}} \tag{10.41}$$

From the similarities of charge expressions in Equations 10.40 and 10.41, a unified expression is used to calculate the inversion charge density for a wide range of devices as a function of Q_b and is given by [17]

$$Q_i(y) = \sqrt{2qn_i\varepsilon_{si}v_{kT}} \cdot \exp\left(\frac{(\phi_s(y) - \phi_B - V_{ch}(y))}{2v_{kT}}\right)\sqrt{\frac{Q_i(y)}{Q_i(y) + Q_0}}, \qquad (10.42)$$

where:

$Q_0 = 2Q_b + 5C_{si}v_{kT}$; the factor "5" is used for accurate modeling of inversion charge

The reported data show that the unified charge density model agrees very well with inversion charge density calculated using exact equation for wide range of body doping concentration [15]. Now, in order to calculate I_{ds} from Equation 10.32, we calculate $dV_{ch}(y)/dQ_i$ from Equation 10.42 for Q_i [15] as

$$\frac{dV_{ch}}{dy} = \frac{d\phi_s}{dy} + v_{kT}\frac{dQ_i}{dy}\left(\frac{Q_0}{Q_i(Q_i + Q_0)} - \frac{2}{Q_i}\right) \qquad (10.43)$$

Then from Equation 10.33, we can show: $(d\phi_s/dy) = (1/C_{ox}) \times (dQ_i/dy)$. Therefore, we get

$$\frac{dV_{ch}}{dQ_i} = -\frac{1}{C_{ox}} + v_{kT}Q_0\frac{1}{Q_i(Q_i + Q_0)} - \frac{2v_{kT}}{Q_i} \qquad (10.44)$$

Now substituting dV_{ch}/dQ_i from Equation 10.44 into Equation 10.32 we obtain the following basic equation for I_{ds}

$$I_{ds} = \left(\frac{W}{L}\right)\mu(T) \cdot \left[\frac{Q_{is}^2 - Q_{id}^2}{2C_{ox}} + 2v_{kT}(Q_{is} - Q_{id}) - v_{kT}Q_0\ln\left(\frac{Q_0 + Q_{is}}{Q_0 + Q_{id}}\right)\right] \qquad (10.45)$$

Equation 10.45 describes the continuous drain current model for symmetric common multigate FinFET devices. The model equation predicts the drain current in all operation regions: subthreshold, linear, and saturation of both fully depleted and lightly depleted channel symmetric DG-FETs. Figures 10.3 shows the simulated and measured $I - V$ characteristics of bulk-FinFET devices obtained by common multigate drain current model [21].

10.3.2 Modeling Real Device Effects

In Section 10.3.1, the core I_{ds} model for common multigate FinFET devices is derived considering an ideal large geometry FinFET device unaffected by the structural or physical phenomena. However, as described in Chapter 6, different physical phenomena, as well as small geometry effects, significantly affect the performance of real FinFET devices. Thus, in this section a brief overview of modeling real-device effects on common multigate FinFETs is presented highlighting the key physical phenomena and geometrical effects.

10.3.2.1 Short-Channel Effects

As described in Section 6.2, the short-channel effects (SCEs) originate from 2D electrostatics where the drain significantly affects the potential barrier at the source

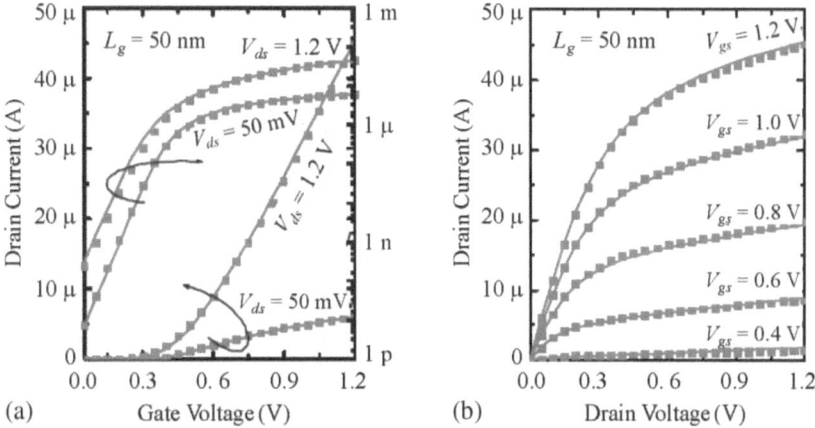

FIGURE 10.3 Drain current modeling for moderately doped symmetric bulk-FinFET devices: (a) $I_{ds} - V_{gs}$ characteristics for different V_{ds}; (b) $I_{ds} - V_{ds}$ characteristics for different V_{gs}. Here, $L = 50$ nm, $t_{fin} = 25$ nm, and TiN gate with equivalent $T_{ox} = 1.95$ nm; symbols are measured data and lines represent compact drain current model [21].

due to its close proximity to source region in short-channel devices. SCEs degrade the device performance through V_{th} roll-off, drain-induced barrier lowering (DIBL), and subthreshold slope (S) degradation. There are several approaches to model SCEs [22–26]. However, as discussed in Chapter 6, the approach assuming a parabolic potential function perpendicular to the silicon/insulator interface to solve 2D Poisson's equation is shown to maintain a balance between the model accuracy and model computation time [24,25].

In Section 6.2.2, we showed that to model SCEs, 2D Poisson's equation is solved in the x-direction along the fin thickness and in the y-direction along the length of the channel, assuming that the inversion charge is negligible and the electric field E_x is independent of y, whereas the electric field E_y is independent of x. Then assuming a parabolic potential distribution along the x-direction, the minimum potential at the center of the channel $\phi_c(y)$ is determined [26]. Then the minimum potential ϕ_{csl} [17] is expressed in terms of the terminal voltages V_{gs} and V_{ds}, L, and the characteristic field-penetration length λ given in Equation 6.20 as

$$\lambda \equiv \sqrt{\frac{\varepsilon_{si}}{2\varepsilon_{ox}}\left(1+\frac{\varepsilon_{ox}t_{fin}}{4\varepsilon_{si}T_{ox}}\right)t_{fin}T_{ox}} \tag{10.46}$$

As discussed in Chapter 6, the parameter λ is known as the scale length that defines the extent of penetration of the electric field from the drain into the fin channel as a function of physical parameters T_{ox} and t_{fin} and, therefore, the amount of SCE in a transistor as discussed below.

V_{th} *roll-off*: Threshold voltage V_{th} roll-off due to SCE as discussed in Section 6.2.3 is modeled by Equation 6.31 and is given by

$$\Delta V_{th,SCE} = -\frac{\left(V_{bi} - \phi_{st}\right)}{\cosh\left(\dfrac{L}{2\lambda}\right) - 1} \tag{10.47}$$

where:

V_{bi} is the built-in potential of the source-drain to body pn-junction described in Section 2.3.2

ϕ_{st} is the surface potential at the source end in the subthreshold region

L is the channel length of the device

From the discussions in Chapter 3.2.2, we can show the value of $\phi_{st} \cong E_g/2$, E_g the energy gap of the silicon fin of the FinFET devices. The term $\Delta V_{th,SCE}$ is further modified with additional fitting parameters to improve the modeling accuracy [27–30].

Figure 10.4 shows the dependence of ΔV_{th} on the gate oxide thickness and silicon body thickness. As the oxide thickness and body thickness decrease the gate control on the body increases, thus suppressing SCE as expected [21].

DIBL effect on V_{th}: Threshold voltage V_{th} degradation due to DIBL as discussed in Section 6.2.3 is modeled by Equation 6.32 given by

$$\Delta V_{th,DIBL} = -\frac{V_{ds}}{2\left[\cosh\left(\dfrac{L}{2\lambda}\right) - 1\right]} \tag{10.48}$$

where:

V_{ds} is the applied drain voltage to the FinFET devices

Again, the term $\Delta V_{th,DIBL}$ can be further modified with additional fitting parameters to improve modeling accuracy [27–30].

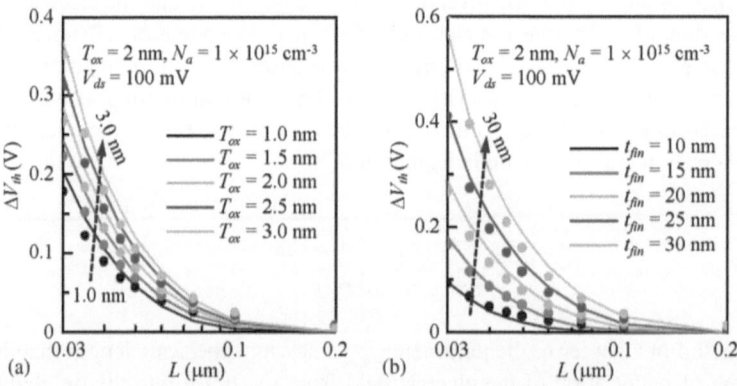

FIGURE 10.4 Drain current model used to simulate SCEs in lightly doped DG-FinFETs showing threshold voltage roll-off for different: (a) oxide thickness, T_{ox}; and (b) fin channel thickness t_{fin} of DG-FinFETs; symbols represent TCAD and lines represent compact model [21].

Subthreshold slope degradation: The planar-MOSFET subthreshold swing model can be applied to modeling S for FinFET devices and is given by [2]

$$S \equiv \left(\frac{d\left[\log\left(I_{ds} \right) \right]}{dV_{gs}} \right)^{-1} \cong \ln\left(10 \right) v_{kT} \left(1 + \frac{C_d}{C_{ox}} + \frac{C_{IT}}{C_{ox}} + \frac{C_{dsc}(\lambda)}{C_{ox}} \right) \qquad (10.49)$$

where:
C_d is the depletion capacitance associated with the depletion region
C_{IT} is the capacitance due to interface states
C_{dsc} is the coupling capacitance between source-drain to channel as described in Section 1.2.2

It can be shown that C_{dsc} depends on L, λ, and V_{ds} similar to ΔV_{th}. Then defining

$$n \equiv \left(1 + \frac{C_d + C_{IT} + C_{dsc}(\lambda)}{C_{ox}} \right) \qquad (10.50)$$

We can write Equation 10.49 as

$$S \cong 2.3 n v_{kT} \qquad (10.51)$$

Thus, the degradation in the subthreshold swing can be modeled through the (L, λ, V_{ds})-dependent $n v_{kT}$ term in Equation 10.51.

10.3.2.2 Quantum Mechanical Effects

Quantum mechanical (QM) confinement of inversion carriers is discussed in Section 6.3 and has been well known in bulk-MOSFETs for a long time [31–33]. The large electric field due to gate voltage V_{gs} perpendicular to the channel leads to a large band bending at the silicon surface and the inversion carriers are confined to dimensions along the thickness t_{fin} of a DG-FinFET as shown in Figure 10.5(a). As described

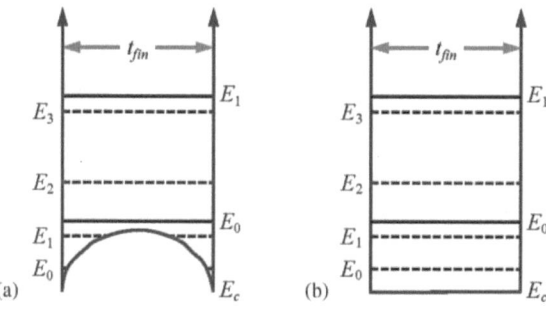

FIGURE 10.5 Energy-band diagrams showing the carrier confinement and associated quantization of electronic energy levels in DG-FinFETs: (a) electrical confinement due to band bending at the top and bottom gate silicon/SiO$_2$ interface; and (b) structural confinement due to ultrathin-body bounded by gate dielectric on the sidewalls.

in Section 6.3.1, this carrier confinement, also known as the *electrical confinement* (EC), leads to splitting of the energy bands into discrete sub-bands which reflects as an increase in the threshold voltage of the transistor and a decrease in the gate capacitance, both of which act to reduce the current drive of the transistor [17,31].

In the case of DG-FinFETs, unlike bulk-MOSFETs, there is strong carrier confinement even at low electric fields making the QM effect even more complex [34]. The carriers are bounded by gate insulator on two sides and are similar to the carriers confined in a rectangular well [17,35–37]. This is referred to as the *structural confinement* (SC) since it arises from the very physical structure of DG-FinFET as shown in Figure 10.5(b). In order to account for the QM effect, it is necessary to model the effect of both EC and SC (Figure 10.5) on the performance of DG-FinFETs. Several groups have reported different analytical and numerical approaches to capture the QM effect in DG-FinFETs [35–37].

The QM confinement of the inversion carriers increases device V_{th}, degrades the gate capacitance, and reduces the effective width of the device due to a shift in the inversion charge centroid as discussed in Section 3.5 (Figure 3.18) along the depth [17,18,31]. A shift in the bottom/top of the conduction/valence band due to SC is used to modify $V_{ch}(y)$ at the source and drain surface potential equations and model QM effect due to SC [17]. In order to model EC, the bias-dependent charge centroid thickness is used to modify the T_{ox} [Equation 3.101] and calculate the reduction in the width of the device [36].

QM effect on threshold voltage: As discussed in Section 6.3.3, the threshold voltage increase due to QM effect is modeled by Equation 6.49 given by

$$\Delta V_{th,QM} = \frac{\pi^2 \hbar^2}{2qm^* t_{fin}^2} \tag{10.52}$$

where:
$\hbar$ is the reduced Planck's constant
m^* is the effective mass of electrons
q is the electronic charge
t_{fin} is the thickness of fin channel

QM effect on drain current: As discussed in Section 6.3.4, QM effect affects the drain current models of FinFET devices due to the increase in the effective gate dielectric thickness given by [38]

$$\delta t_{inv} = \left(\frac{7\varepsilon_{si} \hbar^2}{qm^* Q_i} \right)^{1/3} \tag{10.53}$$

Due to δt_{inv}, the structural parameter of DG-FinFETs, $r \equiv \dfrac{\varepsilon_{si} T_{ox}}{\varepsilon_{ox} t_{fin}}$, defined in Equation 10.31 changes the boundary condition for β and the surface potential function given by Equation 10.30. In order to establish an appropriate QM correction, the boundary condition in Equation 10.30 must be reformulated with the effective oxide thickness due to QM effects given by

$$r^{QM} \equiv \frac{\varepsilon_{si}\left(T_{ox}+\delta t_{inv}\right)}{\varepsilon_{ox}t_{fin}} \qquad (10.54)$$

Thus, I_{ds} degradation due to QM effects can be modeled by modified structural parameters and therefore, surface potential function as described in Section 6.3.4.

Again, the volume inversion [17,20,21,39] due to QM effect, as described in Section 6.3.1, affects the subthreshold performance of FinFET devices. A unique behavior of lightly doped FinFETs with thin fin channel is that the inversion charge is no longer confined at the Si/SiO_2 interface and the entire fin is inverted. For any gate voltage, the electrostatic potential increases at the interfaces and in the fin volume from depletion to weak inversion and in the strong inversion. As a result, the potential shift or total band bending exceeds $2\phi_B$ in every region and in the entire fin. Due to volume inversion: (1) the potential as well as the inversion carrier density is nearly independent of the position inside the body because of negligible potential drop between the surface and the center of the body as shown in Figure 10.6(a); (2) the potential, as well as the inversion charge density, is weakly dependent on body thickness; any small increase in gate voltage in subthreshold increases the potential through the entire body causing inversion in the entire body; and (3) since the electronic potential is virtually independent of body thickness, the total integrated charge inside the body is proportional to body thickness. Thus, as a result of volume inversion, the subthreshold region drain current is also proportional to t_{fin} as shown in Figure 10.6(b).

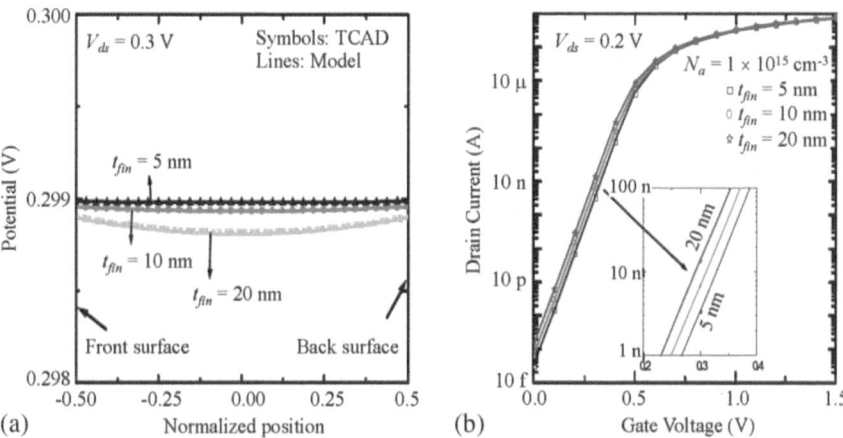

FIGURE 10.6 Volume inversion in lightly doped FinFETs: (a) potential profile in the body between the front and back surfaces in volume inversion; (b) subthreshold $I_{ds} - V_{gs}$ plots for different body thicknesses t_{fin} showing volume inversion (flat potential profile) simulated by the drain current model and numerical device simulation (TCAD); symbols represent TCAD and lines represent compact model; here, body doping $N_a = 1 \times 10^{15}$ cm^{-3} and gate oxide thickness $T_{ox} = 2$ nm [21].

10.3.2.3 Mobility Degradation

Similar to surface mobility degradation in planar-MOSFETs [2], the degradation of carrier mobility in FinFETs also occurs due to four main scattering mechanisms: coulomb scattering, acoustic phonon scattering, surface roughness scattering, and optical phonon scattering. The first three scattering mechanisms have vertical (transverse) field dependency and they are each dominant at different regions of device operation: coulomb scattering at weak inversion, acoustic phonon scattering at mid-inversion, and surface roughness scattering at strong inversion [2]. Similar to planar-MOSFETs, these mechanisms together are modeled through a model called *low-field mobility degradation* and used to define the *effective* mobility [2,27] as described in Chapter 6.

The low-field mobility of the inversion carriers in FinFETs is modeled using Equation 6.70 and is given by

$$\mu_{eff} = \frac{\mu_0}{1+\mu_a\left(E_{eff}\right)^{eu} + \mu_d\left[\frac{1}{2}\left(1+\frac{q_{ia}}{q_b}\right)\right]^{-ucs}} \tag{10.55}$$

where:

μ_0 is concentration-dependent surface mobility

μ_a is a technology-dependent parameter that describes mobility degradation

μ_d is a technology-dependent parameter that describes the second order effect on mobility

eu is a technology-dependent parameter that describes the primary effect of E_{eff} on mobility

ucs is a technology-dependent parameter that describes the secondary effect of E_{eff} on mobility

In order to account for the body bias V_{bs}, dependence of mobility for FinFETs on bulk substrate is modeled by Equation 6.70 and is given by

$$\mu_{eff} = \frac{\mu_0}{1+\left(\mu_a + \mu_c V_{bs}\right)\left(E_{eff}\right)^{eu} + \mu_d\left[\left(1+\eta\frac{q_{ia}}{q_b}\right)\right]^{-ucs}} \tag{10.56}$$

where:

μ_c is a technology-dependent parameter describing mobility degradation due to body bias V_{bs}

η is a constant defining electrons ($\eta = 1/2$) and holes ($\eta = 1/3$)

The mobility Equations 10.55 and 10.56 have been derived assuming strong inversion condition. In the strong inversion regime, the inversion carrier mobility is a function of V_{gs}. In the subthreshold region, the accuracy of the mobility is not critical since Q_i varies with V_{gs} and cannot be modeled accurately. Therefore, in the subthreshold regime, the mobility is usually modeled as a constant concentration-dependent mobility.

The technology-dependent parameter set $\{\mu_a, \mu_c, \mu_d, eu, ucs\}$ are extracted from the measured I_{ds} *versus* V_{gs} characteristics of FinFET devices at low drain bias V_{ds}.

10.3.2.4 Velocity Saturation

As discussed in Section 6.5.1, the influence of the lateral electric field due to the applied V_{ds} on device performance is modeled in drain current calculation by considering the *velocity saturation* in FinFET devices under high lateral electric field. At a high lateral field due to a high applied V_{ds}, the dominant scattering mechanism is optical phonon scattering since the electrons are able to gain enough energy to emit optical phonons. This high lateral field scattering causes the carrier velocity saturation. The velocity saturation causes degradation in I_{ds} [27] as described in Section 6.5.1 and is modeled using Equation 6.78 given by

$$I_{ds} = \frac{I_{ds0}}{\left[1 + \left(\frac{V_{ds}}{E_c L_{eff}}\right)^\alpha\right]^{\frac{1}{\alpha}}} \tag{10.57}$$

where:
I_{ds0} is the drain current without velocity saturation
L_{eff} is the effective channel length of the device
E_c is the field at which carriers are velocity saturated
α is an empirical parameter used for accurate prediction of velocity saturation effect in drain current

The models for channel length modulation (CLM) and output resistance of FinFETs is described in Sections 6.5.2 and 6.6, respectively, for an accurate prediction of real device effects.

10.3.2.5 Source-Drain Series Resistance

The detailed discussions and mathematical formulation to model source-drain series resistance are presented in Chapter 8. In this section, the highlights of modeling are briefly overviewed. In thin-body transistors, the source-drain series resistance is high. In order to reduce the parasitic resistances in FinFETs, selective epitaxial growth (SEG) raised source-drain (RSD) regions are formed in device fabrication as described in Section 8.2 [40]. Thus, the parasitic source-drain resistance model includes distributed contact resistance R_{con}, a spreading resistance, and a bias-dependent source-drain extension (SDE) resistance R_{sde}.

The detailed model formulation for the components of contact resistance R_{con}, spreading resistance R_{sp}, and the SDE resistance R_{sde} is discussed in Chapter 8 [5,41]. Thus, the expressions for modeling the total source resistance and drain resistance, R_s and R_d, respectively are given by

$$R_s = \frac{R_{S1}}{1 + R_{S2} \times \left(V_{gs} - V_{fbsd}\right)} + R_{S3}$$

$$R_d = \frac{R_{D1}}{1 + R_{D2} \times \left(V_{gd} - V_{fbsd}\right)} + R_{D3} \tag{10.58}$$

where:

{RS_1, RS_2, RS_3, RD_1, RD_2, RD_3} is a technology-dependent set of model parameters obtained from the measurement data of a target technology.

10.4   DYNAMIC MODEL

10.4.1   COMMON MULTIGATE $C - V$ MODEL

This section presents the dynamic model of the common multigate DG-FETs for transient analysis of the devices in circuit CAD. The intrinsic capacitance model that describes the transient behavior of the transistors is derived from the terminal charges.

For DG-FETs, the total charge in the body is given by the charges on the top and bottom gate electrodes. The total charge is computed by integrating the charge along the channel. Thus, considering two gates with fin height H_{fin} are electrically interconnected, we get the expression for the total gate charge Q_G as

$$Q_G = WC_{ox} \int_0^L \left(V_{gs} - V_{fb} - \phi(y) \right) \cdot dy \qquad (10.59)$$

where:

L is the channel length of the device

$W \cong 2H_{fin}$ is the channel width of the device

C_{ox} is the gate oxide capacitance of the electrically interconnected common gates

V_{gs} is the applied gate voltage

V_{fb} is the flat band voltage

$\phi(y)$ is the potential at any point y in the direction of current flow

Now, in order to determine the terminal charges, the inversion charge in the body is divided between the source and drain terminals using the Ward-Dutton charge partitioning approach [42,43]. Then the total charge on the source terminal (Q_S) is given by

$$Q_S = -WC_{ox} \int_0^L \left(1 - \frac{y}{L}\right) \cdot \left(V_{gs} - V_{fb} - \phi(y) - \frac{Q_b}{C_{ox}} \right) \cdot dy \qquad (10.60)$$

Now, using the charge conservation principle, the total charge on the drain terminal (Q_D) can be expressed as

$$Q_D = -WC_{ox} \int_0^L \frac{y}{L} \cdot \left(V_{gs} - V_{fb} - \phi(y) - \frac{Q_b}{C_{ox}} \right) \cdot dy \qquad (10.61)$$

In order to evaluate Equations 10.59–10.61, the surface potential $\phi_s(y)$ as a function of the position y along the length of the transistor is obtained using current continuity condition. Since current continuity states that the current is conserved along the length of the transistor, we can write

$$I_{ds}(L) = I_{ds}(y) \quad \text{where} \quad 0 \le y \le L \qquad (10.62)$$

However, the expression for I_{ds} in Equation 10.45 is very complex and is not practical for applying current continuity condition. Thus, for the simplicity of mathematical formulation of $\phi_s(y)$, Equation 10.45 is simplified as shown below [17,28]

$$I_{ds}(y) = \mu(T) \cdot \left(\frac{W}{y}\right) \left[g(Q_{is}) - g(Q_{iy}) \right] \qquad (10.63)$$

where:

$$g(Q_i) \cong \frac{Q_i^2}{2C_{ox}} + 2v_{kT}Q_i \qquad (10.64)$$

Q_{is} is the inversion charge density at the source end
Q_{iy} is the inversion charge density at any point y along the channel

Now, from Equation 10.33, we can show

$$Q_i(y) = -C_{ox}\left(V_{gs} - V_{fb} - \phi_s(y) - \frac{Q_b}{C_{ox}} \right) \qquad (10.65)$$

The function $g(Q_{iy})$ in Equation 10.64 is obtained after neglecting the third term on the right-hand side within the square brackets of Equation 10.45. The approximate Equations 10.63 and 10.64 retain good accuracy in the strong inversion regime, however, overestimate I_{ds} in the subthreshold regime. The advantage of using a mathematically simple analytical expression for terminal charges outweighs the resulting error in the accuracy of $C - V$ model in the subthreshold regime. Thus, using Equation 10.63, the current continuity Equation 10.62 can be written as

$$\frac{g(Q_{is}) - g(Q_{id})}{L} = \frac{g(Q_{is}) - g(Q_{iy})}{y} \qquad (10.66)$$

Thus, using Equations 10.65 and 10.66, $\phi_s(y)$ can be expressed as

$$\frac{(B - \phi_{s0} - \phi_{sL}) \cdot (\phi_{sL} - \phi_{s0})}{L} = \frac{(B - \phi_{s0} - \phi_s(y)) \cdot (\phi_s(y) - \phi_{s0})}{y} \qquad (10.67)$$

where:
   ϕ_{s0} represents the surface potential at the source end
   ϕ_{sL} represents the surface potential at the drain end
   B is defined by the following expression

$$B = 2\left(V_{gs} - V_{fb} - \frac{Q_b}{C_{ox}} + 2v_{kT} \right) \qquad (10.68)$$

The terminal charges are obtained by substituting $\phi_s(y)$ in Equations 10.60–10.62 and evaluation of the integrals [29,30]

$$Q_G = 2H_{fin}LC_{ox}\left[V_g - V_{fb} - \frac{\phi_{s0} + \phi_{sL}}{2} + \frac{\left(\phi_{sL} - \phi_{s0}\right)^2}{6\left(B - \phi_{sL} - \phi_{s0}\right)}\right]$$

$$Q_D = -2H_{fin}LC_{ox}\left[\begin{array}{l} \dfrac{V_g - V_{fb} - \dfrac{Q_b}{C_{ox}}}{2} - \dfrac{\phi_{s0} + \phi_{sL}}{4} + \dfrac{\left(\phi_{sL} - \phi_{s0}\right)^2}{60\left(B - \phi_D - \phi_S\right)} \\[3mm] + \dfrac{\left(5B - 4\phi_{sL} - 6\phi_{s0}\right)\cdot\left(B - 2\phi_{sL}\right)\cdot\left(\phi_{s0} - \phi_{sL}\right)}{60\left(B - \phi_{sL} - \phi_{s0}\right)^2} \end{array}\right] \qquad (10.69)$$

$$Q_S = -\left(Q_G + Q_D + Q_B\right)$$

where:
Q_B = total bulk charge given by $H_{fin}.L.Q_b$ where Q_b is the bulk charge density

The expressions for terminal charges are continuous and valid over subthreshold, linear, and saturation regimes of operation.

Equation 10.69 forms the $C - V$ model for circuit CAD [28]. The terminal charges are used as state variables in the circuit simulation. All the capacitances are derived from the terminal charges to ensure charge conservation. The capacitances are defined as

$$C_{ij} = \frac{\partial Q_i}{\partial V_j} \qquad (10.70)$$

where:
i and j denote the multigate FET terminals

Note that C_{ij} satisfies

$$\sum_i C_{ij} = \sum_j C_{ij} = 0, \qquad (10.71)$$

due to charge conservation.

The capacitances from $C - V$ model are plotted as a function of gate voltage and drain voltage in Figure 10.7(a) and 10.7(b), respectively.

10.5 THRESHOLD VOLTAGE VARIABILITY

In Section 9.2.2.6, different sources of process-induced device performance variability are overviewed. Although the channel doping can be avoided in FinFETs, random discrete doping (RDD) from the source-drain doping gradient causes dopant fluctuations and variability for devices with $L_g < 10$ nm. In a fin with channel doping

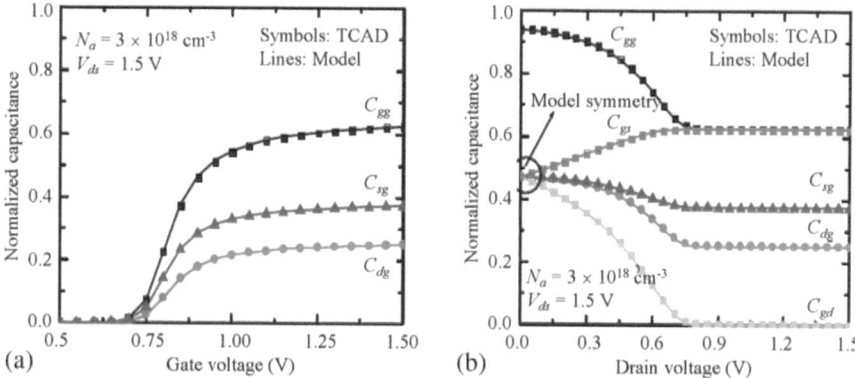

FIGURE 10.7 FinFET dynamic model: modeling transcapacitances as a function of: (a) gate voltage; and (b) drain voltage; model symmetry is seen at $V_{ds} = 0$ where $C_{dg(gd)} = C_{sg(gs)}$; symbols represent TCAD and lines represent compact model [21].

concentration of 5×10^{17} cm^{-3}, about a third of the threshold voltage V_{th}-variability is due to RDD as observed in planar-MOSFETs [2]. Therefore, modeling V_{th}-variability is important even for FinFET devices. The V_{th}-variability model developed for planar-MOSFETs is applicable to FinFETs with appropriate considerations for device width and channel length [44]. In this section, mathematical formulation for V_{th}-variability due to RDD is presented following the reported procedure [44–47].

In order to develop an analytical model for V_{th}-variability in FinFET devices due to RDD in the fin channel, let us consider an ideal FinFET device structure with channel doping $\geq 5 \times 10^{17}$ cm^{-3} as shown in Figure 10.8. If we assume t_{fin} is thicker than the gate-induced channel depletion width, then at the biasing condition $V_{gs} = 0 = V_{ds}$, the p-body below the silicon (Si)/gate-dioxide interface is depleted and the device is in the off-state. When $V_{gs} > V_{th}$ at $V_{ds} = 0$, an n-type inversion layer from source to drain is formed near the substrate at the Si/gate-oxide interface as shown in Figure 10.8. Under a sufficiently strong gate bias $V_{gs} > V_{th}$, referred to as the *strong inversion* condition, the Si surface near the Si/gate-oxide interface

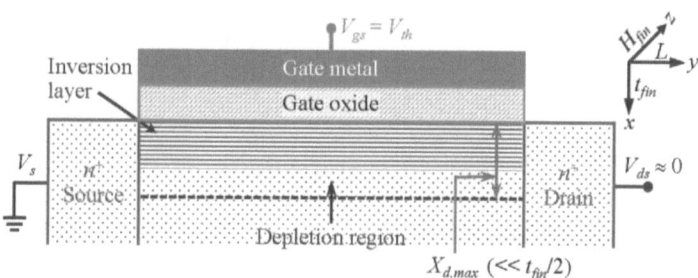

FIGURE 10.8 2D cross-section of an ideal *n*FinFET device structure with n^+ source/drain regions and p-type substrate; (x, y, z) is the coordinate system with x along the fin thickness t_{fin}, y along the channel length L, and z along the width H_{fin} perpendicular to the surface; $x = 0$ at the Si/gate-dielectric interface; here, RDD is shown for top gate for illustration only.

consists of: (1) an inversion layer; and (2) a bulk depletion region beyond the inversion layer as depicted in Figure 10.8.

As shown in Figure 10.8, we consider x, y, and z as the spaces along the fin thickness t_{fin}, channel length L, and fin height H_{fin} of the device, respectively, and $x = 0$ at the Si/gate-oxide interface. At strong inversion, $x = X_{d,max}$ is the depth of the maximum depletion region corresponding to the surface potential $\phi_s = 2\phi_B$; where ϕ_B is the bulk potential. For the simplicity of mathematical formulation, let us make the following simplifying assumptions:

1. The distribution of dopants near the surface of the silicon substrate can be represented by any arbitrary one-dimensional (1D) profile along the x-direction with concentration N_{CH} per unit volume, considering $t_{fin} > X_{d,max}$
2. The silicon surface under the gate consists of a number of thin sheets of dopants of thickness dx and area WL stacked consecutively from the Si/gate-oxide interface to any arbitrary depth beyond $X_{d,max}$, where $W = 2H_{fin} + t_{fin}$
3. Any fluctuations in N_{CH} cause corresponding variation in V_{gs} and therefore in V_{th} to maintain the condition for strong inversion at $\phi_s = 2\phi_B$; and $x = X_{d,max}$
4. V_{th} of a FinFET is defined as the applied gate bias V_{gs} at $V_{ds} \cong 0$ and $\phi_s = 2\phi_B$ similar to planar-MOSFETs
5. Dopant fluctuations in a sheet of dopants imply that the probability of finding the number of dopants in the sheet is "1" or "0." In other words, the variance of dopant fluctuations $\left(\sigma_{V_{th}}^2\right)$ in a sheet is the total number of dopants $(N_{CH} \times WLdx)$ in the sheet

Again, let us consider $V_{ds} \cong 0$, then using *assumption 1* for 1D channel doping profile along the x-direction only, the number of dopants per unit depth of a sheet of dopants (*assumption 2*) is $N_{sh} = N_{CH} \times WL$; where $N_{CH} = N_a$ is the acceptor doping concentration in the fin channel for an nFinFET device.

If $dN_{sh}(x)$ is the fluctuations in N_{sh} per unit depth at a depth x, then to maintain the strong inversion condition (*assumption 3*), the corresponding change in the vertical electric field, $dE_{inv}(x)$, in the fin channel due to V_{gs} is given by Gauss' law as

$$dE_{inv}(x) = -\frac{q}{\varepsilon_{si}} \frac{dN_{sh}(x)}{WL} \qquad (10.72)$$

where:
q is the electronic charge
ε_{si} is the permittivity of silicon fin

Similarly, the change in the vertical electric field $dE_{dep}(x)$ in the semiconductor due to dopant fluctuations $N_{CH}(x)dX_d$ in the depletion region of thickness dX_d is given by

$$dE_{dep}(X_d) = -\frac{q}{\varepsilon_{si}} N_{CH}(X_d)dX_d \qquad (10.73)$$

Thus, the total change in the vertical electric field $dE_s(x)$ in the semiconductor due to dopant fluctuations in a sheet of channel and in depletion regions is given by

$$dE_s = -\frac{q}{\varepsilon_{si}}\left(\frac{dN_{sh}(x)}{WL} + N_{CH}(X_d)dX_d\right) \tag{10.74}$$

In order to relate Equation 10.74 with the fluctuation in gate voltage, we express dX_d in terms of dopant fluctuations in a sheet of dopants in the channel region. Since the change in the electric field at a depth x near the silicon surface is given by $dE(x) = -d\phi_s/x$, then from Equation 10.72 the variation in ϕ_s due to dopant fluctuations in the inversion layer is given by

$$d\phi_s(inv) = \frac{q}{\varepsilon_{si}}\frac{dN_{sh}(x)}{WL}x \tag{10.75}$$

Similarly, from Equation 10.73, $d\phi_s$ due to dopant fluctuations in the depletion region beyond the inversion layer is given by

$$d\phi_s(dep) = \frac{q}{\varepsilon_{si}}N_{CH}(X_d)X_d dX_d \tag{10.76}$$

Therefore, the total change in ϕ_s due to dopant fluctuations in the inversion layer and the depletion layer under the gate is given by

$$d\phi_s = \frac{q}{\varepsilon_{si}}\left(\frac{dN_{sh}(x)}{WL}x + N_{CH}(X_d)X_d dX_d\right) \tag{10.77}$$

Now, in order to maintain the strong inversion condition at $\phi_s = 2\phi_B$ = constant and therefore, $d\phi_s = 0$ (*assumption 3*), we get from Equation 10.77

$$\frac{dN_{sh}(x)}{WL}x + N_{CH}(X_d)X_{d,max}dX_d = 0$$

or, $\tag{10.78}$

$$dX_d = -\frac{1}{WL}\frac{1}{N_{CH}(X_d)}\frac{x}{X_{d,max}}dN_{sh}(x)$$

Then substituting the expression for dX_d from Equation 10.78 into Equation 10.74, we can show

$$dE_s = -\frac{q}{\varepsilon_{si}}\frac{1}{WL}\left(1 - \frac{x}{X_{d,max}}\right)dN_{sh}(x) \tag{10.79}$$

Equation 10.79 shows the fluctuations in the electric field in the body due to the fluctuations in the dopants in the fin channel. In order to find the corresponding fluctuations in the gate voltage to maintain $\phi_s = 2\phi_B$, we use Gauss' law at the Si/gate-oxide interface to get

$$\varepsilon_{si}E_s = \varepsilon_{ox}E_{ox} \tag{10.80}$$

where:
ε_{ox} is the permittivity of gate oxide material

Then differentiating Equation 10.80, we get

$$dE_{ox} = \frac{\varepsilon_{si}}{\varepsilon_{ox}}dE_s$$

$$-\frac{dV_{ox}}{T_{ox}} = \frac{\varepsilon_{si}}{\varepsilon_{ox}}dE_s \tag{10.81}$$

where:
V_{ox} is the voltage drop in the oxide
T_{ox} is the gate oxide thickness

Since for a particular CMOS technology, T_{ox} is a constant, we get from Equation 10.81

$$dV_{ox} = -\frac{T_{ox}}{\varepsilon_{ox}}\varepsilon_{si}dE_s \tag{10.82}$$

Then substituting for dE_s from Equation 10.79 into Equation 10.82, we get the variation in V_{ox} due to the dopant fluctuations in the channel and depletion regions under the gate as

$$dV_{ox} = \frac{T_{ox}}{\varepsilon_{ox}}\frac{q}{WL}\left(1-\frac{x}{X_{d,\max}}\right)dN_{sh}(x) \tag{10.83}$$

We know that for a FinFET device, the gate voltage V_{gs} is given by

$$V_{gs} = V_{fb} + 2\phi_B + V_{ox} \tag{10.84}$$

Here, V_{fb} is the flat band voltage and is a constant for any given technology, therefore, from Equation 10.84, the fluctuations in the gate voltage to maintain the strong inversion condition with $2\phi_B = \text{constant}$ (*assumption 3*) is given by

$$dV_{gs} = dV_{ox} = \frac{T_{ox}}{\varepsilon_{ox}}\frac{q}{WL}\left(1-\frac{x}{X_{d,\max}}\right)dN_{sh}(x) \tag{10.85}$$

Therefore, using *assumption 4*, the variation in V_{th} is given by

$$dV_{th} = \frac{T_{ox}}{\varepsilon_{ox}}\frac{q}{WL}\left(1-\frac{x}{X_{d,\max}}\right)dN_{sh}(x) \tag{10.86}$$

where:

dV_{th} is the total fluctuations in V_{th} corresponding to the total dopant fluctuations in the channel and depletion region under the gate

Since $dN_{sh}(x)$ is a random variable in the area WL at a depth x, therefore, $dV_{th}(x)$ is a random variable due to dopant fluctuations in any sheet of thickness dx. Then the total variation in V_{th} is the random dopant variations over the entire depth, $X_{d,max}$. Therefore, we can write from Equation 10.86

$$dV_{th}(X_d) = \int_0^{dV_{th}} dV_{th}(x) = \frac{T_{ox}}{\varepsilon_{ox}} \frac{q}{WL} \int_0^{X_{d,max}} \left(1 - \frac{x}{X_{d,max}}\right) dN_{sh}(x) \qquad (10.87)$$

From Equation 10.87, the variance of V_{th} is given by

$$Var(dV_{th}(X_d)) = \sigma_{V_{th}}^2 = \left(\frac{T_{ox}}{\varepsilon_{ox}} \frac{q}{WL}\right)^2 Var\left(\int_0^{X_{d,max}} \left(1 - \frac{x}{X_{d,max}}\right) dN_{sh}\right) \qquad (10.88)$$

Now, we know that the variance of a sum is the sum of variances. Therefore, we can express Equation 10.88 as

$$\sigma_{V_{th}}^2 = \left(\frac{T_{ox}}{\varepsilon_{ox}} \frac{q}{WL}\right)^2 \int_0^{X_{d,max}} Var\left[\left(1 - \frac{x}{X_{d,max}}\right) dN_{sh}\right]$$

$$= \left(\frac{T_{ox}}{\varepsilon_{ox}} \frac{q}{WL}\right)^2 \int_0^{X_{d,max}} \sigma_{N_{sh}}^2 \left(1 - \frac{x}{X_{d,max}}\right)^2 \qquad (10.89)$$

Since $\sigma_{N_{sh}}^2 = Var(dN_{sh}) = N_{CH}WL$ is the variance in a sheet of channel dopants per unit depth (*assumption 5*), therefore, for a sheet of thickness dx, we can express Equation 10.89 as

$$\sigma_{V_{th}}^2 = \left(\frac{T_{ox}}{\varepsilon_{ox}} \frac{q}{WL}\right)^2 \int_0^{X_{d,max}} (N_{CH}WLdx) \cdot \left(1 - \frac{x}{X_{d,max}}\right)^2$$

$$= \left(\frac{T_{ox}}{\varepsilon_{ox}}\right)^2 q^2 \frac{N_{CH}}{WL} \int_0^{X_{d,max}} \left(1 - \frac{x}{X_{d,max}}\right)^2 dx \qquad (10.90)$$

$$= \left(\frac{T_{ox}}{\varepsilon_{ox}}\right)^2 q^2 \frac{N_{CH}}{WL} \frac{1}{3} X_{d,max}$$

From Equation 10.90, the variance of V_{th} due to RDD in the channel region of a FinFET device is given by

$$\sigma_{th,RDD} = \frac{1}{\sqrt{3}} \frac{T_{ox}}{\varepsilon_{ox}} q \frac{\sqrt{N_{CH}}}{\sqrt{WL}} \sqrt{X_{d,\max}} \tag{10.91}$$

Now, for a doped FinFET device, the depth of the zero-bias depletion region is given by [2]

$$X_d = \sqrt{\frac{2\varepsilon_{si}}{qN_{CH}} \phi_s} \tag{10.92}$$

where:

$\phi_s = 2\phi_B$ at $X_d = X_{d,\max}$

Therefore, from Equations 10.91 and 10.92, we get

$$\sigma_{th,RDD} = \frac{1}{\sqrt{3}} \frac{T_{ox}}{\varepsilon_{ox}} q \frac{\sqrt{N_{CH}}}{\sqrt{WL}} \sqrt[4]{\frac{2\varepsilon_{si}}{qN_{CH}} 2\phi_B}$$

$$= \frac{1}{\sqrt{3}} \frac{T_{ox}}{\varepsilon_{ox}} \frac{\sqrt[4]{N_{CH}}}{\sqrt{WL}} \sqrt[4]{4q^3 \varepsilon_{si} \phi_B} \tag{10.93}$$

$$= \sqrt{\frac{2}{3}} \left(\sqrt[4]{q^3 \varepsilon_{si} \phi_B} \right) \frac{T_{ox}}{\varepsilon_{ox}} \cdot \left(\frac{\sqrt[4]{N_{CH}}}{\sqrt{WL}} \right)$$

Now, replacing W and L by the effective device width W_{eff} and channel length L_{eff} respectively, in Equation 10.93 to represent the real device dimensions, we can express the V_{th}-variance in FinFET devices as

$$\sigma_{V_{th},RDD} = C \cdot \left(\sqrt[4]{q^3 \varepsilon_{si} \phi_B} \right) \frac{T_{ox}}{\varepsilon_{ox}} \cdot \left(\frac{\sqrt[4]{N_{CH}}}{\sqrt{W_{eff} L_{eff}}} \right) \tag{10.94}$$

where:

$C \cong 0.8165$ and is a number

In Equation 10.94, the bulk potential is given by [Equation 2.36]

$$\phi_B = v_{kT} \ln \left(\frac{N_{CH}}{n_i} \right) \tag{10.95}$$

where:

v_{kT} is the thermal voltage at any ambient temperature T
n_i is the intrinsic carrier concentration in the fin channel

If we ignore the dopant fluctuations in the depletion region, then the second term in Equation 10.89 is zero ($x << X_{d,\max}$) so that we get the expression for V_{th}-variance as

$$\sigma_{V_{th}}^2 = \left(\frac{T_{ox}}{\varepsilon_{ox}}\frac{q}{WL}\right)^2 \int_0^{X_{d,\max}} \sigma_{N_{sh}}^2 \tag{10.96}$$

Again, since $\sigma_{N_{sh}}^2 = Var(dN_{sh}) = \overline{N_{CH}}WLdx$ (*assumption 5*), where $\overline{N_{CH}} = \dfrac{N_{CH}}{2}$ is the average doping concentration in the inversion layer, therefore, we can show from Equation 10.96

$$\sigma_{V_{th}}^2 = \left(\frac{T_{ox}}{\varepsilon_{ox}}\frac{q}{WL}\right)^2 \cdot \frac{1}{2} \int_0^{X_{d,\max}} \left(N_{CH}WLdx\right).$$

$$= \frac{1}{2}\left(\frac{T_{ox}}{\varepsilon_{ox}}\right)^2 q^2 \frac{N_{CH}}{WL} \int_0^{X_{d,\max}} dx \tag{10.97}$$

$$= \frac{1}{2}\left(\frac{T_{ox}}{\varepsilon_{ox}}\right)^2 q^2 \frac{N_{CH}}{WL} X_{d,\max}$$

We know that the surface inversion layer is formed when the intrinsic energy level E_i is pulled below the Fermi level. For a fully depleted fin, we assume an average value of bulk potential to consider the dopant fluctuations in the inversion layer only. Then substituting Equation 10.92 into Equation 10.97, we can show

$$\sigma_{th,RDD} = \frac{1}{\sqrt{2}}\frac{T_{ox}}{\varepsilon_{ox}}q\frac{\sqrt{N_{CH}}}{\sqrt{WL}}\sqrt[4]{\frac{2\varepsilon_{si}}{q(N_{CH})}\phi_B/2}$$

$$= \frac{1}{\sqrt{2}}\frac{T_{ox}}{\varepsilon_{ox}}\frac{\sqrt[4]{N_{CH}}}{\sqrt{WL}}\sqrt[4]{q^3\varepsilon_{si}\phi_B}$$

$$= \frac{1}{\sqrt{2}}\cdot\left(\sqrt[4]{q^3\varepsilon_{si}\phi_B}\right)\frac{T_{ox}}{\varepsilon_{ox}}\cdot\left(\frac{\sqrt[4]{N_{CH}}}{\sqrt{WL}}\right) \tag{10.98}$$

$$= C\left(\sqrt[4]{q^3\varepsilon_{si}\phi_B}\right)\frac{T_{ox}}{\varepsilon_{ox}}\cdot\left(\frac{\sqrt[4]{N_{CH}}}{\sqrt{WL}}\right)$$

Note the value of $C = 1/\sqrt{2} = 0.7071$, if we neglect the dopant fluctuation in the depletion region. Thus, Equation 10.98 is applicable for fully depleted FinFETs with $t_{fin} < X_{d,\max}$. Now, if we define C_{vt} as a technology-dependent parameter given by

$$C_{vt} = C \cdot \left(\sqrt[4]{q^3\varepsilon_{si}\phi_B N_{CH}}\right)\frac{T_{ox}}{\varepsilon_{ox}} \tag{10.99}$$

Then for any particular CMOS technology, Equation 10.94 and Equation 10.99 can be generalized as

$$\sigma_{V_{th,RDD}} = C_{vt} \frac{1}{\sqrt{W_{eff} L_{eff}}}, \qquad (10.100)$$

Note that in Equation 10.99, $C = 0.8165$ models dopant fluctuations both in the inversion layer and depletion region under the gate whereas, $C = 0.7071$ models dopant fluctuations in the inversion layer only applicable to fully depleted FinFETs. Thus, Equation 10.100 can be used to characterize the effect of random dopant fluctuations on V_{th} of FinFET devices of a non-planar CMOS technology at a nanometer node.

10.6   SUMMARY

This chapter presented an overview of the present state-of-the-art compact models for thin-body common multigate FinFET devices. The device model consists of a core model for large geometry devices and models for real devices to analyze the physical and geometrical effects on the device performance. The basic features of the model include capturing the important physics of thin-body multigate transistors such as the volume inversion and the dynamic V_{th} shift for body bias in ultrathin-body transistors. The models are valid for digital as well as analog circuit analysis with the $C - V$ models that simulate the transcapacitances. This chapter is intended to provide readers the present state-of-the-art modeling activities in thin-body FET devices.

REFERENCES

1. L. Nagel and D. Pederson, "Simulation program with integrated circuit emphasis," University of California, Berkeley, Electronics Research Laboratory Memorandum No. UCB/ERL M352, 1973.
2. S.K. Saha, *Compact Models for Integrated Circuit Design: Conventional Transistors and Beyond*, CRC Press, Taylor & Francis Group, Boca Raton, FL, 2015.
3. M.S. Lundstrom and D.A. Antonidis, "Compact models and the physics of nanoscale FETs," *IEEE Transactions on Electron Devices*, 61(2), pp. 225–233, 2014.
4. C.C. McAndrew, "Practical modeling for circuit simulation," *IEEE Journal of Solid-State Circuits*, 33(3), pp. 439–448, 1998.
5. Y.S. Chauhan, D.D. Lu, S. Venugopalan, *et al.*, *FinFET Modeling for IC Simulation and Design: Using the BSIM-CMG Standard*, Academic Press, San Diego, CA, 2015.
6. S.K. Saha, "Modeling process variability in scaled CMOS technology," *IEEE Design & Test of Computers*, 27(2), pp. 8–16, 2010.
7. N. Arora, *MOSFET Models for VLSI Circuit Simulation: Theory and Practice*, Springer – Verlag, Vienna, 1993.
8. S.K. Saha, N.D. Arora, M.J. Deen, and M. Miura-Mattausch, "Advanced compact models and 45-nm modeling challenges," *IEEE Transactions on Electron Devices*, 53(9), pp. 1957–1960, 2006.
9. S.K. Saha, M.J. Deen, and H. Masuda, "Compact interconnect models for gigascale integration," *IEEE Transactions on Electron Devices*, 56(9), pp. 1784–1786, 2009.
10. Y.S. Chauhan, S. Venugopalan, M.-A. Chalkiadaki, *et al.*, "BSIM6: Analog and RF compact model for bulk MOSFET," *IEEE Transactions on Electron Devices*, 61(2), pp. 234–244, 2014.

11. L. Chang, M. Ieong, and M. Yang, "CMOS circuit performance enhancement by surface orientation optimization," *IEEE Transactions on Electron Devices*, 51(10), pp. 1621–1627, 2004.

12. H.C. Pao and C.T. Sah, "Effects of diffusion current on characteristics of metal-oxide (insulator)-semiconductor transistors," *Solid-State Electronics*, 9(10), pp. 927–937, 1966.

13. J. Sallese, F. Krummenacher, F. Pregaldiny, *et al.*, "A design oriented charge-based current model for symmetric DG MOSFET and its correlation with the EKV formalism," *Solid-State Electronics*, 49(3), pp. 485–489, 2005.

14. Y. Taur, X. Liang, W. Wang, and H. Lu, "A continuous, analytic drain current model for DG MOSFETs," *IEEE Electron Device Letters*, 25(2), p. 107–109, 2004.

15. M.V. Dunga, C.-H. Lin, X. Xi, *et al.*, "Modeling advanced FET technology in a compact model," *IEEE Transactions on Electron Devices*, 53(9), pp. 1971–1978, 2006.

16. C. Auth, C. Allen, A. Blattner, *et al.*, "A 22-nm-high performance and low-power CMOS technology featuring fully-depleted tri-gate transistors, self-aligned contacts and high density MIM capacitors." In: *Symposium on VLS Technology*, pp. 131–132, 2012.

17. M.V. Dunga, "Nanoscale CMOS modeling," Ph.D. dissertation, Electrical Engineering and Computer Science, University of California, Berkeley, CA, 2008.

18. N. Paydavosi, S. Venugopalan, Y.S. Chauhan, *et al.*, "BSIM – SPICE models enable FinFET and UTB IC design," *IEEE Access*, 1, pp. 201–215, 2013.

19. Y. Taur, "Analytic solutions of charge and capacitance in symmetric and asymmetric double-gate MOSFETs," *IEEE Transactions on Electron Devices*, 48(12), pp. 2861–2869, 2001.

20. S. Venugopalan, D.D. Lu, Y. Kawakami, *et al.*, "BSIM-CG: A compact model of cylindrical/surround gate MOSFET for circuit simulations," *Solid-State Electronics*, 67(1), pp. 79–89, 2012.

21. M.V. Dunga, C.-H. Lin, D.D. Lu, *et al.*, "BSIM-MG: A versatile multi-gate FET model for mixed-signal design." In: *Symposium on VLSI Technology*, pp. 60–61, 2007.

22. X. Liang and Y. Taur, "A 2-D analytical solution for SCEs in DG MOSFETs," *IEEE Transactions on Electron Devices*, 51(9), pp. 1385–1391, 2004.

23. Q. Chen, E.M. Harrell, and J.D. Meindl, "A physical short-channel threshold voltage model for undoped symmetric double-gate MOSFETs," *IEEE Transactions on Electron Devices*, 50(7), pp. 1631–1637, 2003.

24. K. Suzuki, T. Tanaka, Y. Tosaka, and H. Horie, "Scaling theory for double-gate SOI MOSFETs," *IEEE Transactions on Electron Devices*, 40(12), pp. 2326–2329, 1993.

25. K. Suzuki, Y. Tosaka, and T. Sugii, "Analytical threshold voltage model for short channel n+/p+ double-gate SOI MOSFETs," *IEEE Transactions on Electron Devices*, 43(5), pp. 732–738, 1996.

26. K.K. Young, "Short-channel effect in fully depleted SOI MOSFET's," *IEEE Transactions on Electron Devices*, 36(2), pp. 399–402, 1989.

27. W. Liu, *BSIM4 and MOSFET Modeling for IC Simulation*, World Scientific, Singapore, 2011.

28. M.V. Dunga, C.-H. Lin, A.M. Niknejad, and C. Hu, "BSIM-CMG: A compact model for multi-gate transistors." In: *FinFETs and Other Multi-Gate Transistors*, J.-P. Colinge, (ed.), pp. 113–153, Springer, New York, 2008.

29. D. Lu, C.-H. Lin, A.M. Niknejad, and C. Hu, "Multi-gate MOSFET compact model BSIM-MG." In: *Compact Modeling: Principles, Techniques and Applications*, G. Gildenblat, (ed.), pp. 395–429. Springer, New York, 2010.

30. S. Khandelwal, J. Duarte, A.S. Medury, *et al.*, *BSIM-CMG 110.0.0 Multi-Gate MOSFET Compact Model Technical Manual*, University of California, Berkeley, CA, 2015.

31. S. Saha, "Effects of inversion layer quantization on channel profile engineering for nMOSFETs with 0.1 μm channel lengths," *Solid-State Electronics*, 42(11), pp. 1985–1991, 1998.

32. F. Stern, "Electronic properties of two-dimensional systems," *Reviews of Modern Physics*, 54(2), pp. 437–672, 1982.

33. R. Rios, N.D. Arora, C.-L. Huang, *et al.*, "A physical compact MOSFET model, including quantum mechanics effects, for statistical circuit design applications." In: *IEEE International Electron Devices Meeting Technical Digest*, pp. 937–940, 1995.

34. L.D. Landau and E.M. Lifshitz, *Quantum Mechanics*, Addison-Wesley, Reading, MA, 1990.

35. G. Baccarani and S. Reggiani, "A compact double-gate MOSFET model comprising quantum-mechanical and nonstatic effects," *IEEE Transactions on Electron Devices*, 46(8), pp. 1656–1666, 1999.

36. L. Ge and J.G. Fossum, "Analytical modeling of quantization and volume inversion in thin Si-film double gate MOSFETs," *IEEE Transactions on Electron Devices*, 49(2), pp. 287–294, 2002.

37. S. Venugopalan, M.A. Karim, S. Salahuddin, *et al.*, "Phenomenological compact model for QM charge centroid in multi gate FETs," *IEEE Transactions on Electron Devices*, 60(4), pp. 480–484, 2013.

38. F. Stern and W.E. Howard, "Properties of semiconductor surface inversion layers in the electric quantum limit," *Physical Review*, 163(3), pp. 816–835, 1967.

39. F. Balestra, S. Cristoloveanu, M. Benachir, J. Brini, and T. Elewa, "Double-gate silicon-on-insulator transistor with volume inversion: A new device with greatly enhanced performance," *IEEE Electron Device Letters*, 8(9), pp. 410–412, 1987.

40. Y.-K. Choi, D. Ha, T.-J. King, and C. Hu, "Ultra-thin body PMOSFETs with selectively deposited Ge source/drain." In: *Symposium on VLS Technology*, pp. 19–20, 2001.

41. D. Lu, "Compact models for future generation CMOS," Ph.D. dissertation, Electrical Engineering and Computer Science, University of California, Berkeley, CA, 2011.

42. D.E. Ward and R.W. Dutton, "A charge-oriented model for MOS transistor capacitances," *IEEE Journal of Solid-State Circuits*, 13(5), pp. 703–710, 1978.

43. S.Y. Oh, D.E. Ward, and R.W. Dutton, "Transient analysis of MOS transistors," *IEEE Journal of Solid-State Circuits*, 15(4), pp. 636–643, 1980.

44. P. Stolk, F. Widdershoven, and D. Klaassen, "Modeling statistical dopant fluctuations in MOS transistors," *IEEE Transactions on Electron Devices*, 45(9), pp. 1960–1971, September 1998.

45. T. Mizuno, J.-I. Okamura, and A. Toriumi, "Experimental study of threshold voltage fluctuation due to statistical variation of channel dopant number in MOSFETs," *IEEE Transactions on Electron Devices*, 41(11), pp. 2216–2221, 1994.

46. T. Mizuno, "Influence of statistical spatial-nonuniformity of dopant atoms on threshold voltage in a system of many MOSFET's," *Japanese Journal of Applied Physics*, 35(2B), pp. 842–848, 1996.

47. S.K. Saha, "Modeling statistical dopant fluctuations effect on threshold voltage of scaled JFET devices," *IEEE Access*, 4, pp. 507–513, 2016.

Index